Phase Transformation Kinetics in Thin Films

MATERIALS RESEARCH SOCIETY SYMPOSIUM PROCEEDINGS   VOLUME 230

Phase Transformation Kinetics In Thin Films

Symposium held April 29-May 1, 1991, Anaheim, California, U.S.A.

EDITORS:

M. Chen
IBM Almaden Research Center, Almaden, California, U.S.A.

M.O. Thompson
Cornell University, Ithaca, New York, U.S.A.

R. B. Schwarz
Los Alamos National Laboratory, Los Alamos, New Mexico, U.S.A.

M. Libera
Stevens Institute of Technology, Hoboken, New York, U.S.A.

|M|R|S| MATERIALS RESEARCH SOCIETY
Pittsburgh, Pennsylvania

CAMBRIDGE UNIVERSITY PRESS
Cambridge, New York, Melbourne, Madrid, Cape Town,
Singapore, São Paulo, Delhi, Mexico City

Cambridge University Press
32 Avenue of the Americas, New York NY 10013-2473, USA

Published in the United States of America by Cambridge University Press, New York

www.cambridge.org
Information on this title: www.cambridge.org/9781107409828

Materials Research Society
506 Keystone Drive, Warrendale, PA 15086
http://www.mrs.org

First published 1992
First paperback edition 2012

Single article reprints from this publication are available through
University Microfilms Inc., 300 North Zeeb Road, Ann Arbor, MI 48106

CODEN: MRSPDH

ISBN 978-1-107-40982-8 Paperback

Contents

*Invited Paper

*Invited Paper

Preface

This volume contains papers presented at the Materials Research Society Symposium on "Phase Transformations Kinetics in Thin Films" held in Anaheim, California from April 29 through May 1, 1991. This symposium provided a multidisciplinary forum for explorations, on experimental and theoretical levels, of thin film reactions and stability, phase nucleation and growth, and amorphization. The papers in this volume, refereed by the peer review process, are organized according to materials and techniques and do not reflect the order of presentations at the symposium.

Symposium sessions were organized in the areas of thin-film crystallization, solid-state amorphization, interfacial reactions, solid-state transformations, phase-change optical media and ferroelectric thin films. Six internationally recognized invited speakers reviewed some of the important problems in these areas including metal-mediated growth of Si, stress enhanced reactions, solid-state amorphization, phase-change optical recording, and ferroelectric materials for electronic applications.

Contributed papers ranged from theoretical determination of the limits to melt nucleation to commercial concerns of process-ing techniques for specific properties. Despite this breadth, the similarity of experimental techniques and thermodynamic underpinnings for most of the materials provided a common basis for discussions. As a result, a number of common themes arose from the sessions. For example, several papers described the formation of a disordered (amorphous) phase at an interface during annealing, both in traditional solid-state amorphizing metal couples such as Ni/Zr, and at metal-semiconductor inter-faces such as Pt/GaAs. Theoretical models for solid state amorphization, based on a "mechanical melting temperature" or an "isentropic melting state," were also presented. Other common themes included first phase determinations, kinetic barriers to transformations, stress enhanced reactions, and point defect reactions. On the materials side, ferroelectric thin films for electronic dielectric applications emerged as a major new topic. Several papers discussed the control and understanding of the phase transformations (amorphous to pyrochlore and perovskite), as well as new growth and processing techniques (such as sol-gel precursors).

Overall, the symposium and these manuscripts reflect our rapidly advancing, and sometimes changing, understanding of thin film reactions. The abundance of new and unresolved questions, however, ensures that these topics will continue to be of considerable interest and importance.

M. Chen
M. Thompson
R.B. Schwarz
M. Libera

January 1992

Acknowledgments

We first acknowledge and thank all of the speakers and poster presenters, contributing authors and participants who made this symposium a success. Special appreciation ia also due the referees who worked to review the manuscripts in a timely fashion. We especially thank the invited speakers whose excellent reviews of the fields established the baseline for the sessions. They were:

F. Spaepen	K.N. Tu
D. Wolf	K.A. Rubin
W.L. Johnson	J.F. Scott

We are also grateful for the efforts of the session chairs who successfully adhered to schedules and managed discussions of the technical program. They were:

K. Kavanaugh	M.O. Thompson
M. Nastasi	R.B. Schwarz
Y. Tyan	M. Chen
M. Libera	

Finally, we gratefully acknowledge financial support provided to this symposium by Los Alamos National Laboratory, Eastman Kodak, IBM Corporation, Mitsubishi Kasei Corporation, and Computer Graphics Service.

MATERIALS RESEARCH SOCIETY SYMPOSIUM PROCEEDINGS

*Prior Materials Research Society Symposium Proceedings
available by contacting Materials Research Society.*

PART I

Solid State Amorphization

MOLECULAR DYNAMICS SIMULATION OF THE EFFECT OF INTERFACES IN MELTING AND SOLID-STATE AMORPHIZATION

DIETER WOLF* and SIDNEY YIP**
*Materials Science Division, Argonne National Laboratory, Argonne, IL 60439
**Department of Nuclear Engineering, Massachusetts Institute of Technology, Cambridge, MA 02139.

ABSTRACT

A newly developed molecular dynamics code was used to study the effect of free surfaces, grain boundaries and voids in the process of melting. It was found that conventional "thermodynamic melting" occurs via nucleation of the liquid at the extended defects with subsequent growth into the crystal. In the absence of interfaces, or when this transition is kinetically hindered, however, a second type of melting transition can be triggered by an elastic instability first described by Born ("mechanical melting"). It is suggested that the distinct characteristic features associated with the two types of melting are actually observed in solid-state amorphization experiments. A unified thermodynamics-based description, in the form of an extended phase diagram, of melting and solid-state amorphization is proposed which brings out the parallels between these two phenomena and suggests that their underlying causes are apparently the same. By investigating the effect of surface stresses on the structure and elastic behavior of free-standing thin films, we discuss how these concepts need to be modified in thin-film and small-grained materials.

1. INTRODUCTION

The crystalline-to-amorphous (C-A) phase transformation is currently receiving renewed attention due to new experimental evidence that amorphous alloys can be produced by a variety of irradiation-, chemically-, and mechanically-driven processes [1]. Given that the transformation can be induced by many different mechanisms, the question naturally arises as to what is the underlying nature of the transition that is common to all these processes. In the same context it can be asked what is the connection between melting and amorphization since both phenomena are concerned with the transition from an ordered to a disordered phase.

That amorphization is analogous to melting in certain respects has been recognized recently by a number of workers [1-5]. In particular, Cahn and Johnson [2] have pointed out parallels which exist in the processes involving the heterogeneous nucleation of disorder and Okamoto et al. [5] have discussed the similarity in the volume dependence of the shear modulus during irradiation-induced amorphization and heating-induced melting [6]. In the present work we suggest that the analogy between melting and amorphization may be taken further. By focussing on the role of the two thermodynamic state variables of temperature and volume in the destruction of crystalline order, we propose a unifying description which appears to be a natural extension of basic equilibrium thermodynamics. This representation brings out clearly the essential thermodynamic parallels between these two phenomena without being encumbered by specific mechanistic aspects. It also elucidates the important role played by self diffusion, a manifestation of the equilibrium dynamical behavior of a statistical-mechanical system. A more detailed and comprehensive account of this work has appeared recently [7].

2. MOLECULAR DYNAMICS SIMULATION OF MELTING

The fundamental concept of melting is based on the coexistence of the solid with the liquid when the free energies of the two phases are equal. It is implied that at temperatures above this coexistence the solid is unstable; but neither the mechanism of melting nor the kinetics of the process are considered in the thermodynamic definition. In reality, melting occurs in the presence of external or internal surfaces and over a finite time interval [8]. Despite a wealth of experimental data [9,10], it is not clear conceptually how the observed kinetic behavior is to be interpreted in the context of the thermodynamic basis of the transition.

Three basic physical scenarios of melting have been proposed. The first treats the phenomenon as a homogeneous, bulk process involving a lattice instability (see, for example, Refs. 11-13) in which the (temperature-dependent) normal modes of the lattice become unstable at sufficiently high temperature. The second involves a mechanical instability occurring when the concentration of intrinsic (i.e., thermally generated) defects reaches a critical concentration (see, for example, Ref. 14). The third, originating from experimental observation [8,10,15-18] describes melting as nucleating at extrinsic defects, such as free surfaces, grain boundaries, etc. Several recent experiments demonstrate that when the surface conditions are modified, the melting point can be depressed [15] or the solid can be substantially superheated [8,10,18,19]. The implications are that (a) melting is basically a heterogeneous process, and (b) the mechanism of nucleation at extrinsic defects generally determines the kinetics.

In any study of melting, knowledge of the thermodynamic melting point, T_m, is of primary importance. Hence, at the outset free-energy calculations should be performed for the crystalline and liquid phases to determine T_m. If this is not done, then one does not know the true melting point of the model system described by the particular interatomic interaction potential function adopted for the simulation. The interatomic potential used in the present work is an embedded-atom-method (EAM) potential [20] for copper parameterized for Cu/Ni alloys [21]. The zero-temperature equilibrium lattice parameter for this potential is $a_0 = 3.6208$ Å. An analysis of the free energies of the (undefected, perfect-crystal) solid phase and of the liquid phase then yields a coexistence temperature at zero pressure of $T_m = 1171 \pm 30$ K. The details of that calculation have been described elsewhere [22].

2.1 Thermodynamic Melting

In order to investigate the role of extrinsic defects on melting, we have investigated the high-temperature behavior of a bicrystal containing a symmetrical grain boundary (GB). Far from the interface in the direction of the GB normal the GB is embedded in perfect-crystal blocks which are allowed to slide parallel and perpendicular to the interface plane, thus enabling both GB migration and a volume expansion at the boundary. Details of this 2-d periodic simulation model are given elsewhere [23].

The particular GB we have studied is the so-called Σ29 (001) twist GB. This GB is created by rotating one half of the bicrystal about the <001> axis by an angle of 43.60° relative to the other half. The resulting GB is periodic in the (001) plane with a square planar unit cell containing $\Sigma=29$ atoms, and with sides of length 3.808 a_0 at zero temperature. This boundary was chosen because of (i) the large interplanar spacing of the (001) planes compared to the vibrational amplitudes of the atoms and (ii) its relatively large planar unit cell. The latter allows us to consider this a "generic" high-angle boundary as opposed to boundaries with small planar unit cells, such as symmetric tilt GBs, for which the GB energy is known to be unusually sensitive to relative translations of the two halves of the bicrystal [24]. The simulation cell contains 32 (001) planes, 16 in each half of the bicrystal, with a total of 928 atoms.

To investigate the breakdown of crystalline order upon melting, we define the squared magnitude of the static structure factor, $S(\underline{k})$, which for brevity we denote simply as S(k),

$$S(k) = [1/N \, \Sigma_i \cos(\underline{k} \cdot \underline{q}_i)]^2 + [1/N \, \Sigma_i \sin(\underline{k} \cdot \underline{q}_i)]^2 \quad , \tag{1}$$

where $\underline{q}_i$ is the position of atom i. For the overall S(k), all atoms in the simulation cell are included in the sums in Eq.(1), whereas for the *planar* structure factor, $S_p(k)$, only atoms in a given lattice plane are considered. For an ideal-crystal lattice at zero temperature, $S_p(k)$ then equals unity for any wave vector, $\underline{k}$, which is a reciprocal lattice vector in that plane. By contrast, in the liquid state, without long-range order in the plane, $S_p(k)$ fluctuates near zero.

For the twist GB considered here, the two halves of the bicrystal are rotated relative to one another. A reciprocal-lattice vector lying in a (001) plane in one half of the bicrystal is not a reciprocal-lattice vector in the other half. Two different wave vectors, $\underline{k}_1$ and $\underline{k}_2$, are thus required to monitor planar order in the two halves. They are related by the relative rotation of the two halves of the bicrystal. For a well-defined lattice plane, say in crystal 1, $S_p(k_1)$ then fluctuates about a finite value, somewhat less than unity, which is appropriate for the temperature of the crystal, while $S_p(k_2)$ fluctuates about a value that is essentially zero. In the interface region, due to local disorder, one expects somewhat lower values of $S_p(k_1)$. By monitoring $S_p(k_1)$ and $S_p(k_2)$, every plane may be characterized as (a) belonging to crystal 1 (if $S_p(k_1)$ is near unity and $S_p(k_2)$ near zero), (b) belonging to crystal 2 (if $S_p(k_1)$ is near zero and $S_p(k_2)$ is near unity), or (c) disordered (if $S_p(k_1)$ and $S_p(k_2)$ are both near zero). The vectors $\underline{k}_1$ and $\underline{k}_2$ were chosen, in the present case, to be reciprocal-lattice vectors in the <100> directions.

Figures 1(a) and (b) show the instantaneous planar structure-factor profile for the GB at 1300 K after 5000 and 10,000 time steps, respectively. At zero time steps, immediately after the temperature has been raised from 1200 to 1300 K, the GB is sharply defined by the crossing of $S_p(k_1)$ and $S_p(k_2)$. The intrinsic disorder due to the GB is evidenced by the lower values of the structure factors in the GB region. As time progresses, it is clear from Figs. 1(a) and (b) that a region of disorder forms at the GB and spreads outward. From the linear increase with time of the mean-square displacement of atoms within the disordered region, the diffusion constant is determined to be 4.1 x 10⁻⁹ m²/s; this value is typical of a molten metal and agrees with the value obtained from an independent simulation of the liquid at that temperature. Further evidence of melting is the overall volume expansion when the entire system becomes disordered. This was investigated at 1400 K. Upon complete disordering, the time-averaged overall volume is found to be 26.69 a_0^3. This compares well with the value of 26.75 a_0^3 for a liquid at this temperature, simulated by ordinary 3-d borders, and is significantly greater than the volume of the defect-free ideal crystal at 1400 K (25.29 a_0^3).

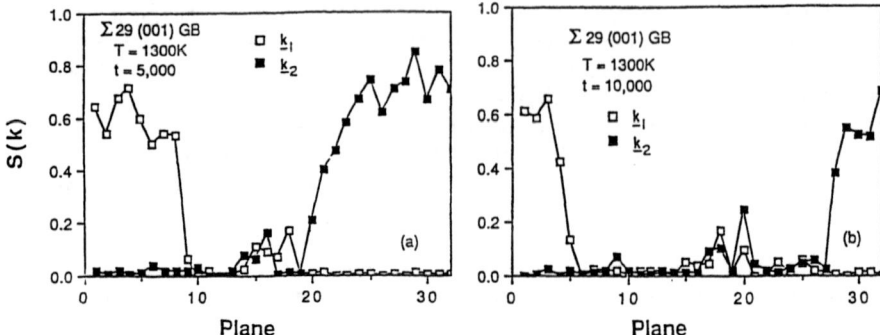

Figure 1. Instantaneous values of $S_p^2(k_1)$ and $S_p^2(k_2)$ defined in Eq. (1) for the 32 slices parallel to the (001) Σ29 twist GB (a) after 5000 time steps and (b) after 10,000 time steps at 1300 K.

From this evidence we conclude that above T_m a GB can nucleate the liquid phase which then propagates through the crystal. One might ask whether other extended defects can also act as nucleation centers. To investigate this question, we have studied the effects of voids of various sizes and of a free (001) surface. As discussed in more detail in Ref. 22, both the insertion of a 13-atom void into the simulation cell and the simulation of a thin slab terminated by free (001) surfaces yields the same behavior, namely, the nucleation at temperatures above T_m of the liquid phase at the extended defect, with subsequent spreading of the solid-liquid interface through the crystal. A planar 5-atom void was also investigated, with the result, however, that the void broke up into highly mobile mono- and divacancies without inducing melting during a 20,000 time-step MD run [22].

From the results of our simulations at several temperatures above T_m, propagation velocities for the spreading of the solid-liquid interfaces into the crystalline regions, v, can be extracted. In Fig. 2 these velocities are plotted as function of temperature for the case of the Σ29 GB. (Similar plots are obtained for our void and free-surface simulations [22].) Extrapolation of these temperature-dependent velocities to zero velocity should yield an estimate of the coexistence temperature, T_m. The temperature so obtained from Fig. 2 is 1179±20 K, in remarkable

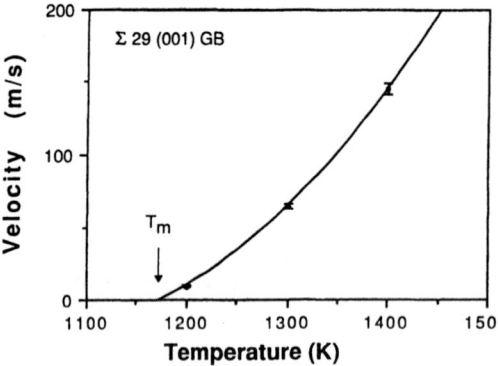

Figure 2. Velocity of propagation of the two crystal-liquid interfaces nucleated at the GB as function of temperature above the thermodynamic coexistence temperature T_m. The solid line shows a quadratic fit to the data points [7,22]. The data obtained for the free surface and the 13-atom void fall on the same curve [22].

agreement with the temperature of 1171±30 K obtained from the free-energy analysis [22]. Equally good agreement was obtained from our simulation involving nucleation of the liquid at the free surface or at the 13-atom void. The time required for nucleation of the liquid at the free surface was found to be much longer, however, than at the GB and at the void [22].

From these results we draw the conclusion that thermodynamic melting of a "real" crystal (i.e., one containing extended defects) occurs via nucleation of disorder at the extrinsic defects with subsequent thermally-activated [7,8,25,26] growth of the liquid phase into the crystal. For a typical propagation velocity of 100 m/s at ~150 K above the melting point (see Fig. 2), a single crystal 1mm in diameter, bordered by free surfaces, would thus require a time of the order of 10^{-5} s to melt completely.

2.2 Mechanical Melting

In order to induce the melting transition, some degree of superheating into a range of metastability is obviously needed to drive the phase transformation at finite rate, because exactly at the thermodynamic melting point a solid-liquid interface, nucleated at the extended defects, cannot propagate. With both the nucleation and growth of the liquid phase requiring mobile atoms near extended defects, however, thermodynamic melting may be hindered by either eliminating the nucleation centers or lowering the atomic mobility. The latter may be achieved by lowering T_m, for example, through a hydrostatic expansion of the crystal. As discussed in more detail in Sec. 3, this kinetic hindrance of thermodynamic melting appears to play an important role in solid-state amorphization.

In computer simulations thermodynamic melting is easily suppressed by elimination of extended defects, i.e., by the simulation of a perfect crystal with 3-d periodic borders [11]. Experimentally superheating is extremely difficult even in the most favorable cases due to the presence of dislocations [8,10,18,19]. Over half a century ago Born [12] pointed out the existence of an absolute limit to superheating for any crystalline structure. By considering the volume dependence of the normal modes of a crystal lattice he demonstrated the existence of a phonon instability at a certain critical volume expansion, V_s. By couching the discussion in terms of the elastic response of a crystal lattice under isotropic tension, Born's phonon instability can be shown to correspond to an elastic instability in the minimal shear constant, C', defined by (for cubic crystals)

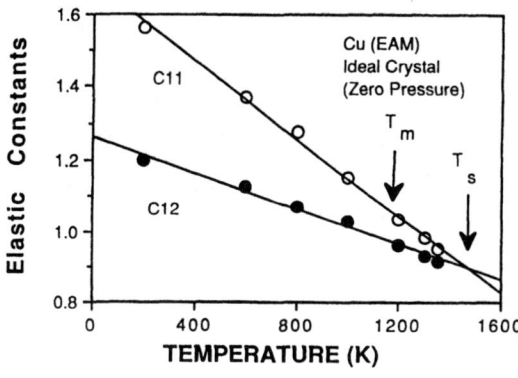

Figure 3. Elastic constants C_{11} and C_{12} (in 10^{12} dyn/cm^2) vs. temperature for a perfect crystal containing no extended defects. The solid lines are straight-line fits to the data points. The "stability temperature" is denoted by T_s.

$$C' = (C_{11} - C_{12})/2 \qquad\qquad (2)$$

i.e., C_{44} expressed in a coordinate system with z axis parallel to <001> and the x-y axes parallel to the <110> and <1$\bar{1}$0> directions in which C_{44} assumes its smallest value for any cubic crystal. Born's criterion, when applied to a superheated crystal lattice, establishes the existence of a maximum volume, V_s, coupled with a maximum superheating temperature, T_s, above which the crystal becomes mechanically unstable and therefore has to undergo some kind of phase transformation (to the liquid state or some other crystal structure). The temperature associated with the maximum thermal expansion is referred to as the mechanical melting point, T_s.

Figure 3 illustrates a method for determining T_s from the elastic constants of a superheated perfect crystal. From a least-squares fit to the decrease of the elastic constants C_{11} and C_{12} with increasing temperature, a value of $T_s = 1432\pm12$ K is obtained, which is about 260 K higher than the thermodynamic melting point, $T_m = 1171\pm30$ K, obtained for the Cu(EAM) potential. The elastic constants in Fig. 3 were obtained from constant-pressure molecular dynamics simulations [27] with evaluation of both the Born and fluctuation contributions [28].

In practice it is very difficult, even in simulations, to reach the maximum superheating temperature, T_s, because of statistical fluctuations in the volume and temperature of the sample. By gradually stepping up the simulation temperature, we were able to superheat a perfect crystal (with the same unit-cell geometry as that used above, however without the GB) to within about 80 K of T_s; beyond this temperature the crystal could not be stabilized. Figure 4 illustrates how rapidly a perfect crystal melts above T_s. After a step increase of the simulation temperature above T_s to 1700 K, only about 500 MD time steps (or about 10-20 lattice vibration periods) are required to destroy the long-range order within the (001) planes. This evidence suggests that the liquid phase is formed *homogeneously*, as one would expect from a phonon or elastic instability.

In the above discussion we have emphasized the role of the lattice instability in establishing a maximum superheating *temperature*, T_s, at zero external pressure (see Fig. 3). Because of the thermal expansion accompanying any temperature change, this temperature is related to a maximum *volume*, $V_s(T_s)$, above which the crystal cannot exist. To see the variation with volume alone, in Fig. 5 the results of Fig. 3 are replotted accordingly, from which the critical stability volume, $V_s(T_s)$, can be extracted. As expected, this volume expansion exceeds that associated with the thermodynamic melting point, $V_m(T_m)$.

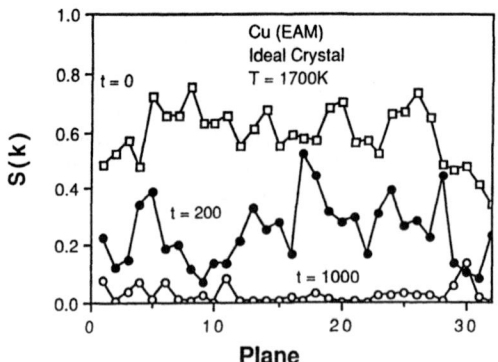

Figure 4. Mechanical melting of a perfect crystal, simulated by use of 3-d periodic borders, after suddenly increasing the temperature from 1300 to 1700 K. The three instantaneous slice-by-slice profiles of the planar static structure factor after 0, 200, and 1000 time steps show that planar order is lost simultaneously in all parts of the crystal.

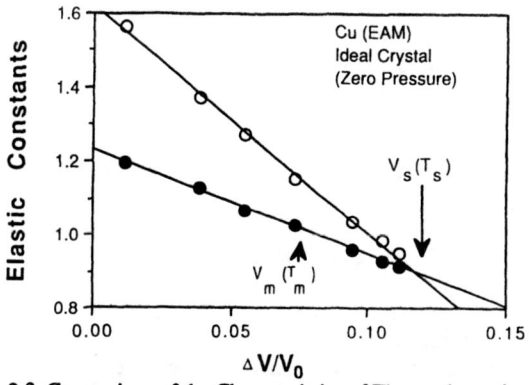

Figure 5. Elastic constants of Fig. 3 replotted vs. relative volume change, $\Delta V/V_0$ [with $V_0=V(T=0)$]. $V_s(T_s)$ is the maximum volume up to which, at $T=T_s$, crystalline order can be sustained. This volume exceeds that associated with the thermodynamic melting point, $V_m(T_m)$.

2.3 Comparison of the Characteristics of Thermodynamic and Mechanical Melting

The above simulations illustrate that every crystal, in principle, has two melting points, T_m and T_s. Conceptually the two transitions have distinct physical origins: whereas thermodynamic melting is governed by the free energies of the liquid and the solid phases, mechanical melting is triggered by a phonon instability. Since the volume expansion required for mechanical melting is always larger than that associated with thermodynamic melting [29], the free energy always favors thermodynamic over mechanical melting; i.e., $T_m < T_s$. However, as illustrated above, the former requires thermally activated atomic mobility and therefore may be kinetically hindered by slow atomic diffusion in the liquid phase. If a crystal is melted under atmospheric conditions (i.e., by conventional heating to melting), the thermodynamic state variables usually will be such that high atom mobility in the liquid enables the nucleation and growth of the liquid phase at extended defects. However, if a crystal is disordered at lower temperature (for example, by uniformly expanding the crystal), the consideration of limited atom mobility as a possible hindrance to phase change by thermodynamic processes may be of significant importance, particularly because atomic mobilities decrease *exponentially* with decreasing temperature but increase only approximately *linearly* with increasing volume. The crystal may therefore not be able to disorder at the volume specified by equilibrium thermodynamics until a larger volume is reached. The largest possible volume is the instability volume, V_s', associated with the ultimate ("mechanical") stability limit, where the crystal structure breaks down without change in volume.

To conclude, we summarize the four main distinguishing characteristics of thermodynamic and mechanical melting.

(a) Whereas thermodynamic melting, characterized by (T_m, V_m), is based on the free energies of both the crystalline and liquid states, mechanical melting is triggered by a phonon instability in the crystal lattice at a critical volume, V_s, associated with T_s.

(b) Thermodynamic melting requires the existence of extended defects at which the liquid phase can nucleate. By contrast, mechanical melting can occur with and without the presence of such defects, although one would expect that the combination T_s, V_s of thermodynamic state variables needed to trigger the phonon instability depends somewhat on whether or not the system contains extended defects and on the nature of these defects.

(c) The growth of the liquid phase into the crystal (by propagation of solid-liquid interfaces) requires thermally-activated diffusion in the liquid. Mechanical melting, by contrast, is caused by a phonon instability, and therefore happens typically within a few dozen lattice vibration periods; i.e., typically within about 10^{-12} sec, without requiring thermal atom mobility.

(d) As a consequence of (c), thermodynamic melting involves relatively slow kinetic processes in contrast to mechanical melting. Also, thermodynamic melting is a *heterogeneous* process, involving nucleation and growth of the liquid phase, whereas mechanical melting takes place *homogeneously*, without the need for the presence of lattice defects.

3. THE T-V PHASE DIAGRAM AND ITS LOW-TEMPERATURE EXTENSIONS

We can summarize our discussion of thermodynamic and mechanical melting by referring to a typical phase diagram of a monatomic system in the T-V plane as shown in Fig. 6. The phase-diagram representation is useful not only for expressing the relation between the thermodynamic variables at the two melting transitions, but also for discussing the underlying thermodynamic basis of the connection between melting and solid-state amorphization.

In Fig. 6 we have drawn the usual phase boundaries delineating the single-phase regions of crystal (C), liquid (L), and vapor (Vap) which, with one exception, require little comment. The condition for thermodynamic melting is expressed by the melting curve, $T_m(V)$, which terminates at the triple-point temperature, T_t. This is the lowest temperature, according to equilibrium thermodynamics, at which the crystal, at a volume V_t^C, can coexist with the liquid (and also with the vapor). The freezing curve, $T_f(V)$, which is in near-coincidence with the mechanical stability curve, $T_s(V)$, and lying more or less parallel to the melting curve, also terminates at T_t, where the liquid has a volume V_t^L.

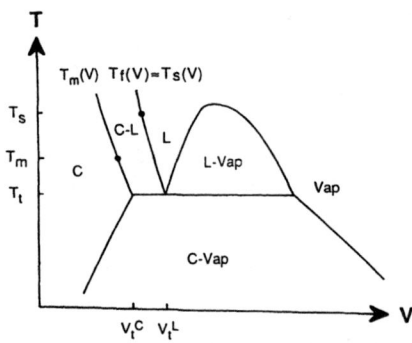

Fig. 6. Schematic temperature-volume phase diagram of a monatomic substance showing the single-phase regions of crystal (C), liquid (L), and vapor (Vap), and the various two-phase regions. On the horizontal triple line, at temperature T_t, the crystal (at volume V_t^C) and the liquid (at volume V_t^L) coexist with the vapor. The points on the thermodynamic melting line, $T_m(V)$, and the freezing curve, $T_f(V)$, indicate conditions of ambient pressure. To a good approximation, the freezing curve and the mechanical-stability line, $T_s(V)$, coincide.

While it is not clear whether the stability and freezing curves, $T_s(V)$ and $T_f(V)$, are the same, we believe they should lie rather close to each other. There also exists experimental evidence pointing to similar values for V_s and V_f [6]. Since in the present discussion it is not necessary to differentiate between the stability and freezing curves, we assume throughout that the two curves coincide.

It has been noted by Tallon [6] that an empirical expression for the variation of C' with volume change, ΔV, for isobaric heating can be written as

$$C'(T) = C'(0) \exp[-\gamma_G \Delta V(T)/V(0)] \quad , \qquad (3)$$

where γ_G is an effective Gruneisen parameter which is independent of temperature. Expansion of the exponential, and truncation after the linear term, according to

$$C'(T) = C'(0) [1-\gamma_G \Delta V(T)/V(0)] \quad , \qquad (4)$$

yields a vanishing $C'(T)$ at the critical volume expansion $\Delta V = V(0)/\gamma_G$, which corresponds approximately to the total volume expansion, $V_s(T) \sim V_s(0)$, needed to induce the mechanical instability. We will see below (see Fig. 7) that the variation of C' over the volume range of interest is not quite linear. Linearly extrapolated values [6] are therefore likely to be underestimates. While this does not affect our conclusion that the limiting volume for mechanical stability should be similar to the liquid volume at melting, it is, in fact, evidence supporting the argument that V_f cannot exceed V_s.

By definition, our discussion of melting in Sec. 2 pertains to the part of the phase diagram above the triple-point temperature. Although for a number of substances T_t is extremely close to T_m, the melting point at ambient pressure, in Fig. 6 we nevertheless show T_m separately for clarity of discussion. Correspondingly, T_s indicates the temperature for mechanical melting at ambient pressure.

3.1 Low-Temperature Extension of the Mechanical-Melting Curve

Although the equilibrium phase diagram in Fig. 6 indicates that there is no crystal-to-liquid transition below the triple-point temperature, one can define for $T < T_t$ a critical volume $V_s(T)$ at which the crystal, upon uniform volume expansion, becomes mechanically unstable. This means that the instability curve for mechanical melting in Fig. 6 can be extended to arbitrarily low temperature. To demonstrate this extension, we have calculated C' at zero temperature, using lattice dynamics. The results are shown as open circles in Fig. 7 in which C' is seen to vanish at a volume $V_s(0)$. This is the volume obtained when the stability curve is extended to zero temperature. Since the variation of C' with volume expansion can be appropriately determined at any temperature, whether by lattice dynamics or molecular dynamics, it follows that at arbitrary temperature $V_s(T)$ is a well-defined quantity, thus justifying the extension of the stability curve, $T_s(V)$, in Fig. 6 down to zero temperature.

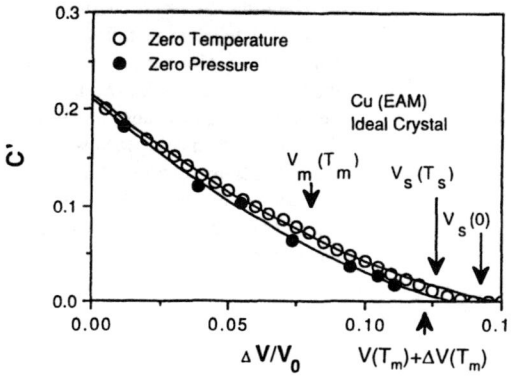

Fig. 7. Decrease of the minimal shear constant (in 10^{12} dyn/cm^2) defined in Eq. (3) as function of a relative volume increase, $\Delta V/V_0$, introduced either by heating towards mechanical melting (full circles; MD results) or by isotropic hydrostatic-stress induced volume expansion at T=0 (open circles; lattice-dynamics results).

In Fig. 7 we also show molecular dynamics results (filled circles) for heating at zero pressure. These data give a vanishing C at $V_s(T_s)$ which is somewhat smaller than $V_s(0)$ but slightly larger than the volume of the melt, $V_f(T_m)$. The close proximity of these values indicates that the freezing and stability curves are, indeed, close together, with the latter on the side of higher temperature at equal volume, or conversely, higher volume for equal temperature.

The extension of the mechanical-melting curve in Fig. 6 to zero temperature implies that the system can be maintained in its crystalline phase when its volume is less than $V_s(T)$. Yet on the basis of equilibrium thermodynamics, the region to the right of the sublimation curve in Fig. 6 is the two-phase region C-Vap in which the crystal and vapor coexist. The interpretation of our extended curve, $V_s'(T)$, therefore implies the existence of a metastable crystalline phase in which sublimation, a thermally activated process, is kinetically suppressed. Such a state can be produced, for example, by a relatively rapid isothermal volume expansion of the crystal. However, as we will argue below, this picture is incomplete because one also has to consider the extension of the thermodynamic melting curve.

3.2 Low-Temperature Extension of the Melting Curve

In our discussion of thermodynamic melting in Sec. 2, we have emphasized the importance of nucleation-and-growth kinetics. In considering a similar extension below T_t of the related melting curve, kinetics plays an equally crucial role. In Fig. 8 we show schematically the extended curve $T_m'(V)$ along with the previous extension $T_s'(V)$. Notice that neither the triple line separating the C-L region from the C-Vap region nor the sublimation curve appear in this diagram. The reason is that both phase separations can be suppressed when kinetics is taken into account. As a result, the region lying to the left of $T_m'(V)$ is the effective single-phase region for the crystalline state; however, it is important to keep in mind that the region between the original subli- mation curve and the extended part of the melting curve represents a crystal existing as a metastable overexpanded solid. Similarly, the extended two-phase region below the triple-point temperature represents the region where the metastable overexpanded solid can co-exist with the metastable supercooled liquid .

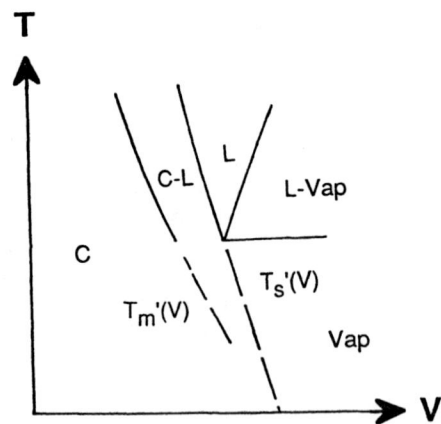

Fig. 8. Generalized T-V phase diagram showing the extensions of both the thermodynamic-melting and the mechanical-stability (or freezing) curves below the triple-point temperature. Along the extension of the former, an expanded crystal becomes unstable against the disor- dered phase by heterogeneous amorphization; by contrast, along the extension of the mechanical-melting curve, homogeneous amorphization can occur.

3.3 Implications for Solid-State Amorphization

The effective phase diagram we have proposed in Fig. 8 has immediate implications for solid-state amorphization. In this picture, the C-A transformation is viewed as an isothermal melting process driven by volume expansion. For solid-state amorphization, volume expansion driven by external forces at constant temperature plays the same role as that of the thermal expansion during heating to melting at constant pressure. The extended melting and freezing curves depicted in Fig. 8, therefore expose the essential physics behind the parallels between the two disordering processes in a particularly clear and simple way. When the expansion is large enough to cross the curve $T_m'(V)$, the crystal becomes unstable against structural disordering, and heterogeneous amorphization may take place. Further, if one crosses the $T_s'(V)$ curve, disordering will occur through the process of homogeneous amorphization, and the two processes can compete with each other.

Finally, we mention that there is considerable experimental evidence, discussed in detail in Refs. 7 and 30, that solid-state amorphization, the process in which the long-range crystalline order is destroyed by external means (such as mechanical or chemical means, or by irradiation), can proceed by the same two distinct mechanisms as melting and that, in contrast to conventional melting, both types of transition can actually be observed. In a typical solid-state amorphization experiment, the temperature is held fixed at some relatively low value, well below T_m and T_t. The role of the irradiation, or of the mechanical or chemical means, in inducing the crystal-to-amorphous transition is to expand the crystal lattice to the coexistence point in phase space where the thermodynamic transition can, in principle, occur. However, relatively low atom mobility gives rise to a competition between the heterogeneous and homogeneous processes, a competition governed by the level of atomic mobility at that point in phase space.

4. OUTLOOK FOR THIN-FILM MATERIALS

The elastic behavior of thin films and superlattices may differ dramatically from that of bulk materials, as evidenced by the existence of the "supermodulus effect", i.e., the anomalous strengthening at small modulation wavelengths or film thicknesses, Λ, of some elastic moduli coupled with the softening of others [31-33]. Based on this experimental evidence, and on insights gained from computer simulations in which the distinct effects of the interfacial stresses and the atomic-level structural disorder at the interfaces were separated [34,35], one would expect both the thermodynamic and mechanical melting points of these materials, as well as their thermodynamic and thermo-elastic properties, to depend strongly on Λ (or, in nanophase materials, on the grain size). Although to date the above ideas have not been applied to thin-film materials or to two-component systems, such extensions appear highly desirable and timely, particularly since all the basic theoretical tools and concepts seem to be in place.

Acknowledgments. We have benefitted from discussions with P. R. Okamoto, S. R. Phillpot, J. F. Lutsko and M. Kluge. This work was supported by the U.S. Department of Energy, BES Materials Sciences, under Contract W-31-109-Eng-38.

References

1. For recent reviews, see W. L. Johnson, Prog. Mater. Sci. **30**, 81 (1986); *Solid-State Amorphizing Transformations*, R. B. Schwartz and W. L. Johnson, eds., Elsevier Sequoia, Netherlands, 1988; J. Less-Common Met., **140** (1988).
2. R. W. Cahn and W. L. Johnson, J. Mater. Res. **1**, 724 (1986).
3. P. Richet, Nature **331**, 56 (1988).
4. H. J. Fecht and W. L. Johnson, Nature **334**, 50 (1989).
5. P. R. Okamoto, L. E. Rehn, J. Pearson, R. Bhadra, and M. Grimsditch, J. Less-Common Met. **140**, 231 (1988).
6. J. L. Tallon, Phil. Mag. **39**, 151 (1979); J. L. Tallon and W. H. Robinson, Phil. Mag. **36**, 741 (1977); J. L. Tallon, J. Phys. Chem. Solids **41**, 837 (1984).
7. D. Wolf, P. R. Okamoto, S. Yip,. J. F. Lutsko and M. Kluge, J. Mat. Res. **5**, 286 (1990).
8. R. L. Cormia, J. D. Mackenzie and D. Turnbull, J. Appl. Phys. **34**, 2239 (1963).
9. A. R. Ubbelohde, *Molten State of Matter: Melting and Crystal Structure*, Wiley, Chichester, 1978.
10. J. Daeges, H. Gleiter, and J. H. Perepezko, Phys. Lett. **A119**, 79 (1986); R. W. Cahn, Nature **323**, 668 (1986).
11. R. M. J. Cotterill, J. Cryst. Growth **48**, 582 (1980).
12. M. Born and K. Huang, *Dynamical Theory of Crystal Lattices* , Oxford, 1962.
13. L. L. Boyer, Phase Transitions **5**, 1 (1985).
14. R. W. Cahn, Nature **273**, 491 (1978).
15. R. W. Cahn, Nature **323**, 668 (1986).
16. N. G. Ainslie, J. D. MacKenzie and D. Turnbull, J. Phys. Chem. **65**, 1718 (1961).
17. P. Buffat and U.-P. Borel, Phys. Rev. **A13**, 2287 (1976).
18. J. B. Boyce and M. Stutzmann, Phys. Rev. Lett. **54**, 562 (1985).
19. C. J. Rossouw and S. E. Donnelly, Phys. Rev. Lett. **55**, 2960 (1985).
20. M. S. Daw and M. I. Baskes, Phys. Rev. Lett. **50**, 1285 (1983); Phys. Rev. B **29**, 6443 (1984).
21. S. M. Foiles, Phys. Rev. B **32**, 7685 (1985).
22. J. F. Lutsko, D. Wolf, S. R. Phillpot and S. Yip, Phys. Rev. B **40**, 2841 (1989).
23. J. F. Lutsko, D. Wolf, S. Yip, S. R. Phillpot, and T. Nguyen, Phys. Rev. B **38**, 11572 (1988).
24. D. Wolf, J. de Phys. Colloq. C4 **46**, C4-197 (1985).
25. J. Frenkel, Phys. Z. Sowjetunion **1**, 498 (1932).
26. J. Q. Broughton, G. H. Gilmer and K. A. Jackson, Phys. Rev. Lett. **49**, 1496 (1982).
27. M. Parrinello and A. Rahman, J. Appl. Phys. **52**, 7182 (1981).
28. J. Ray and A. Rahman, J. Chem. Phys. **80**, 4423 (1984), and Phys. Rev. B **32**, 733 (1985).
29. F. F. Abraham, Adv. Phys. **35**,1 (1986).
30. W. J. Meng, P. R. Okamoto, L. J. Thompson, B. J. Kestel, and L. E. Rehn, Appl. Phys. Lett. **53**, 1820 (1988); P. R. Okamoto and M. Meshii, in *Science of Advanced Materials*, H. Wiedersich and M. Meshii, eds., ASM, Metals Park, 1990.
31. W. M. C. Yang, T. Tsakalakos, and J. E. Hilliard, J. Appl. Phys. **48**, 876 (1977).
32. B. M. Clemens and G. L. Eesley, Phys. Rev. Lett. **61**, 2356 (1988).
33. See, for example, *Science of Composite Interfaces*, R. G. Brandt and S. Fishman, eds., Mat. Sci. Eng. B **126** (1990) and references therein.
34. D. Wolf and J. F. Lutsko, J. Mat. Res. **4**, 1427 (1989); Phys. Rev. Lett. **60**, 1170 (1988) and J. Appl. Phys. **66**, 1961 (1989).
35. J. A. Jaszczak, S. R. Phillpot, and D. Wolf, J. Appl. Phys. **68**, 4573 (1990).

KINETICS OF SOLID-STATE REACTIONS IN Ni-Zr THIN FILMS

R.B. SCHWARZ and J.B. RUBIN
Center for Materials Science and MST-7, Los Alamos National Laboratory, MS K765, Los Alamos, NM 87545

ABSTRACT

We have studied the kinetics of the solid-state amorphizing reaction in thin film multilayers of Ni and Zr. Crystalline Ni and Zr films were deposited in ultra-high vacuum onto platinum resistance thermometers embedded in alumina. An electronic feedback circuit controls the temperature of the substrata by adjusting the power dissipated by the platinum resistors. We find that structural relaxation in the as-deposited Ni and Zr films affects the initial stages of the reaction. For long reaction times there is a discontinuous change in the reaction rate. The time to reach this transition increases with film thickness and depends exponentially on $1/T$, with an apparent activation energy of 3 eV atom^{-1}.

INTRODUCTION

A solid-state amorphizing reaction (SSAR) has been shown to occur [1,2] at the interfaces of two pure metals A and B having the following characteristics: (1) metals A and B have a large negative heat of mixing in the amorphous state and (2) metals A and B have vastly different diffusivities in each other and in the amorphous alloy to be formed. The first condition provides the thermodynamic driving force for the reaction, whereas the second condition ensures that the SSAR is kinetically favored over the nucleation and growth of crystalline phases.

Measurements of the SSAR kinetics in early and late transition metal multilayers [3,4,5,6] have identified three regimes. For short annealing times the reaction rate is low. Data in this region are sparse because of the difficulty in bringing the multilayers to the reaction temperature within a short time interval. However, Rutherford backscattering (RBS) [7,8] and resistivity measurements [9,10] suggest that there is a time delay for the initiation of the reaction and this delay has been taken as evidence for an interfacial reaction barrier. Other possible explanations for the initial delay in the SSAR are based on grain growth [11] and the release of mechanical stresses [12] in the films. Both mechanisms would produce changes in the film properties during the initial stage of annealing.

For intermediate times the amorphous layer thickness increases proportionally to $t^{1/2}$. It is well established that for this regime the SSAR rate is limited by the diffusion of the smaller late transition atom in the amorphous layer already formed [13,14,15].

For long reaction times, the kinetics change again, showing a decreased reaction rate. This third stage has been attributed to the exhaustion of crystalline material [16], formation of Kirkendall voids [11], relief of mechanical stress [17,18], and to the establishment of the terminal solubility in the early transition metal [8].

We present measurements of the SSAR kinetics in vacuum-deposited thin film multilayers of nickel and zirconium. To study the early stages of the reaction we developed a substrate system of fast thermal response and high stability. The reaction kinetics are deduced from resistivity measurements obtained during isothermal and constant-heating-rate annealings.

Mat. Res. Soc. Symp. Proc. Vol. 230. ⌐1992 Materials Research Society

EXPERIMENTAL

The Ni/Zr multilayers were prepared by electron-beam evaporation in ultra-high vacuum onto alumina substrata. The substrata incorporate a platinum thin film resistor. These resistance thermometers [19] are 0.125" x 0.400" x 0.030" and weigh 90 mg. The substrate surface on which the films are deposited is polished with $0.25\,\mu$m diamond paste. A pad of tungsten, 2.5 mm long and 150 nm thick, is then vacuum deposited onto either end of this surface. These pads are used as electrical contacts to measure the resistance of the Ni/Zr films using a 4-point AC technique. Tungsten was chosen because of its low diffusivity in both crystalline zirconium and amorphous NiZr alloy [20,21]. The current and voltage leads are 0.001" diameter platinum wires, which are laid across the tungsten pads and attached with a thin coating of colloidal graphite. Low thermal conductivity fiberglass straps are used to hold the substrate in the vacuum chamber. Each assembly is baked in situ for at least 20 h at 700°C to remove adsorbed water.

The platinum substrate resistor is connected as one arm of a Wheatstone bridge, the other three arms being resistors with a near-zero temperature coefficient of resistance. For the platinum resistor, $(1/R)(dR/dT) = 0.00385\ C^{-1}$. An electronic feedback circuit [22] detects any imbalance in the bridge and increases or decreases the voltage applied to the bridge accordingly. An increase in the applied voltage increases the power dissipated by the platinum resistor and thus increases its temperature and resistance. Because of the low thermal mass of the resistor, thermal equilibrium following a change in the set point is achieved within a few seconds. Notice that there is no need for a separate heater and thermometer because the platinum resistor fulfills both requirements.

The substrata were heated repeatedly to 700°C without hysteresis error. The accuracy of the temperature measurement was verified by measuring the resistance of a vacuum-deposited 85 nm-thick nickel film during heating at a constant rate of 10 K min^{-1}. A clear change in dR/dT occurs at 358°C which corresponds to the Curie temperature of nickel [23]. The present thermal system represents a substantial improvement in the accuracy of surface temperature measurements. To demonstrate this point, we also measured the temperature of the nickel film by attaching a 0.001" diam. Pt/Pt-10% Rh thermocouple (the thinnest we could work with) to the nickel film using the colloidal graphite. For a substrate temperature of 300°C the thermocouple measured only 289°C, with the error increasing for higher set temperatures. Errors of this magnitude cannot be neglected when studying reaction kinetics, which depend exponentially on temperature. Even larger errors are expected when heating, in vacuum, ceramic or glass substrata which are held by simple mechanical contact against a metal block of known temperature.

The base pressure before thin film deposition was below $1x10^{-9}$ torr, and during deposition rose to $5x10^{-8}$ torr. The films were deposited with the substrata kept at ambient temperature. Following deposition, the set point of the substrate temperature controller was increased in one step and a stable reaction temperature was reached within 20 s.

RESULTS

The curves in Figs. 1 - 3 identify three kinetics regimes labeled (a), (b), and (c) [Notice the different scales for the abscissa]. The temperature dependence of regime (a) changes with annealing temperature; whereas for 300°C (Fig. 1) 1/R in regime (a) has a $t^{1/2}$ dependence, at 350°C (Fig. 3) 1/R versus $t^{1/2}$ has a clear convex curvature. One possible explanation for this convex dependence is that the resistance of the unreacted Ni and Zr films changes concurrently with the SSAR taking place at the Ni/Zr interface. To check this, we annealed a 400 nm-thick zirconium film at 330 °C. The change in 1/R of the as-deposited film is shown in Fig. 4 as a function of $t^{1/2}$, where t is the time from the end of the deposition. Although a measurement has not been made, a similar change in 1/R is expected to occur in the nickel films following deposition.

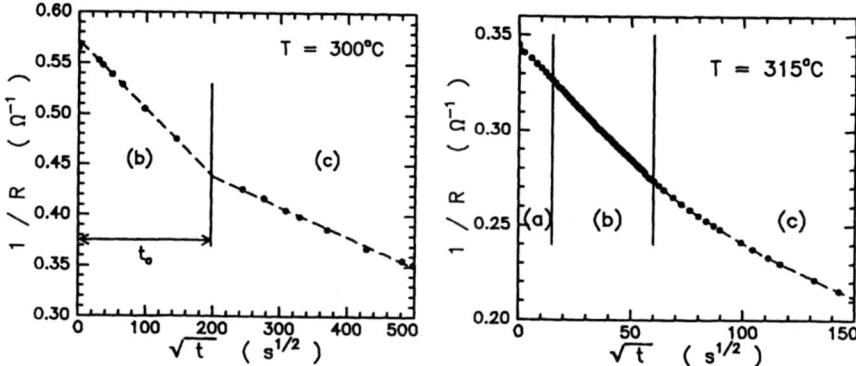

Fig. 1. (left) Change in 1/R of a Ni-Zr bilayer annealed isothermally at 300°C.

Fig. 2. (right) Change in 1/R of a Ni-Zr bilayer annealed isothermally at 315°C.

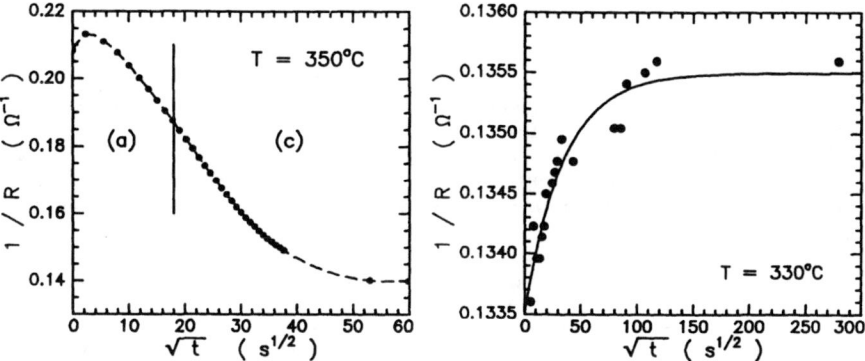

Fig. 3. (left) Change in 1/R of a Ni-Zr bilayer annealed isothermally at 350°C.

Fig. 4. (right) Change in 1/R of a 400 nm-thick Zr film annealed isothermally at 330°C.

The intermediate regime, (b), is characterized by a linear $t^{1/2}$ dependence which can be clearly seen in Figs. 1 and 2. This time dependence has been discussed earlier [13-15]. At the highest reaction temperature of 350°C (Fig. 3), regime (b) becomes narrow and there seems to be a continuous transition from regime (a) to (c).

The transition between regimes (b) and (c) is defined as the annealing time, t_0, at which 1/R deviates from a $t^{1/2}$ dependence. We investigated the dependence of t_0 on film thickness and reaction temperature. For the former, we reacted three bilayers, having thicknesses in the ratios of 1:2:4, at 330°C. Although the data in Fig. 5 are not sufficient to discern the analytical dependence of $t_0(x)$, they do show that t_0 increases with increasing x. The dependence of t_0 on temperature was investigated on bilayers of constant thickness. Figure 6 shows that t_0 is linearly dependent on 1000/T, suggesting that t_0 is determined by a thermally activated process with an activation energy close to 3 eV atom^{-1}.

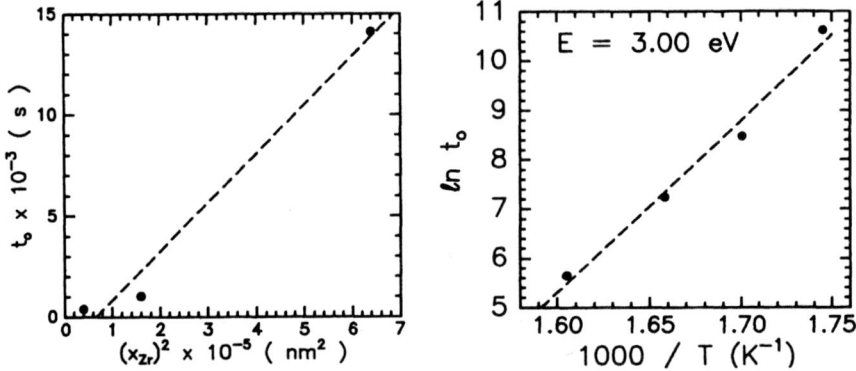

Fig. 5. (left) Time t_0 denoting the transition between annealing regimes (b) and (c) at 330 C, as a function of the square of the Zr layer thickness.

Fig. 6. (right) Time t_0 denoting the transition between annealing regimes (b) and (c) as a function of reciprocal temperature.

DISCUSSION

Platinum resistance thermometers embedded in high purity alumina, and an electronic feedback circuit to control the power dissipated by these resistors, provides an ideal system for studying the kinetics of the solid-state reactions in thin films at temperatures between 300 and 1000 K. Because these devices serve both as heaters and thermometers, only two electrical leads into the vacuum chamber are necessary for their operation.

The as deposited Zr film is structurally unstable and heating it to the SSAR reaction temperature enables it to relax, thereby decreasing it resistance. The data in Fig. 4 shows that at 330°C, 1/R increases rapidly, approaching a saturation value after approximately 3 h. It is this rapid rate of change that makes it difficult to study the initial kinetics of the SSAR. Possible causes for the change in 1/R include (a) recovery and/or recrystallization of the films, (b) relaxation of internal stresses in the films, and (c) annihilation of defects, such as mono- or di-vacancies, which are likely to be trapped into the films during condensation from the vapor phase. The total change in 1/R for the zirconium film is approximately 1.5% whereas the initial increase in 1/R for the

Ni/Zr bilayer annealed at 350°C is approximately 2%. If the short-lived increase in 1/R due to a structural relaxation in the zirconium film is superimposed onto the 1/R change due to the SSAR, which is linear in $t^{1/2}$, the resultant 1/R vs $t^{1/2}$ curve would be similar to that shown in Fig. 3, which shows a maximum at very short reaction times. If the increase in 1/R is weaker, as appears to be the case for lower reaction temperatures, then the measured 1/R curve may not have a maximum but may instead appear as if there is a time delay for the onset of the reaction. The structural relaxation in the as-deposited films affects not only the initial stage of the SSAR; notice in Fig. 4 that at 330°C the relaxation continues for several hours, and thus should also affect, to a lesser degree, annealing stages (b) and (c).

The present measurements show that recovery affects the kinetics of SSARs in thin films. Researchers [9] have tried to circumvent structural relaxation effects in SSARs by depositing the thin films at the reaction temperatures. Because the time required for structural recovery is significantly longer than the deposition time, it is unlikely that these effects can be entirely eliminated. A comparison of the shape of the 1/R curve within region (a) in Figs. 1 - 3 suggests that to minimize the recovery effects, experiments which purport SSAR kinetics measurements should be made at the lowest possible temperature.

The transition from regimes (b) to (c) at time t_o, clearly seen in Fig. 2 as a discontinuity in the rate of change of 1/R, corresponds to a slowing of the SSAR rate. The time to reach this transition depends on the thickness of the as-deposited films and on temperature. Notice, however, that the thicknesses of the present nickel and zirconium films greatly exceeds that which can be reacted by a SSAR [24,25], so that the exhaustion of one or both materials cannot be the cause for the abrupt decrease in the reaction rate. The temperature dependence of t_o suggests a diffusion-controlled process with activation energy of 3 eV atom^{-1}. This large activation energy rules out any possible participation of the nickel atoms, which have activation energies for diffusion in crystalline Zr and amorphous NiZr close to 1 eV atom^{-1}. The magnitude of the activation energy suggests a recovery process in the zirconium film controlled by self diffusion, which has an activation energy of 2.77 eV [26].

Although the present measurements show that recovery effects are important and can explain some of the features of the 1/R versus $t^{1/2}$ curves, it appears that additional mechanisms are required to explain the time delay for the initiation of the SSAR observed by van Rossum et al. [4] in Ni/Hf films using Rutherford backscattering techniques. This is because Rutherford measurements should be insensitive to the relaxation state of the films.

ACKNOWLEDGEMENTS

We thank the Los Alamos Ion Beam Materials Laboratory for assistance in the characterization of the films by Rutherford backscattering. This work was supported by the U.S. Department of Energy, Office of Basic Energy Sciences, and by Laboratory Directed Research and Development funds.

REFERENCES

1. R. B. Schwarz and W. L. Johnson, *Phys. Rev. Lett.* 51, 415 (1983).
2. R. B. Schwarz, *MRS Bulletin May/June*, 55 (1986).
3. H. Schroder and K. Samwer, *J. Mater. Res.* 3, 461 (1988).

4. M. Van Rossum, M.-A. Nicolet, and W. L. Johnson, *Phys. Rev.* B 29, 5498 (1984).
5. R. J. Highmore, J. E. Evetts, A. L. Greer, and R. E. Somekh, *Appl. Phys. Lett.* 50, 566 (1987).
6. K. M. Unruh, W. J. Meng, and W. L. Johnson, *Mater. Res. Soc. Symp. Proc.* 37, 551 (1985).
7. W.L. Johnson, M. Atzmon, M. Van Rossum, B.P. Dolgin, and X.L. Yeh, in *Rapidly Quenched Metals*, eds. S. Steeb and H. Warlimont, (North Holland, Amsterdam, 1985), Vol. 5, p. 1515.
8. K. Pampus, J. Boettinger, B. Torp, H. Schröder, and K. Samwer, *Phys. Rev.* B 35, 7010 (1987).
9. K. Samwer, H. Schröder, and K. Pampus, *Mat. Sci. Eng.* 97, 63 (1988).
10. P. Guilmin, P. Guyot, and G. Marchal, *J. de Physique* 46, C8-485 (1985).
11. H. Schröder, K. Samwer, and U. Köster, *Phys. Rev. Lett.* 54, 197 (1985).
12. K. Pampus, K. Samwer, J. Boettinger, H. Schröder, and B. Torp, *Z. Phys. Chem.* 157, 251 (1988).
13. J. J. Highmore, J. E. Evetts, A. L. Greer, and R. E. Somekh, *Appl. Phys. Lett.* 50, 566 (1987).
14. W. J. Meng, E. J. Cotts, W. L. Johnson, *Mat. Res. Soc. Symp. Proc.* 77, 223 (1987).
15. J.B. Rubin and R.B. Schwarz, *Appl. Phys. Lett.* 55, 36 (1989).
16. M. Van Rossum, M.-A. Nicolet, and W.L. Johnson, *Phys. Rev.* B 29, 5498 (1984).
17. H.U. Krebs and K. Samwer, *Europhys. Lett.* 2, 141 (1986).
18. K.P. Unruh, W.L. Johnson, A.P. Thakoor, and S.K. Khanna, *Mater. Res. Soc. Symp. Proc.* 37, 551 (1985).
19. Platinum Resistance Thermometers, type TFD, from Omega Engineering, Stamford, CT.
20. G. M. Hood and R. J. Schultz, *Acta Metall.* 22 459 (1974).
21. J. Bottiger, K. Dyrbye, K. Pampus, and B. Torp, *Int. J. Rapid Solidification* 2, 191 (1986).
22. R. B. Schwarz and J. Rubin, *J. Sci. Instrum.* (submitted, 1991).
23. B. D. Cullity, *Magnetic Materials* (Addison-Wesley, Reading, MA, 1972), p.122.
24. W. J. Meng, C. W. Nieh, and W. L. Johnson, *Appl. Phys. Lett.* 51, 1693 (1987).
25. R. B. Schwarz, J. B. Rubin, and T. J. Tiainen, in *Science of Advanced Materials*, edited by H. Wiedersich and M. Meshii (ASM Intl., Materials Park, OH, 1990), p.1.
26. *Diffusion and Defect Data*, eds. F. Wohlbier and D. J. Fischer (Trans. Tech., Aedermannsdorf).

METASTABLE PHASE EQUILIBRIA IN CO-DEPOSITED $Ni_{1-x}Zr_x$ THIN FILMS

J. B. RUBIN and R. B. SCHWARZ

Center for Materials Science and MST-7, Los Alamos National Laboratory, MS K765, Los Alamos, NM 87545

ABSTRACT

We determine the glass forming range (GFR) of co-deposited $Ni_{1-x}Zr_x$ (0 < x < 1) thin films by measuring their electrical resistance during in situ constant-heating-rate anneals. The measured GFR is continuous for 0.10 < x < 0.87. We calculate the GFR of Ni-Zr melts as a function of composition and cooling rate using homogeneous nucleation theory and a published CALPHAD-type thermodynamic modeling of the equilibrium phase diagram. Assuming that the main competition to the retention of the amorphous structure during the cooling of the liquid comes from the partitionless crystallization of the terminal solid solutions, we calculate that for $dT/dt = 10^{12} \, K \, s^{-1}$, the GFR extends to x = 0.05 and x = 0.96. Better agreement with the measured values is obtained assuming a lower 'effective' cooling rate during the condensation of the films.

INTRODUCTION

The formation of amorphous alloys by rapid solidification techniques requires that crystallization be suppressed during the undercooling of the liquid from its melting temperature, T_m, to the glass transition temperature, T_g. The GFR, or range of compositions over which this undercooling is possible, is determined by a combination of kinetic and thermodynamic factors. For alloys between an early- and a late-transition metal and for conventional quenching methods such as melt-spinning, where the cooling rates are on the order of $10^7 \, K \, s^{-1}$, the GFR is usually discontinuous, being composed of distinct composition ranges centered near deep eutectics. As the cooling rate is increased, however, these individual composition ranges widen and, for extremely high cooling rates, may overlap to include all intermediate compositions, excluding only the terminal solution regions [1]. The formation of metallic films by condensation from the vapor phase represents an extreme case of rapid quenching, with an effective cooling rate on the order of $10^{12} \, K \, s^{-1}$. It is expected, therefore, that the GFR for thin films of $Ni_{1-x}Zr_x$ produced by co-evaporation will be continuous over a wide region $x_1 < x < x_2$, bounded by narrow regions of crystalline solid solutions for $0 < x_1$ and $x_2 < x < 1$.

In the present work we investigate the GFR of $Ni_{1-x}Zr_x$ (0 < x < 1) thin films prepared by the simultaneous condensation of nickel and zirconium vapors in ultra-high vacuum. We use in-situ electrical resistance measurements during constant heating-rate annealing to detect amorphous → crystalline transformations in the as-deposited films. This data is used to deduce the GFR. We also calculate the GFR by a model to be described in the next section.

CALCULATION OF THE GFR

Theoretical modeling of the GFR is important for understanding the basic mechanism for glass formation, and continues to be an area of active research. The basic ideas behind the calculations of the GFR are explained with the help of Figure 1 which shows the composition-temperature diagram for a hypothetical A-B binary alloy.

The equilibrium phases are the terminal solutions α and β, the congruent-melting compound γ, and the liquid λ. Also shown are the four T_o curves (dashed lines), defined as the composition-temperature loci at which the free energy of the liquid equals that of the solid phase of the same composition. Therefore, below each T_o curve there exists a thermodynamic driving force for partitionless crystallization. The importance of the T_o curves derives from the observation that under conditions of rapid cooling, partitionless solidification is almost always kinetically favored over solidification involving solute partitioning. Thus, at those compositions where partitionless crystallization can be avoided while the melt is cooled to below T_g, it should be possible to kinetically trap the alloy in the glassy (amorphous) state. Figure 1 also shows an assumed, composition-invariant T_g (dotted line). The T_o criterion predicts that the GFR for our hypothetical binary alloy will consist of the two narrow regions which solidify into the λ phase, as shown at the bottom of the figure. At all other compositions the undercooled liquid crystallizes partitionlessly into either α, β, or γ.

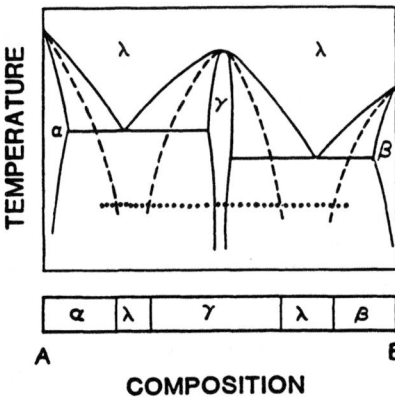

Fig. 1. Temperature vs. composition diagram for a hypothetical metallic A-B binary system.

The prediction of glass formation based on T_o curves, however, does not consider kinetics; it assumes that partitionless crystallization occurs instantaneously once it is thermodynamically favored. In practice, a finite undercooling is required to produce an observable nucleation rate and thus the GFR must necessarily depend on the cooling rate. To include the kinetics of solidification in the calculations of the GFR, several authors [2,3] have used isothermal transformation kinetics to calculate Time-Temperature-Transformation (TTT) curves or Continuous-Cooling-Transformation (CCT) curves [4]. Nash and Schwarz [5] have extended the method by combining homogeneous nucleation theory [6] with a CALPHAD modeling of the binary phase diagram to derive CCT curves for NiTi alloys as a function of cooling rate, dT/dt, and alloy composition, x. The results are used to calculate the temperature $T'(x, \varsigma, dT/dt)$ for the formation of a volume fraction ς of crystalline material during cooling at the constant rate dT/dt. Finally, defining a critical value of ς below which the solid can be defined as amorphous, they determined the GFR for rapidly solidified Ni-Ti melts by the intersection of the T' curves with T_g. The GFR found by this T' criterion is wider than that predicted from purely thermodynamic arguments.

Figure 2 shows the T_o curves (dashed lines) and T' curves (solid lines) for terminal solutions of Ni-Zr alloys, calculated by the method described in the last paragraph. The expressions for the excess free energy of mixing for the liquid, FCC, and HCP phases were obtained from Saunders [7], while the lattice stabilities were taken from de Boer et al. [8]. The glass transition temperature T_g was approximated by the crystallization temperature T_x, which is plotted based on literature values [9-13]. The compositions of congruent-melting and high-temperature peritectic compounds for this system are shown by vertical dotted lines. The equilibrium melting- or decomposition-temperature for each compound is indicated by the filled square at the top of the dotted line. For each intermetallic phase we may calculate a pair of T_0 and T' curves similar to those shown for the terminal solutions. However, experience has shown [1] that for cooling rates of the order of 10^{12} K s^{-1}, the main competition to glass formation comes from the terminal solutions. Extrapolating a curve drawn through the T_g values (dash-dot line in Fig. 2) to intersect the two T' curves of the terminal solutions for a cooling rate of 10^{12} K/s gives a GFR which extends from $0.05 < x < 0.96$.

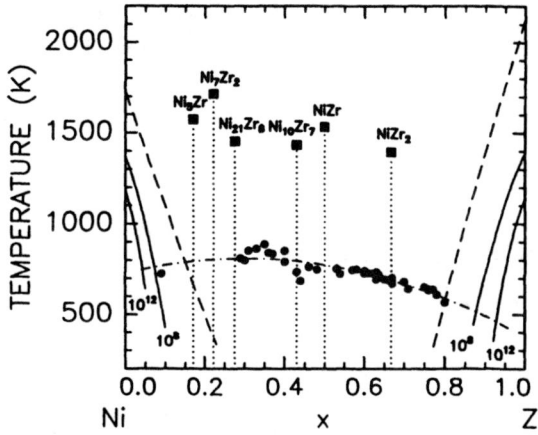

Fig. 2. Calculated T_0 curves for the terminal solutions (dashed) and T' curves at cooling rates of 10^8 and 10^{12} K s^{-1} (solid). Filled circles are crystallization temperatures for melt-quenched $Ni_{1-x}Zr_x$ alloys.

EXPERIMENTAL

$Ni_{1-x}Zr_x$ films were produced by co-evaporation from two electron-beam sources in a UHV system at deposition pressures below 5×10^{-8} torr. Each source was controlled by a thickness-rate monitor, enabling us to control the alloy composition during deposition. The alloy deposition rates varied from 0.3 nm s^{-1} for the Ni-rich films, to 1.0 nm s^{-1} for Zr-rich compositions. All films had a total thickness of 400 nm. The depositions were made onto alumina substrata which were polished with 0.25 μm diamond paste. Embedded in these substrata were calibrated platinum thin-film resistors which were used both to heat the substrata and to measure the film temperature. The experimental design and measurement circuit are described in the previous paper of this symposium [14]. Following deposition, the films were annealed in-situ at a rate of 10 K min^{-1} at pressures below 10^{-10} torr. During the anneals, the resistance of the films was measured by an AC four-point probe method. Afterwards, the films were removed from the vacuum chamber and the compositions and thickness determined by Rutherford Backscattering Spectrometry (RBS).

RESULTS

The resistivities as a function of temperature for the various alloy show distinct features. Fig. 3 shows the resistivity versus temperature curve (normalized to its value at 60°C) for a sample of approximately equiatomic composition. Initially, the resistivity decreases linearly with temperature, a to b, which is characteristic of an amorphous structure. Crystallization occurs at about 400°C, b to c, resulting in an abrupt drop in resistivity. This is followed by a second small drop in resistivity at about 495°C. At d, the film is fully crystalline, and further temperature cycling, d ↔ e, gives a reproducible resistance vs. temperature curve of positive slope.

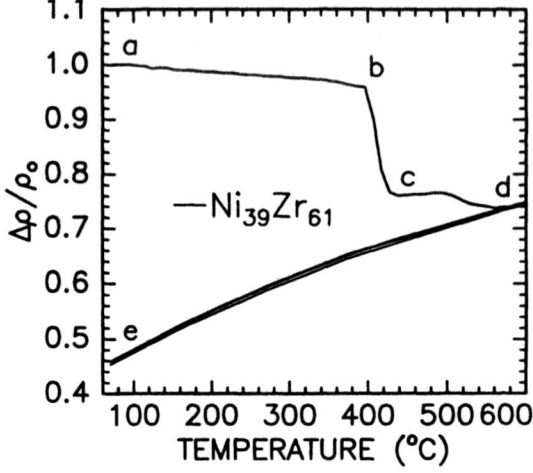

Fig. 3. Electrical resistivity vs temperature curve for a 400-nm thick amorphous $Ni_{39}Zr_{61}$ film annealed at 10 K min^{-1}.

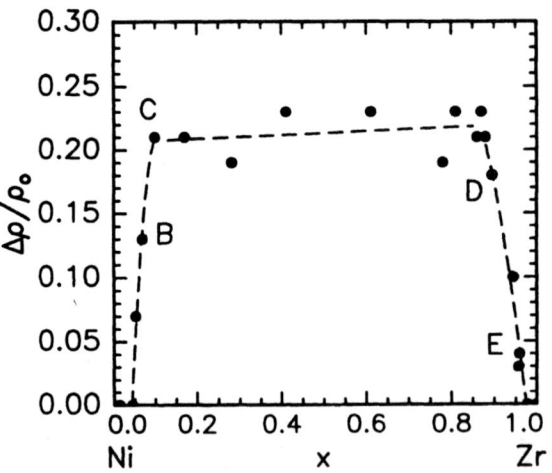

Fig. 4. Change in electrical resistivity caused by crystallization of $Ni_{1-x}Zr_x$ films as a function of composition. The intersection of the dashed lines with $\Delta\rho/\rho$ gives the limits of the two metastable terminal solutions.

Fig. 4 show the resistivity drop accompanying crystallization as a function of alloy composition. This resistivity drop should be proportional to the amount of amorphous phase present. This Figure identifies five composition regions. For $0 < x < 0.04$ and $0.97 < x < 1$, the as-deposited films are crystalline and $\Delta\rho/\rho = 0$. For $0.10 < x < 0.87$ the films are single-phase amorphous. The two narrow regions $0.04 < x < 0.10$ and $0.87 < x < 0.97$ represent transition regions where the films most likely contain both crystalline and amorphous phases. This coexistence of phases may reflect local compositional fluctuations.

DISCUSSION

A comparison of the calculation (Fig. 2) with experiment (Fig. 4) allows us to estimate an 'effective' cooling rate for the formation of amorphous Ni-Zr thin films by co-deposition in vacuum. Extrapolating the T_g values to intersect the T' curves for a cooling rate of 10^{12} K s^{-1} gives a GFR of $0.05 < x < 0.96$, while for dT/dt = 10^8 K s^{-1}, the predicted GFR is $0.08 < x < 0.92$. The experiments give a range of $0.10 < x < 0.87$. Better aggreement would require assuming an 'effective' cooling rate lower than 10^8 K s^{-1}, which seems unrealistic. Although the comparison of the calculated and measured GFRs enables us to define and 'effective' cooling rate, it is not clear that this rate is meaninfull in the context of the formation of amorphous alloys by the RS of melts.

The GFR predicted by our model assuming an 'effective' cooling rate of 10^{12} K s^{-1} overestimates the actual composition range for glass formation. We discuss next possible reasons for the discrepancy.

The volume fraction of crystal, ς, used to define the GFR, is calculated by integrating the homogeneous nucleation rate equation [4,5],

$$I_h = A \cdot \exp(-K \, \sigma^3 / \, \Delta G^2) . \tag{1}$$

The calculated GFR is therefore quite sensitive to the values chosen for the free energy difference between the liquid and crystal, ΔG, and the interfacial free energy, σ. Errors in ΔG may arise from the uncertainties in the assessed equilibrium phase diagram, particularly in the position of the liquidus lines. Any such errors would be magnified by the large temperature range over which the thermodynamic modeling of these phases must be extrapolated to reach the temperature regimes used in the present calculations. Another possible source of error may lie in our estimate of σ for Ni-Zr alloys. These are calculated by a simple weighted average of the σ values for the pure elements, which in turn are determined by [16]

$$\sigma = 0.45 \, \Delta H_f \, N^{-1/3} V^{-2/3} , \tag{2}$$

where ΔH_f is the molar enthalpy of fusion, N is Avogadro's number, and V is the molar volume of the liquid. The value for σ of an alloy was assumed to be independent of temperature.

Existing experimental assessments of the Ni-Zr system [17] permit an adequate modeling of equilibrium phases, but the reliability of these models in providing thermodynamic data far into metastable regions must be viewed with caution [18]. Our experimental results, along with similar studies of other systems, could provide a basis for improving modeling techniques. For σ, a more realistic estimate would incorporate a compositional and temperature dependence based on thermodynamic data, but few empirical studies have been made, and the problem would remain of determining non-equilibrium values.

ACKNOWLEDGEMENTS

We thank the Los Alamos Ion Beam Materials Laboratory for assistance in the characterization of the films by Rutherford backscattering. This work was supported by the U.S. Department of Energy, Office of Basic Energy Sciences, and by Laboratory Directed Research and Development funds.

REFERENCES

1. C.-J. Lin and F. Spaepen, *Acta Metall.* 34, 1367 (1986).
2. D. R. Uhlman, *J. non-cryst. Solids* 7, 337 (1972).
3. L. E. Tanner and R. Ray, *Acta metall.* 27, 1727 (1979).
4. D. G. Morris, *Acta metall.* 31, 1479 (1983).
5. P. Nash and R.B. Schwarz, *Acta Metall.* 36, 3047 (1988).
6. J. H. Hollomon and D. Turnbull, in *Prog. Metall. Phys.*, Vol. 4 (Interscience, New York, 1953), p. 333.
7. N. Saunders, CALPHAD 9, 297 (1986).
8. F. R. de Boer, R. Boom, W. C. M. Mattens, A. R. Miedema, and A. K. Niessen, *Cohesion in Metals - Transition Metal Alloys*, (North Holland, Amsterdam, 1988).
9. Y. D. Dong, G. Gregan, and M. G. Scott, *J. Non-Cryst. Solids* 43, 403 (1981).
10. K. H. J. Buschow, B. H. Verbeek, and A. G. Dirks, *J. Phys. D: Appl. Phys.* 14, 1087 (1981).
11. K. Jansson and M. Nygren, *Mat. Sci. Eng.* 97, 373 (1988).
12. Z. Altounian, T. Guo-hua, and J. O. Strom-Olsen, *J. Appl. Phys.* 54, 3111 (1983).
13. K. H. J. Buschow, *J. Phys. F : Met. Phys.* 14, 593 (1984).
14. R. B. Schwarz and J. B. Rubin, paper in this symposium.
15. B. Predel, in *Calculation of Phase Diagrams and Thermochemistry of Alloy Phases*, eds. Y. A. Chang and J. Smith (TMS-AIME, Metals Park, Ohio, 1980), p. 72.
16. D. Turnbull, *J. Appl. Phys.* 21, 1022 (1950).
17. M. E. Kirkpatrick and W. L. Larsen, *Trans. ASM* 54, 580 (1961).
18. R. B. Schwarz, P. Nash, and D. Turnbull, *J. Mater. Res.* 2, 456 (1987).

ON THE DIFFUSION BEHAVIOUR IN STRESSED NI-ZR COUPLES

G. MAZZONE, A. MONTONE and M. VITTORI ANTISARI
ENEA, Divisione Scienza dei Materiali, CRE Casaccia, C.P. 2400, 00100 ROMA A.D.
ITALY

ABSTRACT

Solid state reactions at the interface of bulk Ni-Zr couples have been induced at several temperatures by compressive plastic deformation. The reaction product is an amorphous phase whose thickness increases with applied load and sample temperature. In addition, similar samples have been thermally reacted in order to measure the thermal interdiffusion coefficent in the same conditions. Measurements on stressed couples show that the interdiffusion coefficent (several orders of magnitude larger than the corresponding thermal value) follows a dual regime Arrhenius behaviour. The activation energy is independent of load and of the order of 0.2 eV in the low temperature regime extending up to 550 K. A different behaviour characterized by a single value of the activation energy in the whole temperature range has been observed in thermally treated samples.

INTRODUCTION

In previous papers (1)(2) we have proposed a novel approach to the study of the kinetics of solid state reactions (SSR) induced by plastic deformation in elemental diffusion couples. The experimental method is based on the compressive plastic deformation of bulk trilayers prepared by placing an elemental foil between two sheets made of the second element of the couple. The main advantage of this method over mechanical alloying or cold rolling is that the temperature of the interface during deformation can be measured and varied independently of applied load. In fact, good thermal contact between sample and pistons of the fatigue machine used to deform the trilayer keeps the temperature increase due to mechanical deformation as low as a few tens of K. The temperature can then be varied by heating sample and pistons with a suitable RF coil. The experimental work was performed on Ni-Zr couples which show good adhesion when deformed with a high enough load. For temperatures lower than about 700 K the reaction product was an amorphous layer, whose thickness was used to derive an approximate interdiffusion coefficient under the hypothesis that diffusion in the already grown glass is the rate limiting step of the reaction.

As a result of these experiments, it was possible to measure the interdiffusion coefficient under different processing conditions. The width of the glassy layer was consistent with values of the diffusivities several order of magnitude larger than those observed in thermal experiments and increased with both the value of applied load and the rate at which the load was applied. The temperature dependence of the interdiffusion coefficient was described by a dual regime Arrhenius behaviour with a change in the apparent activation energy at about 570 K.

The purpose of this paper is to present further experimental data which may contribute to the understanding of this behaviour.

EXPERIMENTAL

Square samples of 25 mm^2 were prepared by placing a 0.5 mm thick 99.9 % electropolished Ni foil between two 1.6 mm thick 99.7 % lapped Zr sheets. The samples were deformed with loads of 75 or 150 kN, both applied in about 50 ms, at temperatures ranging from 330 to 730 K. The details of the method as well as the sample temperature measurements are reported in ref.(1).

Moreover, specimens deformed at room temperature with a load of 50 kN and having a glassy interlayer about 1 nm thick were thermally treated in a vacuum furnace at temperatures between 370 K and 720 K for times suitable to derive with good accuracy the interdiffusion coefficent.

All the specimens were characterized by TEM observations on cross sections perpendicular to the interface prepared with the usual method of slicing, grinding and dimpling, while the final thinning was performed by Ar or Xe ion beam milling in a liquid N2 cooled stage.

Some TEM specimens were heat treated "in situ" with a heating specimen holder to derive the time exponent of the glassy phase growth law.

Electron microdiffraction and X-ray microanalysis experiments were carried out to characterize the microstructure of the glassy layer with a Jeol 4000 FX electron microscope operated at 400 kV.

It is well known that, with a thermionic electron gun, it is difficult to have good angular resolution in the diffraction pattern if the microbeam is focused on the specimen to define a small analyzed area. This difficulty may be overcome using a beam having a nearly one-dimensional cross section. In order to obtain this condition the circular cross section of a beam with small divergence is deformed using the condenser lens stigmator as a cylindrical lens. As a consequence the cross section changes to a very elongated ellipse oriented in such a way that focusing occurs only in the direction normal to the interface. Accordingly the microdiffraction spatial resolution is kept only in the direction normal to the Ni-Zr interface while good angular resolution in the diffraction pattern is obtained in the other direction.

EXPERIMENTAL RESULTS

If one assumes that Ni diffusion trough the amorphous layer is the rate limiting step of the reaction, an approximate interdiffusion coefficient D can be derived from the square of the half-thickness of the amorphous layer divided by the reaction time for which an upper bound is given by the deformation time (2).

Experimental D values are reported in Fig.1 versus reciprocal temperature in the usual Arrhenius representation. The reaction temperatures have been approximately corrected for the small effect due to plastic deformation by adding to the constant piston temperature the measured peak value of the temperature increase. The experimental results of plastically deformed samples are described by a dual regime Arrhenius behaviour with an activation energy for both loads of 0.2 eV up to about 550 K.

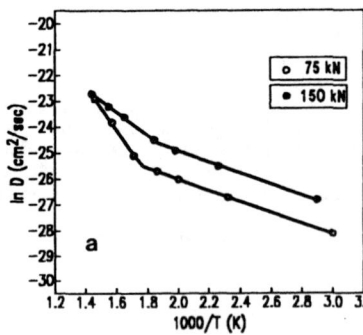
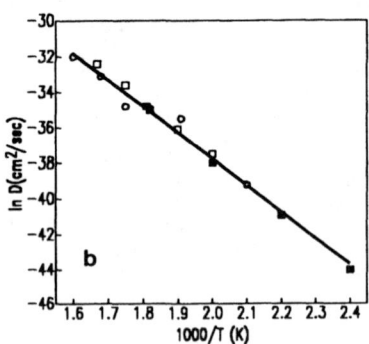

Fig.1. Arrhenius plot of diffusion data. Fig.1a is relative to samples deformed with the loads reported in the inset, while Fig.1b is relative to thermal annealing. Open squares represent "in situ" TEM measurements, closed squares furnace annealings and open circles are from Ref. (3)

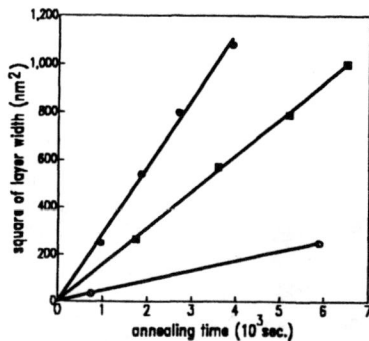

Fig.2. The square of the glassy layer thickness versus the annealing time as derived from "in situ" TEM experiments (open circles 423 K, closed circles 523 K, squares 458 K).

At higher temperatures the activation energy for this type of samples depends on applied load and appears to be lower for larger deformations. Fitting the experimental data with two independent diffusion mechanisms one deduces a high-temperature activation energy of 1.3 eV for a load of 75 kN and 0.5 eV for a load of 150 kN.

For the thermally reacted samples, Fig.1b shows for the whole temperature range an Arrhenius behaviour with a single activation energy of 1.3 eV. It is worth noticing that our data are in very good agreement with those obtained by Schultz (3) on bulk samples also shown in Fig 1. On the other hand, because of the large scatter of values reported by Toma (4) it is difficult to compare our interdiffusion coefficients with those derived from thin film diffusion couples. In any case, our values are in the range defined by previous measurements with good agreement on the activation energy for diffusion (5)(6).

"In situ" heating of electron transparent cross-sections has allowed us to deduce the 1/2 time exponent of the growth law as reported in Fig.2.

X-ray microanalysis was used to obtain the chemical composition of the reacted layers. Because of beam broadening effects (7) this analysis was restricted only to specimens with an interface layer larger than a few tens of nm. The results reported in Table 1 have been obtained keeping the beam entirely inside the glassy layer.

Table 1 shows that the composition of our thermally grown samples is in agreement with previous measurements (8) within experimental uncertainties. Mechanical deformation seems to increase the Zr concentration at the Zr-rich side. Concerning sample contamination, no oyigen or nitrogen was detected in our samples using an ultra-thin window detector.

Typical results of electron microdiffraction are shown in the micrographs in Figs. 3 and 4 which show specimens deformed in the low and in the high temperature range respectively. In Fig.3 one observes the diffuse haloes typical of

TABLE 1
Ni content (at%) of amorpous phase at the a-crystal interfaces

Applied Load (kN)	Temperature (K)	Ni at a-Ni interf.	Ni at a-Zr interf.
0	500	66	47
75	570	66	39
150	620	67	38

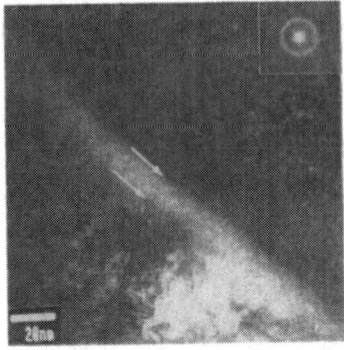

Fig.3. Dark field image of the glassy layer (marked by arrows) observed in the sample deformed with 150 kN at 480 K. The inset shows the corresponding diffraction pattern.

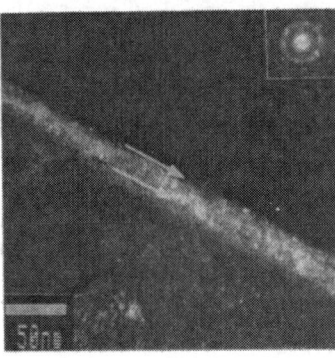

Fig.4. Dark field image of the glassy layer (marked by arrows) observed in the sample deformed with 150 kN at 630 K. The inset shows the corresponding diffraction pattern.

the diffraction pattern of an amorphous phase and a contrast in the micrograph consistent with a homogeneous microstructure. On the contrary, in Fig.4 sharp diffraction spots are superimposed on the diffuse haloes and the contrast appears to be generated by crystal particles. The interplanar spacings derived from Fig.4 are not consistent with the crystal structure of any stable Ni-Zr intermetallic compound or Ni and Zr oxides. However the spectrum shows good agreement both in peak positions and intensities with the X-ray diffraction results obtained from a Ni-Zr intermetallic compound already observed in a Ni-Zr alloy with 17 at% Zr (9).

The micrograph of Fig.5 shows the interface layer of the specimen deformed with 150 kN at 680 K. It is interesting to notice that glass crystallization by nucleation of orthorombic NiZr at the Zr-aNiZr interface, occurs at a thickness (37 nm) of the reacted layer well below the critical thickness reported for thermally reacted samples (11) (12). This partial crystallization is also observed on the specimen processed with 75 kN at the same temperature, while for both loads, deformation at 730 K produces crystalline interlayers more than 100 nm thick.

DISCUSSION

The first point which deserves to be mentioned concerns the activation energies involved in the diffusion process leading to SSR between Ni and Zr. The dual regime Arrhenius behaviour has been observed in our case only when the diffusion couples react

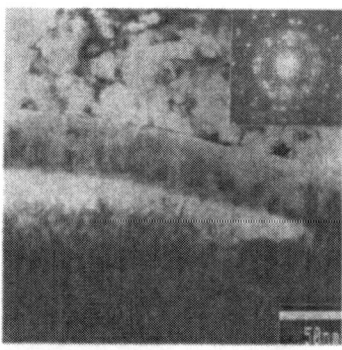

Fig.5. Bright field image of the glassy layer observed in the sample deformed with 150 kN at 630 K. The inset shows the corresponding diffraction pattern.

during plastic deformation, while, contrary to the experimental findings of Ding et al. (13), the process of diffusion in thermally activated samples is described by a single activation energy also for temperatures as low as 400 K. Considering that no clear-cut explanation for the dual behaviour is available, we are not able, at the moment, to explain the discrepancy between our results and those reported in ref. (13). In any case the reaction rates observed in our mechanically deformed samples require a strong enhancement of the interdiffusion coefficent. Following earlier approaches (14) (15) we have ascribed this behaviour to the creation of excess free volume by plastic deformation of the growing glass. The excess free volume would then be the cause of fast diffusion according to the model of Turnbull and Cohen (16).

The results obtained by using two different values of applied load together with the microstructural characterization of the reaction products allow a better description of this phenomenon.

The low value of the activation energy at low temperatures may be interpreted in terms of diffusion through short-circuit paths. The increase of the preexponential term with no change of the activation energy is consistent with this picture if we assume that increasing the applied load one enhances the density of short circuit paths but not their structure. The strain localization in shear bands often observed in metallic glasses deformed at low temperature suggests a possible origin of these fast diffusion paths. However one observes that the Zr-aNiZr interface is quite flat and this fact requires fast transverse diffusion at least at this interface if the atomic flux in the glass is localized along a discrete number of paths.

Another possible explanation of the low temperature behaviour is suggested by molecular dynamics simulations (17) which show how a vacancy-like localized increase of free volume can give rise to high atomic mobility of the surrounding atoms for times of the order of 10^{-13} s and with negligible activation energy. After a few atomic jumps, according to the simulation, the excess free volume spreads out and the effect on diffusivity disappears. One could then speculate that these small regions of low atomic density localized both in space and time and whose concentration may depend on the deformation parameters, might account for our experimental results.

The presence of intermetallic particles associated with the increased slope of the Arrhenius plot could be related to the onset of Zr mobility so that in the high temperature regime the diffusion of both atomic species contributes to the growth of the interface layer. The activation energy appears to depend, in this case, on the deformation strength. This could possibly be related to different size distributions of vacancy-like defects resulting from plastic deformation in different conditions.

The incipient crystallization observed in this regime seems to be the result of homogeneous nucleation of the intermetallic particles while the usually observed crystallization by nucleation of orthorombic NiZr at the Zr-aNiZr interface is found in our specimens only for temperatures as high as 680 K. Since, in general, the energy barrier for homogeneous nucleation is higher than that for heterogeneous nucleation, this unusual behaviour can be accounted for on the basis of the critical thickness approach of Johnson and coworkers (11)(12). Actually, according to this model, heterogeneous nucleation at a moving interface is possible only if the interface velocity is so low that the time required for crossing a distance comparable with the critical nucleus thickness in the growth direction is larger than the inverse of the heterogeneous nucleation rate. In our case the atomic mobility of both atomic species could be large enough to allow the nucleation of the crystalline compounds on the time scale of the experiment, but the interface velocity would be still sufficiently high to hinder heterogeneous nucleation.

Finally we remark that the critical thickness observed in the present experiment is much smaller than that measured on thermally reacted samples (12). Even considering that critical thickness is controlled by a delicate balance of the thermally activated processes involved in glass growth and intermetallic nucleation, a possible role can be played by the chemical composition of the glass at the Zr-aNiZr interface, which, in our case, because of its lower Ni content, is less stable against crystallization than thermally grown glasses (16).

If the composition range of the growing glass is determined by the metastable equilibrium with the terminal solid solutions, the observed tendency to a lower Ni content at the Zr-aNiZr interface in mechanically deformed samples can be attributed to a change of the chemical potential of crystalline Zr induced by the plastic deformation. Actually it has been observed that severe mechanical deformation can substantially increase the free energy of hexagonal metals like Zr while it has a negligible effect on fcc Ni (19)(20).

REFERENCES

1) S.Martelli, G.Mazzone, A.Montone and M.Vittori Antisari, J.Phys. (Paris), Colloq. 51, C4-241, (1990)
2) G.Mazzone, A.Montone and M.Vittori Antisari, Phys.Rev.Lett.,65, 2019 (1990)
3) L.Schultz in Science and Technology of Rapidly Quenced Alloys, edited by M.Tenhover, W.L.Johnson and L.E.Tanner (Mater. Res. Soc. Proc. 80,Pittsburgh, PA 1987) pp 97-104
4) A.Toma, J.Phys.:Condens. Matter, 2, 3167 (1990)
5) E.J.Cotts, W.J.Meng and W.L.Johnson, Phys. Rev. Lett.,57, 2295 (1986)
6) R.J.Highmore, J.E.Evett, A.L.Greer and R.E.Somekh, Appl. Phys. Lett., 50,566 (1987)
7) J.I.Goldstein, J.L.Costley, G.W.Lorrimer and S.J.B.Reed in SEM 1977, edited by O.Johari (IITRI, Chicago, ILL 1977) p 315
8) J.C.Barbour, Phys. Rev. Lett.,55, 2872 (1985)
9) K.Kramer, Trans. Met. Soc. (AIME), 215, 256 (1959)
10) S.B.Newcomb and K.N.Tu, Appl. Phys. Lett., 48, 1436 (1986)
11) W.J.Meng, C.W.Nieh and W.L.Johnson, Appl. Phys. Lett., 51, 1693 (1987)
12) W.J.Meng, C.W.Nieh, E.Ma, B.Fultz and W.L.Johnson, Mat. Sci. Eng. 97, 87, (1988)
13) F.Rong Ding, P.R.Okamoto, L.E.Rehn, J. Mater. Res., 4, 1444 (1989)
14) F.Spaepen, Acta Metall., 25, 407 (1977)
15) A.S.Argon, Acta Metall., 27, 47, (1979)
16) D.Turnbull and M.H.Cohen, J. Chem. Phys., 52, 3038 (1970)
17) V.Rosato, private communication
18) J.Eckert, L.Schultz, E.Hellstern, K.Urban, J. Appl. Phys., 64, 3224, (1988)
19) E.Hellstern, H.J.Fecht, Z.Fu, W.L.Johnson, J.Appl. Phys., 65, 305 (1989)
20) H.J.Fecht, E,Hellstern, Z.Fu, W.L.Johnson, Adv. Powder Met.,1, 111 (1990)

AMORPHOUS PHASE FORMATION AND REACTIONS AT Pt/GaAs INTERFACES

DAE-HONG KO and ROBERT SINCLAIR
Department of Materials Science and Engineering, Stanford University, Stanford, CA 94305

ABSTRACT

The amorphous phase formation and initial crystalline reactions at Pt/GaAs interfaces has been investigated via high-resolution electron microscopy (HREM), microdiffraction, and energy dispersive spectroscopy (EDS). A thin amorphous intermixed layer consisting of three elements, platinum, gallium, and arsenic was observed at Pt/GaAs interface in an as-deposited sample. This interlayer grew to 4.5nm in an amorphous state upon low temperature(e.g. 200°C) annealing by a solid-state amorphization reaction. Following the growth of the amorphous interlayer, subsequently, Pt_3Ga and $PtAs_2$ phases nucleated within the amorphous layer and grew at the Pt and GaAs sides, respectively. We also observed the same reaction processes with in-situ annealing HRTEM.

INTRODUCTION

While it has long been suggested that a "glassy" or amorphous layer forms upon the initial reaction at a metal-semiconductor interface[1,2], the existence of this layer has only been demonstrated definitively by recent high-resolution electron microscope (HREM) observations [3-6]. It is now well known that many metals form a 1-2nm amorphous interdiffused layer when deposited on clean silicon surfaces, which in some systems grows upon annealing but in others crystallizes into stable or metastable phases. Such behavior can be interpreted in terms of a solid-state amorphization, driven by a negative heat of mixing of the elements with the amorphous phase produced by kinetic considerations[4, 7].

The situation for compound semiconductors is more complex. As there are three elements involved at even the simplest interface, it is not at all clear that how the constituents will react when they come into contact. However, this issue is important in semiconductor device fabrication, because the electronic properties of a metal-semiconductor contact (such as the Schottky barrier height or ohmic contact resistivity) are strongly influenced by the details of the interfacial chemistry[8-10]. Some studies using electron diffraction have reported an interdiffused amorphous layer produced upon low temperature annealing of nickel and palladium thin films on indium phosphide[11,12] and of cobalt on gallium arsenide[13]. Considering the widespread reports of amorphous phase formation upon deposition in metal-silicon systems, it seems worthwhile to examine metal-compound semiconductor interfaces, to see whether an equivalent reaction occurs for them. Because the expected thickness might only be a few atomic layers, it is clear that HREM is the most powerful technique for the study of such a phase[5]. In this article, we report the formation of an amorphous intermixed layer at the Pt/GaAs interface in an as-deposited condition, and the growth of an amorphous interlayer. We also report the initial crystalline reaction between Pt and GaAs.

EXPERIMENTAL PROCEDURES

Semi-insulating gallium arsenide substrates with (100) orientation were degreased with trichloroethylene and etched in a $H_2O_2:H_2O:H_2SO_4$ (1:1:5) solution for 1 minute and a $H_2O_2:NH_4OH:H_2O$ (1:1:10) solution for 2 minutes. The cleaned wafers were directly loaded into the deposition chamber, after which a 500Å thick platinum layer was deposited by electron beam deposition with a base pressure of 1×10^{-7} torr. The deposition rate for the platinum was maintained at 2Å/second. The materials were treated by rapid thermal annealing (RTA) at 200°C, 300°C, and 350°C, and by furnace annealing at 200°C and 220°C. We prepared cross-

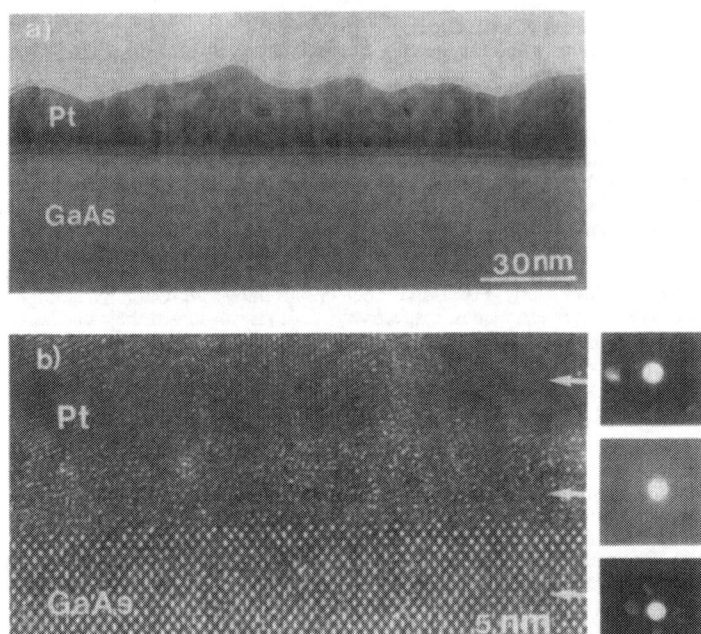

Figure 1. Cross-section TEM micrographs of the as-deposited Pt/GaAs sample. (a)A high-resolution image at lower magnification showing native oxide (white band) and uniform intermixed layer along the interface between platinum and GaAs substrate. (b) A high-resolution image showing amorphous native oxide and amorphous intermixed layer between the crystalline platinum and the crystalline GaAs substrate. Attached microdiffraction pattern taken from the intermixed layer shows a typical amorphous halo.

sectional samples for the TEM by gluing the wafers at room-temperature with a room temperature epoxy, then sectioning, mechanically thinning, polishing, dimpling, and, finally, ion milling in a liquid-nitrogen cooling stage, with an accelerating voltage of 4kV, a total beam current of 0.5mA, and a beam angle of 15°. Samples were thus maintained entirely at room temperature or lower during the TEM sample preparation. They were then analyzed with a Philips EM 430 ST microscope operating at 300kV for high resolution microscopy and with an AKASHI EM-002B microscope operating at 200kV, with an electron beam probe size of 20Å (nominal), for the microdiffraction and the energy dispersive spectroscopy (EDS) analyses. We also performed in-situ annealing in the Philips microscope and observed the reaction with high resolution microscopy.

RESULTS AND DISCUSSION

Figure 1 is a high-resolution TEM micrograph of a cross section of the as-deposited Pt/GaAs sample, which shows a poly-crystalline platinum film on a single crystal GaAs substrate. Between these phases, this micrograph resolves two distinct interfacial layers with a

HRTEM appearance typical of amorphous materials: a thin, somewhat discontinuous bright layer close to the Pt film and a 2-3nm thick uniform interlayer with dark contrast close to the GaAs substrate. The dark contrast of this layer compared to the substrate GaAs and its location (below the native "oxide" layer) suggest that this layer is a Pt-GaAs intermixed layer. Likewise, it is similar in contrast to amorphous metal-silicon intermixed layers (for example, titanium-silicon[6]) that occur when reactive metal is deposited onto a clean silicon surface. In an EDS analysis (not shown here) carried out using a 20Å probe, by moving the probe from the GaAs to the intermixed layer we detected a distinct signal from each of the three elements, which indicates that this layer consists of Ga, As and Pt. It should be noted that clear distinction of the compositions in such thin layers has been demonstrated using such an instrument[9]. Microdiffraction patterns were taken from this intermixed layer, as well as from the adjacent platinum and gallium arsenide phases. The probe size for this microdiffraction analysis was about 20Å, small enough to cover only the intermixed layer. The platinum and GaAs phases reveal [100] and [110] diffraction patterns of the FCC Bravais lattice, respectively. In contrast, the microdiffraction pattern taken from the intermixed layer reveals a typical amorphous halo, even for such a small diffraction column.

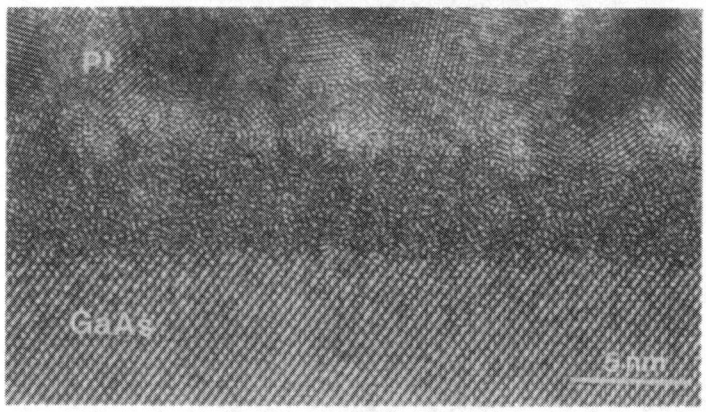

Figure 2. Cross-section HRTEM micrograph of a Pt/GaAs sample annealed at 200°C for 30 min., showing 4.5 nm amorphous interlayer between crystalline Pt film and single crystalline GaAs substrate.

The amorphous interlayer grew during the annealing at 200°C for 30 min. As the HRTEM image of this sample in Figure 2 shows, the interlayer thickened uniformly to about 4.2 nm. There are no lattice fringes or any other evidence of crystallinity in the interlayer, and its contrast level is still between that of the Pt and the GaAs substrate. Hence, it is confirmed that the Pt/GaAs system underwent a solid-state amorphization reaction, and that this reaction resulted in the growth of the amorphous interlayer during the annealing at 200°C.

In Figure 2, the entire amorphous interlayer appears below the oxide layer just like that of the as-deposited sample, which shows that the amorphous layer grew to the GaAs substrate side by the diffusion of Pt to the GaAs. In-situ HRTEM experiments exhibited similar behavior (not shown here). During the in-situ annealing, the crystalline GaAs fringes just beneath the amorphous interlayer disappeared and were replaced with the newly formed amorphous interlayer. Hence, the amorphization reaction occurs at the amorphous interlayer-GaAs interface, consuming the GaAs phase by the diffusion of Pt through the existing amorphous interlayer to the GaAs side

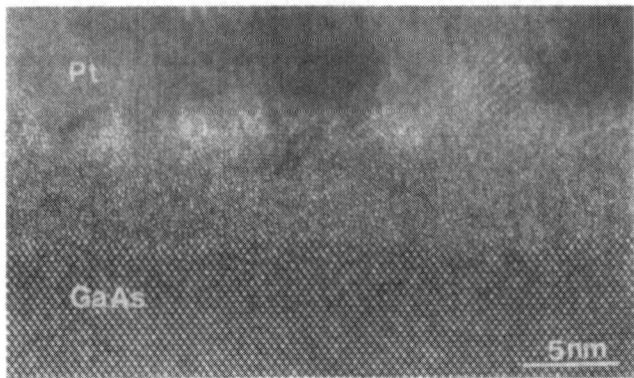

Figure 3. Cross-section HRTEM micrograph of Pt/GaAs sample annealed at 200°C for 90 min., showing crystalline phase of Pt$_3$Ga in the interlayer close to Pt film.

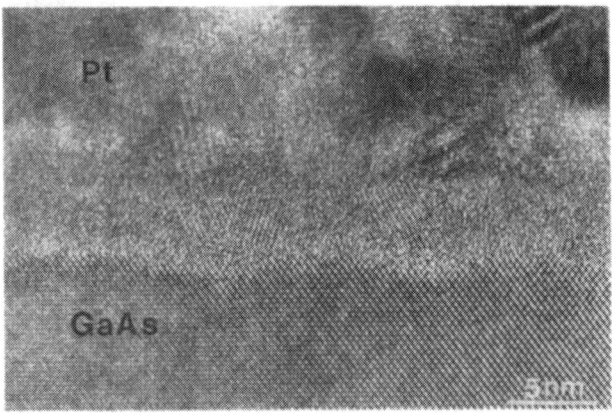

Figure 4. Cross-section HRTEM micrograph of Pt/GaAs sample annealed at 200°C for 12 hours, showing crystalline phase of PtAs$_2$ in the interlayer close to the GaAs substrate.

After 60 min annealing at 200°C, the intermixed layer grew in an amorphous state to a thickness of 5.0nm, after 90 min at 200°C to 5.5nm (see Figure 3). In both cases, however, crystalline phases formed at the very top of the amorphous layer close to the oxide layer. After 60 min at 200°C, a small number of crystalline nuclei were observed to form, and these crystallites were clearly seen after annealing for 90 min at the same temperature. In Figure 3, the HRTEM image of the 90 min annealed sample shows crystalline lattice fringes immediately beneath the native oxide layer. All the crystallites show a lattice spacing of 2.2Å with an interplanar angle of 71°, which corresponds to (111) planes of the Pt$_3$Ga phase. These cross lattice fringes and interplanar angle as well as the intermixed layer's location (below the oxide layer) suggest that the crystallites are the Pt$_3$Ga phase. The layer beneath the newly formed Pt$_3$Ga phase still remains amorphous and does not show any evidence of crystallinity.

During further annealing at 200°C for 12 hours, the intermixed layer grew to a thickness of 7.0 nm (see Figure 4). In addition to this, at this stage, nucleation of another crystalline phase is observed in the lower amorphous layer close to the GaAs substrate (as shown in the (110) HRTEM micrograph in Figure 4). Cross lattice fringes of the crystalline phase in the high-resolution TEM image correspond to those of the $PtAs_2$ phase. Measurement of the lattice spacings gives values of 3.0 Å and 3.4 Å, which correspond to (200) and (111) planes of the $PtAs_2$ phase.

The platinum-GaAs interface, upon annealing for 30 min at 220°C exhibits a reaction behavior similar to that observed after 12 hours of annealing at 200°C. After 30 min at 220°C the intermixed layer has grown to 6.0 nm toward the GaAs substrate, producing a crystalline Pt_3Ga phase in the region close to the crystalline platinum film and a small number of $PtAs_2$ crystallites in the amorphous layer adjacent to the substrate GaAs. An extended annealing for 60min at 220°C produced more $PtAs_2$ crystallites. *In-situ* HRTEM at 220°C showed the same morphology and sequence as those obtained in ex-situ annealed samples. During the *in-situ* experiment, we followed the whole reaction using one single specimen and thereby avoided having to surmise the course of the interfacial reactions.

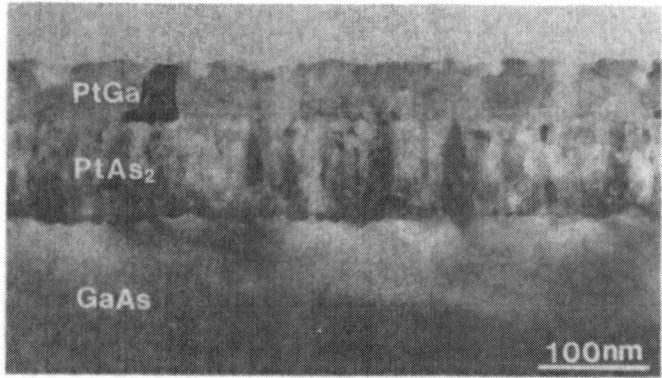

Figure 5. Cross-section Bright Field TEM micrograph of Pt/GaAs sample annealed at 400°C for 20 min., showing layered structure of PtGa/PtAs₂/GaAs after complete reaction.

The Pt_3Ga and $PtAs_2$ phases were observed to grow toward the Pt film and the GaAs substrate respectively during the subsequent higher temperature (e.g. 250°C) annealing. Annealing at 400°C for 20min produced PtGa and $PtAs_2$ as equilibrium phases with the GaAs, and formed a layered structure of PtGa/PtAs₂/GaAs (see Figure 5), which is consistent with the previous results[14-16].

SUMMARY

In summary, we observed a thin amorphous intermixed layer at the platinum-GaAs interface in the as-deposited condition. This interlayer forms below the native oxide layer and consists of the three elements of gallium, arsenic, and platinum. The interlayer grew in an amorphous state upon low temperature annealing by a solid-state amorphization reaction. Following the growth of the amorphous interlayer, we observed within it the nucleation of the Pt_3Ga and $PtAs_2$ phases, which grew at the Pt and GaAs sides, respectively. In-situ annealing HRTEM demonstrated the same reaction processes to those in ex-situ experiments.

ACKNOWLEDGEMENTS

The authors would like to thank Dr. W. S. Lee and Dr. B. G. Park for carrying out the depositions. We also thank Dr. W. Moberly at the Stevens Institute of Technology and our collaborators at ISI in the United States and Mr. M. Moriyama at AKASHI in Japan for their assistance with the microdiffraction and EDS analyses. This work was sponsored by the National Science Foundation, Grant No. DMR 8902232.

REFERENCES

1. R. M. Walser and R. W. Bene, Appl. Phys. Lett. **28**, 624 (1976).
2. A. Hiraki, Surface Science Reports **3**, 357 (1984).
3. K. Holloway and R. Sinclair, J. Less-Common Metals **140**, 139 (1988).
4. K. Holloway, R. Sinclair and M. Nathan, J. Vac. Sci. Techn. **A7**, 1479 (1989).
5. R. Sinclair, K. Holloway, K. B. Kim, D. H. Ko, A.S. Bhansali, A. F. Schwartzman and S. Ogawa, Inst. Phys. Conf. Ser. **100**, 599 (1989).
6. R. Sinclair, Mater. Trans. Jpn. Inst. Met., **31**, 628 (1990).
7. R. B. Schwarz and W. L. Johnson, Phys. Rev. Lett. **51**, 415 (1983).
8. J. Silverman, P. Pellegrini, J. Comer, A. Golvbovic, M. Weeks, J. Mooney, and J. Fitzgerald, Mat. Res. Soc. Proc., **54**, 515 (1986).
9. S. Ogawa, T. Yoshida, T. Kouzaki, and R. Sinclair, J. Appl. Phys., in press (1991).
10. K. B. Kim, M. Kniffin, R. Sinclair, and C. R. Helms, J. Vac. Sci. Techn. **A6**, 1473 (1988).
11. T. Sands, C. C. Chang, A. S. Kaplan, and V. G. Keramidas. K. M. Krishnan, and J. Washburn, Appl. Phys. Lett. **50**, 1346(1987).
12. R. Caron-Popowich, J. Washburn, T. Sands, and A. S. Kaplan, J. Appl. Phys. **64**, 4909 (1988).
13. F. Y. Shiau and Y. A. Chang, Appl. Phys. Lett. **55**, 1510 (1989).
14. C. Fontaine, T. Okumura, and K. N. Tu, J. Appl. Phys. **54**, 1404 (1983).
15. A.K. Sinha and J.M. Poate, in *Thin Films-Interdiffusion and Reactions*, edited by J.W. Mayer (Wiley, New York, 1978), p. 418
16. T. Sands, V. G. Keramidas, A. J. Yu, K-M. Yu, R. Gronsky, and J. Washburn, J. Mater. Res. **2**, 262 (1987).

THERMODYNAMIC ANALYSIS FOR THE SOLID-STATE AMORPHIZATION AND SUBSEQUENT CRYSTALLIZATION OF GaAs/Co

F.-Y. SHIAU*, S.-L. CHEN, M. LOOMANS AND Y. A. CHANG
Department of Materials Science and Engineering, University of Wisconsin, Madison, WI 53706, USA.
*Currently with Tze-Chang Foundation of Science and Technology, Semiconductor Research Center, Hsin-Chu, Taiwan, ROC

ABSTRACT

Phase equilibria along the CoGa-CoAs join were determined by DTA and metallography. On the basis of these data and the phase diagram of Co-Ga-As at 600°C, thermodynamic values for the various phases along the GaAs-Co join were estimated. The Gibbs energy of the amorphous phase is approximated to be that of the supercooled liquid phase. These data were used to rationalize the amorphization process.

1.0 INTRODUCTION

When GaAs/Co thin-film couples are annealed above ambient temperature, amorphization occurs first, followed by crystallization and eventually decomposition to the thermodynamically stable phases[1-3]. In order to provide a thermodynamic rationale for the observed phase transformations, we need to have thermodynamic data for the various competing phases, i.e. the amorphous phase, the GaAs phase (B3), the Co phase (A3), the CoGa phase (B2) and the CoAs phase (B31). In the present study, the Gibbs energy of the amorphous phase is taken to be that of the liquid phase since the difference in the Gibbs energy between these two states is not known. However, this approximation underestimates the stability of the amorphous phase. Since obtaining accurate thermodynamic values of the phases which occur in GaAs/Co couples requires too much experimental effort, we resorted to determining the phase equilibria of the psuedobinary CoGa-CoAs. Using these data and the 600°C isotherm of Co-Ga-As determined by Shiau et al.[4], thermodynamic values of the various phases are estimated. Using these values, phase transformations of GaAs/Co to the various phases are discussed.

2.0 EXPERIMENTAL METHODS AND RESULTS

2.1 Phase Equilibrium Determination

Phase equilibrium samples with an average weight of 0.5 g were prepared from GaAs, Co, Ga and As of purity of 99.99%. The component elements in the appropriate proportions were weighed and sealed in evacuated quartz capsules. The samples were annealed at 1200°C for 2 weeks. Their melting temperatures were determined by DTA using a procedure described elsewhere[5].

2.2 Experimental Results

The melting points for all nine samples were plotted in Fig. 1. These data suggest an eutectic invariant reaction between CoGa and CoAs. The eutectic composition of 47 ± 2 mol%

CoAs was obtained by extrapolating the liquidus data, and the solubility limits at 600°C are taken from Shiau et al.[4]. The curves shown in Fig. 1 were calculated from thermodynamic models as discussed later. The CoGa phase was assumed to melt congruently at 50 at% Ga.

3.0 THERMODYNAMIC MODELLING OF THE CoGa-CoAs PSEUDOBINARY SYSTEM

Using the phase diagram data given in Fig. 1, the thermodynamic parameters for the various phases as well as the Gibbs energies of melting and of phase transformation for CoGa and CoAs were estimated and are given in Table I.

Table I
Thermodynamic Values for the Various Phases Along the CoGa-CoAs Join

$G_1^{\beta \to \ell}$ = 36500 - 25T in J/mol

$G_1^{\beta \to \mu}$ = 35250 - 15T in J/mol

$G_2^{\mu \to \beta}$ = 39231 - 27T in J/mol

$G_2^{\mu \to \beta}$ = 57071 - 22T in J/mol

w_{12}^{ℓ} = 4.3991 - 5704.5/T

w_{21}^{ℓ} = 6.0354 - 7375.5/T

w_{12}^{ℓ} = w_{21}^{ℓ} = 3.2399 - 3930.8/T

w_{12}^{μ} = w_{21}^{μ} = 4.1846 - 5014.8/T

The Gibbs energy for the β-phase is given below.

$$G^{\beta} - (x_1 \, ^{\circ}G_1^{\ell} + x_2 \, ^{\circ}G_2^{\ell}) = - x_1 \, \Delta ^{\circ}G_1^{\beta \to \ell} - x_2 \Delta ^{\circ}G_2^{\beta \to \ell} + \Delta^{id}G$$
$$+ (RT/2) \, x_1 x_2 \, [(w_{12}^{\beta} + w_{21}^{\beta}) + (w_{12}^{\beta} - w_{21}^{\beta}) \, (x_2 - x_1)]$$

The term G stands for the Gibbs energy, x the mole fractions of the component elements, R the gas constant, T the absolute temperature, and w_{12}^{β}, w_{21}^{β} the solution parameters. The subscript 1 denotes CoGa and 2 CoAs. The superscript id stands for the ideal quantity, i.e. $\Delta^{id}G = - T \, \Delta^{id}S$, ° the component elements in their pure states, and β → ℓ the change of state from β to ℓ, i.e. melting of the β-phase. Equations similar to the above equations were used to represent the Gibbs energies of other phases, i.e. ℓ and μ. The superscript symbols β, μ, ℓ denote the Co(Ga,As) phase with the B2 structure, the Co(Ga,As) phase with the B31 structure and the liquid phase respectively.

The computed phase diagram as shown in Fig. 1 using the parameter values given in Table I is in good agreement with the experimental data.

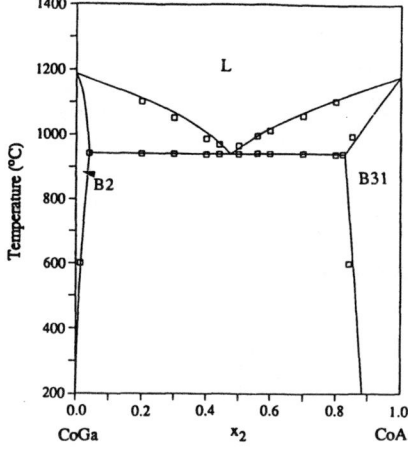

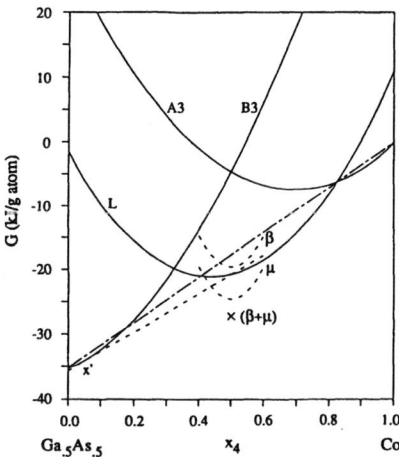

Fig. 1   The pseudobinary CoGa–
CoAs phase diagram,
with $x_2 = x_{CoAs}$.

Fig. 2   The Gibbs energies for
the GaAs-Co join at
573 K, with $x_4 = x_{Co}$.
The dashed-dotted line
represents G of the
mechanical mixture of
$(Ga_{.5}As_{.5}, Co)$.

4.0   DISCUSSION

The amorphization of GaAs/Co couples is believed to be due
to the openness of the structure of GaAs in comparison to the
size of the Co atoms.   The interstitial voids of the GaAs
structure are found to be about the same size as Co atoms.
This condition allows rapid diffusion of Co into GaAs as an
interstitial species.   According to the data reported in the
literature[7,8], $D_{Co}^{GaAs}/D_{Ga}^{GaAs} \sim 10^{11}$ and $D_{Co}^{GaAs}/D_{As}^{GaAs} \sim 10^{14}$ at
573K.   This clearly indicates that for all practical purposes,
Co diffuses rapidly into GaAs with Ga and As remaining
immobile, in support of one of the two criteria suggested by
Schwarz and Johnson[9] for SSAR.

Given sufficient time, the amorphous phase will transform
to a metastable crystalline phase with the formula
μ-Co(Ga$_{.48}$As$_{.52}$) (B31).   It subsequently decomposes to the
stable CoGa and CoAs phases.   The formation of μ is in
accordance with the argument formulated by Lin, Schulz, Hsieh
and Chang[10].   Since the composition of μ falls on the GaAs-Co
join, redistribution of the Ga and As atoms is minimized in
comparison to the formation of CoGa and CoAs.

Let us now discuss the possible phase transformations
using the Gibbs energies for the GaAs-Co join.   The
thermodynamic values for the various phases as given in Tables
II and III are estimated from those given in Table I and the
600°C isotherm of Co-Ga-As[4].   Figure 2 shows the Gibbs energies

of β(B2), μ(B31) and ($\beta+\mu$) in the compositional vicinity of $x_4=0.5$ as well as those of ℓ, A3 and B3 from $x_4=0.0$ to $x_4=1.0$. All of these quantities are referred to Ga(ℓ), Co(A3) and As(c). The symbol (c) denotes the crystalline state of As. The Gibbs energies of the stable phases at compositions greater than 50 mol% Co are not shown in Fig. 2 since they are of no consequence to our discussion for possible phase transformation.

Table II
Thermodynamic Values for Ga, Co, As,
GaAs(B3) CoGa(B2) and CoAs(B31)

$Ga_{(\ell)}+As_{(\ell)}$ = GaAs(B3) $\Delta°G = -116\ 180+52.208T+0.4036T\ell nT^{11}$

$Co(Al)+1/4As_4(g)$ = CoAs(B31) $\Delta°G = -\ 86\ 986+48.468T^{13}$

$Co(Al)+Ga(\ell)$ = CoGa(B2) $\Delta°G = -\ 78\ 252+30.905T^{14}$

$As(c)$ = $As(\ell)$ $\Delta°G = 23\ 814-19.01T+0.4036T\ell nT^{11}$

$As(c)$ = $1/4As_4(g)$ $\Delta°G = 41\ 995-104.92T+8.425T\ell nT^{11}$

$Co(Al)$ = $Co(\ell)$ $\Delta°G = 16\ 192-9.1584T^{12}$

$Co(A3)$ = $Co(Al)$ $\Delta°G = 452-0.6455T^{12}$

Table III
Thermodynamic Values for the Various Phases
Along the $Ga_{.5}As_{.5}$–Co Join

$°G_3^{A3} - °G_3^{B3} = 63\ 690$ J/gatom at 573 K

$°G_4^{B3} - °G_4^{A3} = 63\ 690$ J/gatom at 573 K

$w_{34}^{\ell} = w_{43}^{\ell} = -\ 10\ 649/T$

$A_{34}^{A3} = w_{43}^{A3} = w_{34}^{B3} = w_{43}^{B3} = -\ 7454/T$

$xs_G^{A3} - (x_3°G_3^{\ell} + x_4°G_4^{\ell}) = -\ x_3\ \Delta°G_3^{A3\to\ell} - x_4\ \Delta°G_4^{A3\to\ell}$
$+ (RT/2)\ x_3x_4\ [(w_{34}^{A3} + w_{43}^{A3}) + (w_{34}^{A3} - w_{43}^{A3})\ (x_4 - x_3)]$

where $\Delta°G_3^{A3\to\ell}$ refers to one gram-atom of GaAs, i.e. $Ga_{.5}As_{.5}$. The subscripts 3, 4 denote the $Ga_{.5}As_{.5}$ and Co components.

Let us next discuss the possible phase transformations occurring in a GaAs/Co couple in light of Fig. 2. As mentioned earlier, when a CoAs/Co couple is annealed at 573 K, Co atoms diffuse rapidly into the interstitial voids of GaAs, forming a supersaturated solid solution with the GaAs(B3) structure. The degree of supersaturation of Co in GaAs is not known. Let us suppose that this composition at 573 K is represented by $x_4 \approx 0.05$, as shown in Fig. 2. A tangent is drawn at this composition as shown in this figure. The formation of an amorphous phase with a value of $x_4 = 0.5$ (taking the Gibbs energy of the supercooled liquid to be the same as that of the amorphous phase) is possible, as is the formation of $\mu(x_4=0.5)$ and the stable $\beta+\mu$ two-phase mixture[15]. However, the formation of $\beta(x=0.5)$ is not possible since an increase in the Gibbs energy would result. For the other three cases, transformation from $(Ga_{.5}As_{.5},Co)$ at x_4 would decrease the Gibbs energies. In other words, the transformation of $(Ga_{.5}As_{.5},Co)$ at x_4 to the amorphous phase, the μ phase, or the mixture of $\beta+\mu$ is not restricted by thermodynamics. The initial formation of the amorphous phase is due to kinetics. We can rationalize this phenomenon in the following way. As Co atoms diffuse into the lattice of GaAs, breaking of the Ga-As bond would occur with simultaneous formation of bonds with Co. When the Co concentration reaches a critical value, the basic GaAs structure is no longer stable. The amorphous state is formed since it is thermodynamically allowable. Formation of the μ phase is more difficult in comparison since an entirely different arrangement of Co, Ga and As atoms is needed. An analysis of the crystal structure of B3 and B31 shows that the B31 structure cannot form directly from the B3 structure. Once the amorphous phase is formed, given sufficient time, the atoms of Co, Ga and As will rearrange themselves to nucleate and form the μ phase with the B31 structure. The driving force for this phase transformation is provided by the Gibbs energy difference between the supercooled l and B31 phases as shown in Fig. 2. In essence the amorphous phase provides a bridge (a transitional state) between the two different structures, B3 and B31. Subsequent to forming the μ-phase, given additional time, the Gibbs energy difference between μ and the $\beta+\mu$ mixture (see Fig. 2) provides the driving force for the formation of the stable phase configuration of $\beta+\mu$. In this case, long-range diffusion of Ga and As atoms is required.

It is interesting to note that for the binary systems Co-Zr[16], Cu-Zr[16], Ni-Zr[16] and Ti-Si[17], the calculated Gibbs energies of the supercooled liquid phases at sufficiently low temperatures where SSAR occurs are lower than those of the competing crystalline solution phases, i.e. BCC, FCC and HCP. While the rather high stabilities of the supercooled liquid state are favorable for SSAR, they are clearly not necessary. As shown in the present case of GaAs/Co, the Gibbs energy of the supercooled liquid is clearly more positive than that of the μ-phase (B31). Yet the initial phase formed is the amorphous phase, not the μ-phase. As long as the formation of the amorphous phase is not forbidden by thermodynamics, SSAR may occur even if its Gibbs energy is more positive than that of the competing crystalline solution phases.

A second criterion formulated by Schwarz and Johnson[9] for SSAR is large exothermic enthalpy of formation. This is certainly true for the systems studied by them and others[9,18-22]. For instance, the enthalpy of formation of CoZr is -42.2 kJ/gatom. However, we believe this criterion is not a necessary condition for SSAR. In the present study of GaAs/Co, the enthalpy of forming the stable mixture of $\beta+\mu$ from GaAs and Co is only -8.8 kJ/gatom, with the corresponding term for forming the supercooled liquid to be merely -3 kJ/gatom. The necessary condition is the constraint imposed by thermodynamics as displayed in Fig. 2. In fact, Shiau[1] found amorphization of $CoIn_2$ when the enthalpy of formation for this phase is endothermic.

5.0 ACKNOWLEDGEMENT

We wish to thank the Department of Energy for financial support of this study through Grant No. DEE-FG02-86ER452754 and S. Mohney for reading and criticizing the manuscript.

6.0 REFERENCES

1. F.-Y. Shiau, PhD Thesis, University of Wisconsin, Madison, WI 53706, (1990).
2. F.-Y. Shiau and Y. A. Chang, Presented at this Symposium at Annaheim, CA, April 29-May 1, 1991.
3. F.-Y. Shiau, S.-L. Chen, M. Loomans and Y. A. Chang, J. Mater. Res., (1990), accepted for publication.
4. F.-Y. Shiau, Y. Zuo, J.-C. Lin and Y. A. Chang, Z. Metallk., 80, 544 (1989).
5. S.-W. Chen, C.-H. Jan, J.-C. Lin and Y. A. Chang, Metall. Trans., 20A, 2247 (1989).
6. Y.-Y. Chuang, R. Schmid and Y. A. Chang, Metall. Trans., 15A, 1921 (1984).
7. G. S. Kulikov and I. N. Mikulitsa, Sov. Phys.-Solid State, 14, 2335 (1973).
8. H. D. Plafrey, M. Brown and A. F. W. Willoughby, J Electrochem. Soc., 128, 2224 (1981); J. Electron. Mater., 12, 863 (1983).
9. R. B. Schwarz and W. L. Johnson, Phys. Rev. Lett., 51, 415 (1983).
10. J.-C. Lin, K. J. Schulz, K.-C. Hsieh and Y. A. Chang, J. Electrochem. Soc., 136, 3006 (1989).
11. A. E. Schultz, PhD Thesis, University of Wisconsin, Madison, WI (1988).
12. R. R. Hultgren, P. D. Desai, D. T. Hawkins, M. Gleiser, K. K. Kelley and D. D. Wagman, "Selected Values of the Thermodynamic Properties of the Elements", American Soc. for Metals, Metals Park, Ohio 44073 (1973).
13. M. I. Kochnev, Doklady Akad. Nauk SSSR, 70, 433 (1950).
14. A. Mikula, Y. A. Chang and J. P. Neumann, Trans. Jpn. Inst. Metals, 19, 307 (1978).
15. J. W. Cahn, J. Am. Ceramic Soc., 52, 118 (1969).
16. N. Saunders and A. P. Miodownik, J. Mater., Res., 1, 38 (1986).
17. B. M. Clemens and R. Sinclair, MRS Bulletin, 19 (1990).
18. R. B. Schwarz, K. L. Wong, W. L. Johnson and B. M. Clemens, J. Non-Cryst. Solids, 61-62, 129 (1984).
19. B. M. Clemens, W. L. Johnson, and R. B. Schwarz, J. Non-Cryst. Solid, 61-62, 817 (1984).

20. S. B. Newcomb and K. N. Tu, Appl. Phys. Lett. <u>48</u>, 1437 (1986).
21. J. C. Barbour, F. W. Saris, M. Wastasi, and J. W. Mayer, Phys. Rev., <u>B32</u>, 1363 (1985).
22. H. Schroder K. Samwer, and U. Koster, Phys. Rev. Lett., <u>54</u>, 197 (1985).

GROWTH KINETICS OF AN AMORPHOUS PHASE BETWEEN GaAs AND Co

F.Y. SHIAU* AND Y.A. CHANG
Department of Materials Science and Engineering, University of Wisconsin-Madison, Madison, WI53706, U.S.A. *Currently with Tze-Chiang Foundation of Science and Technology, Semiconductor Research Center, Hsin-Chu, Taiwan, R.O.C.

ABSTRACT

Solid-state amorphization reaction between GaAs and Co thin-films was investigated by transmission electron micorscopy and Auger electron spectroscopy. Upon annealing of GaAs/Co thin-film couples at 260-300 $^\circ$C, an amorphous phase was observed to form. The amorphization was attributed to the openness of the GaAs structure relative to the size of the Co atoms. This allows rapid diffusion of Co into the GaAs lattice and promotes the occurrence of SSAR. Annealing at higher temperatures or for longer times led to the formation of a crystalline phase, designated as the μ-phase which was determined to be a metastable supersaturated solid solution of CoAs exhibiting the B31 structure of the approximate composition of $Co(Ga_{.48}As_{.52})$. The growth kinetics of both the amorphous phase and the μ-phase are parabolic in nature. The parabolic rate constant is higher for the μ-phase than for the amorphous phase. The activation energies are 1.47 and 1.35 eV, respectively.

INTRODUCTION

Solid-state amorphization reaction (SSAR) has been reported in numerous binary metal/metal systems, including Au/La[1], Au/Y[2], Ni/Zr[3-5], Co/Zr[6], Ni/Hf[7], Co/Sn[8] and the metal/silicon systems, Rh/Si[9], Ni/Si[10], Ti/Si[11,12], Hf/Si[13] and Zr/Si[13]. More recently SSAR has also been reported to occur in the III-V compound semiconductor/metal systems, GaAs/Ni[14], InP/Ni[15], InP/Pd[16], GaAs/Co[17,18] and InP/Co[18,19]. However, none of the III-V/metal studies has yet provided kinetic data for the growth of amorphous phases nor offered rationalization for the formation of these phases. The objective of the present study is to provide experimental information on the formation and growth of the amorphous phase between GaAs and Co and to provide a rationalization for the formation of this type of phase. In next paper, the role of thermodynamics in the formation of the amorphous phase will be discussed further, based on the experimentally determined phase equilibria in the Ga-As-Co system.

EXPERIMENTAL PROCEDURE

Si-doped ($2-4 \times 10^{17}$ cm^{-3}) (001) GaAs substrates were degreased in acetone and trichoroethylene (TCE), etched in 5 % HCl for 2 min., rinsed in de-ionized H_2O, and then dried with N_2 gas before being loaded into the evaporation chamber. Cobalt thin films 40-100 nm in thickness were deposited onto the wafers by electron beam evaporation under a vacuum of 10^{-7} torr. The deposition rate was about 0.2 nm/sec. The samples were encapsulated in 7 mm i.d. quartz tubes under a vacuum of 10^{-4} torr for subsequent heat treatments at temperatures up to 340 °C for various periods of time.

Cross-sectional transmission electron microscopy (XTEM) specimens were mechanically thinned and ion milled at 5 keV on a liquid-nitrogen cooled stage. Plan-view TEM specimens were prepared by back-etching from the GaAs side with a 5% bromine-methanol solution. Structural analysis was done using a JEOL 200 CX scanning transmission electron microscope and a Vacuum Generator HB501 scanning transmission electron microscope equipped with energy-dispersive X-ray analyzers (EDS's). A well-focused electron beam (with a probe size of 1 nm) obtained in the HB501 STEM is a particularly powerful tool for characterizing amorphizaton. In the present experiments, the single III-V/M interface precludes the use of the X-ray diffractometer as a tool to investigate the SSAR.

RESULTS

The formation of an amorphous initial phase in the samples heat-treated at 260-280 °C for certain periods of time has been experimentally determined previously using TEM techniques. An example is shown in the XTEM micrograph in Fig.1(a). The reacted layer with no granular appearance indicated that the amorphous phase was still stable in the samples annealed at 280 °C for 4 h. After annealing at 300 °C for 3 h., a recrystallization was observed to initiate from the amorphous phase/GaAs interface, as seen in Fig.1(b). This is not surprising since this fully crystalline phase was found to be highly oriented with respect to GaAs after annealing at 300 °C for periods of time longer than 3 h. or at temperature higher than 300 °C. The columnar crystalline phase displayed in Fig.1(c) was observed from the sample annealed at 300 °C for 9 h. The EDS and TEM diffraction pattern analysis[20,21] have shown that this first crystalline phase exhibits the CoAs structure with a composition of $Co(Ga_{.48}As_{.52})$. Co-Ga-As phase diagram[22] studies indicated it to be a supersaturated solid solution of the CoAs phase, i.e. the μ phase. The results outlined above specify the critical annealing conditions for characterizing the growth kinetics of the amorphous and μ phases.

The kinetic data for the growth of the amorphous and the μ phases as presented in Fig.2 were obtained from two sets of samples with the initial thickness of Co being 40 and 100 nm , , respectively. As shown in this figure, the thickness,X, of the amorphous and supersaturated μ phases varies linearly with $t^{1/2}$ within the scatter of the data. This suggests that the growth kinetics are controlled by diffusion. At 300 °C, the initial phase formed is amorphous until t≈180 mins. ($t^{1/2}$≈13.5 min.$^{1/2}$). Subsequently, it transforms to the supersaturated μ phase. At t = 1100 mins.($t^{1/2}$≈33 min.$^{1/2}$), the supersaturated phase has not yet decomposed to other phases in equilibrium with GaAs. At higher temperatures such as 320 and 340 °C, the growth kinetics of the amorphous phase could not be determined because of the rapid

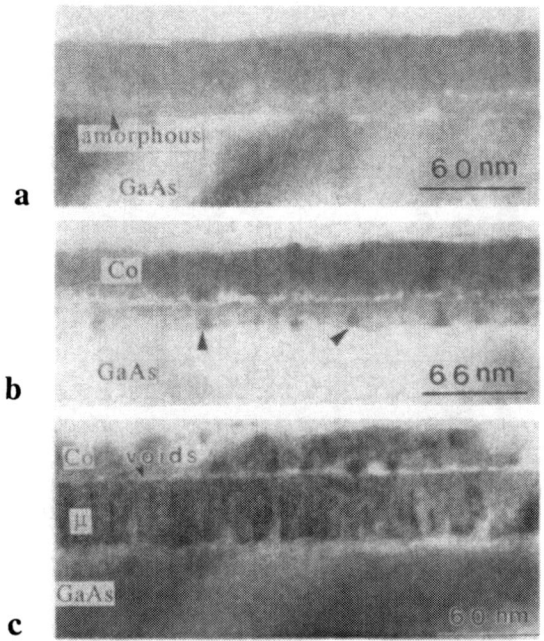

Fig.1 XTEM micrographs for the couples annealed at (a) 280 °C,
4 h., (b) 300 °C, 3 h.,and (c) 300 °C, 9 h.

crystallization of the μ-phases. Values of the parabolic rate
constants,$K=X^2/2t$, obtained for the growths of the amorphous and μ
phases are also displayed in Fig.2a.

The parabolic growth constants are plotted in Fig.2b in the
Arrhenius manner. The activation energies obtained are 130±10
kJ/mol (1.35±0.1 eV) and 142±20 kJ/mol (1.47±0.2 eV), respectively
for the growths of the amorphous and supersaturated μ phases.

DISCUSSION

As has been pointed out by many investigators[1-13], the ability
of one species to diffuse anomalously fast into the other is the
most important factor for SSAR. The crystalline-to-amorphous phase
transformation in Co/GaAs is also attributed to the fast diffusion
of Co through the GaAs. The Kirkendall voids observed at the
Co/amorphous phase interface shown in Fig.1 experimentally
substantiate this suggestion. The diffusion coefficient of Co in
GaAs at 580 °C was reported to be, at least, 10^5 times faster than
that of the self-diffusion of Ga or As in GaAs[23]. At 200-300 °C,
the ratio would be even bigger since the activation energy for the
diffusion of Co in GaAs is much smaller than those for self-
diffusion. At high temperatures, an interstitial-vacancy

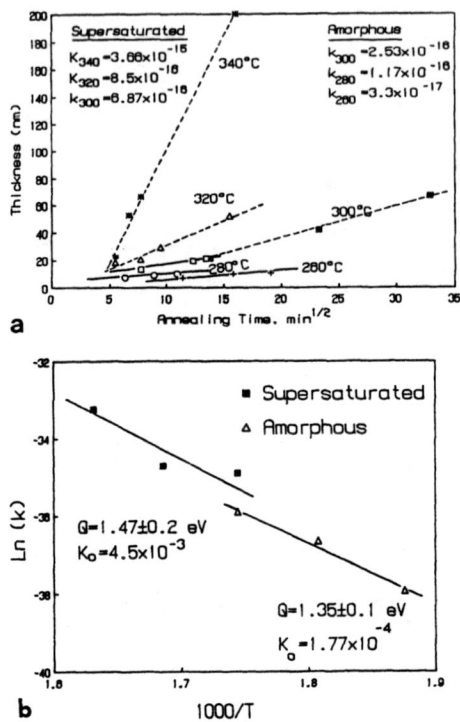

Fig. 2. (a) Parabolic plots and (b) Arrhenius plots for the
amorphous phase and the μ phase.

mechanism[23] was proposed for the fast diffusion while at 200-300 °C
an interstitial mechanism would dominate since the vacancy
concentration would become negligible. Moreover, the interstitial
voids of the GaAs were calculated to be 1.02 times larger than the
size of a Co atom[18]. This crystal structure openness would promote
the fast diffusion of Co in GaAs via an interstitial mechanism.

At 260-300 °C as the Co diffuses into the GaAs lattice, the
GaAs interstitial sites are gradually occupied by the Co atoms and
the coordination number of the Ga and As is significantly altered.
The electrons which are in the Ga-As bonds can not be localized in
the sp^3 state any more, but must be shared with the interstitial Co
atoms. Each of the Ga-As bonds will have less than two electorns.
They become unsaturated bonds. Once the Co concentration reaches a
critical value, the basic GaAs lattice can no longer be sustained.
The Ga-As bonds are then broken. If the Ga-As bond breaking rate
is larger than the rate of rearrangement of the atoms to nucleate
the first crystalline compound, there exists a certain period of
time for some transient state (i.e. an amorphous phase) to
prevail. Realizing that 1) the movements of the Ga and As in a
collapsed and interstitially fully occupied structure could be
significantly retarded, and 2) the first crystalline phase, the μ

phase, has a rather complex orthorhombic structure[24], the nucleation of the first crystalline phase can be kinetically suppressed, leading to the amorphous phase formation.

In order for the amorphous initial layer to continue to grow, Co diffusion through the amorphous interlayer needs to be fast as well. The reaction rate at 260-280 °C ,as displayed in Fig.2a,for the amorphizaton is considerably high, order of 10^{-16} cm^2/sec. The amorphous phase is sufficiently stable kinetically at these temperatures. At 300 °C, the growth rates of the amorphous phase initially and the μ-phase subsequently were measured. The results shown in Fig.2a clearly demonstrate that the rate of growth of the μ-phase, if extrapolated down to 280 °C, is greater than that of the amorphous phase. The Arrhenius plots of the parabolic rate constants for the growth of the amorphous and the μ phases also show that the rate of growth of the μ-phase is greater. To a first approximation, we may state that the interdiffusion coefficients of the μ-phase are higher than those of the amorphous phase. This seems to be surprising at first since the amorphous phase closely resembles the liquid state and should have holes in its structure. However, as has been demonstrated recently by Jan, Swenson, Zheng, Lin and Chang[25], the intrinsic diffusivity of Ni in Ni$_3$GaAs as well as the interdiffusion coefficients are rather high. These results were rationalized in terms of an interstitial diffusion mechanism on the basis of structural analysis. If this were also true for the μ-phase, the interdiffusion coefficients would be expected to be higher in the crystalline phase than in the amorphous phase.

An examination of the data in Fig.2a shows that extrapolations of X for the μ-phase to zero intercept the abscissa at $t^{1/2} \geq 0$. The term X denotes the thickness of the phase formed. On the other hand, extrapolations of X for the amorphous phase intercept the abscissa at $t^{1/2} \leq 0$. Lur and Chen[13]'s data on the formation of an amorphous phase in Ti/Si couples yielded the same results. This phenomenon has not been observed for the formation of crystalline phases in diffusion couples as far as we know. One possible explanation is that the growth kinetics of the amorphous phase initially do not follow a parabolic relationship. Oxidation of alloys such as (Pb, In)[26] sometimes behave in a similar manner. There is an initial fast oxidation rate followed by slower parabolic growth. However, the kinetics of SSAR are not well understood. Undoubtedly, more careful investigations of the early stage of SSAR will lead to better understanding of this phenomenon.

CONCLUSIONS

Solid-state amorphization reactions between GaAs and Co were observed at low temperatures, 260-300 °C. The formation of the amorphous phase initially was attributed to the openness of the GaAs structure relative to the size of the Co atoms. This allows rapid diffusion of Co into the GaAs lattice and promotes the occurrance of SSAR. Annealing at higher temperatures or for longer times led to the formation of a crystalline phase, which was determined to be a metastable supersaturated solid solution of CoAs exhibiting the B31 structure with the approximate composition of Co(Ga$_{.48}$As$_{.52}$). The growth kinetics of both the amrophous phase and the μ-phase are parabolic in nature. The parabolic rate constant is higher for the μ-phase than that for the amorphous phase. The activation energies are 1.47 and 1.35 eV, respectively.

REFERENCES

1 R.B. Schwarz and W.L. Johnson, Phys. Rev. Lett., 51, 415(1983).
2. R.B. Schwarz, K.L. Wong, W.L. Johnson and B.M. Clemens, J. Non-Cryst.Solid, 61-62, 129 (1984).
3. B.M. Clemens, W.L. Johnson, and R.B. Schwarz, J. Non-Cryst. Solid, 61-62, 817 (1984).
4 S.B. Newcomb and K.N. Tu, Appl. Phys. Lett. 48 1437 (1986).
5 J.C. Barbour, F.W. Saris, M. Wastasi, and J.W. Mayer, Phys. Rev., B32, 1363 (1985).
6. H. Schroder, K. Samwer, and U. Koster, Phys. Rev. Lett., 54, 197 (1985).
7. M. Van Rossum, M.A. Nicolet, and W.L. Johnson, Phys. Rev., B29, 5498(1984).
8. P. Guilmin, P. Guyot, and G. Marchal, Phys. Lett., 109A, 174 (1985).
9. S.R. Herd, K.N. Tu, and K.Y. Ahn, Appl. Phys. Lett., 42, 597 (1983).
10. M. Matan, Appl. Phys. Lett., 49, 257 (1986).
11. K. Holloway and R. Sinclair, J. Appl. Phys., 61, 1359 (1987).
12. W. Lur and L.J. Chen, Appl. Phys. Lett., 54, 1217 (1989).
13. J.Y. Cheng and L.J. Chen, Appl. Phys. Lett. 57, 612 (1990).
14. V.A. Uskov, A.B. Fedotov, E.A. Eroteeva, A.I. Rodionov, and D.T.Dzhumakulov, Izv. Akad. Nauk SSSR, Neorgan. Mater., 23, 186 (1987).
15. T. Sands, C.C. Chang, A.S. Kaplan, v.G. Keramidas, K.M. Kirshnan, and J. Washburn, Appl. Phys. Lett., 50 1436 (1987).
16. R. Caron-Popowich, J. Washburm, T. Sands, and A.S. Kaplan, J. Appl.Phys., 64, 4909 (1988).
17. F.Y. Shiau and Y.A. Chang, Appl. Phys. Lett., 55, 1510 (1989).
18. F.Y. Shiau, PhD Thesis, University of Wisconsin, Madison, WI (1990).
19. F.Y. Shiau and Y.A. Chang, in Thin-Film Structures and Phase Stability(Eds.: B.M. Clemens and W.J. Johnson), Mat. Res. Soc. Symp. Proc.,(1990).
20. F.Y. Shiau and Y.A. Chang, Mat. Res. Soc. Symp., 148, 29 (1989).
21. F.Y. Shiau, Y.A. Chang and L.J. Chen, J. Electron. Mater., 17, 433 (1988).
22. F.Y. Shiau, Y. Zuo, J.C. Lin, X.Y. Zheng, and Y.A. Chang, Z. Metallk., 80, 544(1989).
23. G.S. Kulikov and I.N. Nikulitsa, Sov. Phys. -Solid State 14, 2335 (1973).
24. R.W.G. Wyckoff, in Crystal structure, 2nd ed. (Wiley, New York, 1963),Vol. 1, p122.
25. C.H. Jan, D. Swenson and Y.A. Chang, in Fundamentals and Applications of Ternary Diffusion (Ed..: G.R. Purdy), Pergamon Press, New York, 127(1990).
26. M.X. Zhang, Y.A. Chang and V.C. Marcotte, J. Electrochem. Soc. 137, 3158(1990).

PART II

Metal-Metal Thin Film Reactions and Transformations

FIRST PHASE FORMATION KINETICS IN THE REACTION OF Nb/Al

K.R. COFFEY[*], K. BARMAK[**], D.A. RUDMAN[***,+] AND S. FONER[****]
[*]I.B.M., SSPD, 5600 Cottle Rd., San Jose, CA 95193
[**]I.B.M., GTD, East Fishkill Facility, Hopewell Jct., NY 12533
[***]N.I.S.T., 325 Broadway, Boulder, CO 80303
[****]N.M.L., M.I.T., Cambridge, MA 02039

ABSTRACT

Phase formation kinetics in the reaction of Nb/Al multilayered thin films were investigated using scanning calorimetry, x-ray diffraction, and transmission electron microscopy. The first phase to form upon annealing the Nb/Al layered structure is the NbAl$_3$ intermetallic. Its formation is clearly identified by the calorimetry to be a two stage process, which has been modeled as the nucleation and three-dimensional growth to coalescence of the product phase in the plane of the initial interface, followed by the thickening of the product layer by one-dimensional growth perpendicular to the interface plane. For the initial reaction stage the reaction front velocity is higher than can be supported by diffusional transport within the lattice adjacent to the moving interface. Thus diffusion along nonequilibrium interfaces must be the growth mechanism. The large volume fraction consumed during the initial reaction stage indicates a lower nucleation site density than expected at a Nb/Al interface at local equilibrium, suggesting that the interface transport is reducing the driving force for nucleation.

INTRODUCTION

This paper reports the kinetics of first phase formation in the reaction of evaporated multilayer thin films of niobium and aluminum. The Nb/Al system contains three intermetallic compounds in the equilibrium phase diagram.[1] We have previously published[2] the results of a systematic study of the product phase formation sequence in the reaction of Nb and Al and we will now present scanning calorimetry results to provide quantitative information regarding the first phase (NbAl$_3$) formation kinetics in this system and consider the implications of these results to first phase formation in thin film reactions in general.

In agreement with the previous workers[3,4] in this system, we observe the first phase to form upon annealing the Nb/Al layered structure to be the NbAl$_3$ intermetallic. However, the NbAl$_3$ formation is clearly identified by analysis of the calorimetry and x-ray diffraction (XRD) data to be a two stage process, and not simply the thickening of a planar product layer. A model has been previously presented[5] in which the first step is taken to be the nucleation of isolated grains in the initial interface and the growth of these grains until they coalesce and form a planar product phase layer. The second step is the expected thickening of the now planar product layer by uniform growth perpendicular to the interface plane.

EXPERIMENTAL

All the films for this study were deposited by UHV electron beam evaporation. The total film thickness was always 1 μm while the individual layer thickness ranged from 34 nm to 367 nm for Al and 38 nm to 383 nm for Nb. The effect of oxygen contamination in the Nb/Al interface on the subsequent reactions has been studied using this UHV deposition equipment,[6] and the preparation conditions used for these multilayer samples are those found consistent with an oxide-free interface. Transmission electron microscopy (TEM) was used to characterize the as-deposited films.[2] In order to ensure continuous planar Al layers and avoid islanding of the Al, liquid nitrogen was used to cool a copper substrate holder during deposition. For all samples deposited under these conditions, the Nb grain size is < 30 nm independent of layer thickness. The Al grain size is of the order of the Al layer thickness, maintaining a single layer of Al grains in each deposited layer.

The scanning calorimetry technique developed for this work used a differential thermal analysis cell. The experimental process is non-isothermal: starting from room temperature, the cell temperature is increased linearly with time until the maximum desired temperature has been reached. The power evolved from the thermodynamically irreversible reactions is calculated from the raw data as a function of time and temperature. The calorimetry data is conventionally displayed as power versus temperature. Further details of the experimental techniques are provided elsewhere.[7]

RESULTS

Figure 1 is the result of a typical calorimetry run. The multilayer film was of 33 at.% Al overall composition (d_{Al} = 45 nm, d_{Nb} = 98 nm) and was annealed at 20 K/min. The vertical axis is the power evolved due to the exothermic reactions in the film and the horizontal axis is temperature. The horizontal axis is also implicitly time, since the calorimetry was done at a constant heating rate. Three maxima in the reaction rate (peaks), labelled A, B, and C in the figure, are observed. The identification of the phases formed in each of these peaks was accomplished by XRD analysis of a series of samples of the same configuration annealed to different maximum temperatures, as indicated by the lowercase letters in Fig. 1. The as-deposited films contain only Nb and Al. After annealing to a maximum temperature T_M = 708 K (point b in Fig. 1), the presence of $NbAl_3$ along with residual Al and Nb was observed. Further annealing to T_M = 980 K (point c) results in the consumption of the Al and the further formation of $NbAl_3$. The presence of the $NbAl_3$ phase as the only product phase at temperatures less than 980 K indicates clearly that both peaks A and B in Fig. 1 result from reactions in which this single phase is the product. The presence of $NbAl_3$ was confirmed for maximum annealing temperatures as low as 662 K by electron diffraction in a TEM sample.[2] The $NbAl_3$ formation was always completed by 980 K, before the formation of other product phases at peak C.

The area and position of peaks A and B depend systematically on the as-deposited Al layer thickness, as shown in Fig. 2. Peak A shifts only very weakly with Al layer thickness, while peak B shifts to higher temperatures (later times) with thicker initial Al layers. Inspection of the calorimetry curves for different samples also indicates that peak A corresponds to the reaction of 30 nm of the deposited Al layer, independent of initial layer thickness. Thus peak A becomes a smaller fraction of the total $NbAl_3$ formation (A + B) for samples with thicker initial Al layer as can be seen by comparison of Figs. 1 and 3. The increase in magnitude and the shift to later times and temperatures of peak B with as-deposited Al layer thickness identify it as the expected planar thickening of an $NbAl_3$ product layer.

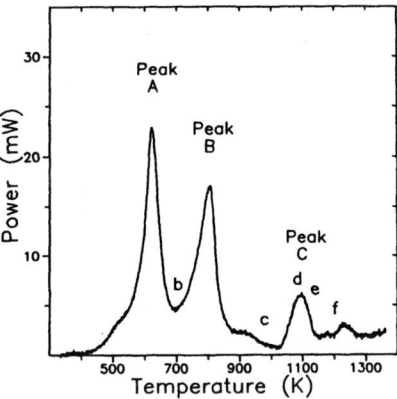

Figure 1 - Scanning calorimetry experiment using a heating rate of 20 K/min. The multilayer film sample was 36 mg of 33 at.% Al overall composition (d_{Al} = 45 nm, d_{Nb} = 98 nm). Peaks A and B are both stages of $NbAl_3$ formation. Peak C is the first stage of the formation of the A15 superconducting phase Nb_3Al. The lowercase letters refer to maximum annealing temperatures used to prepare samples for the XRD analysis.

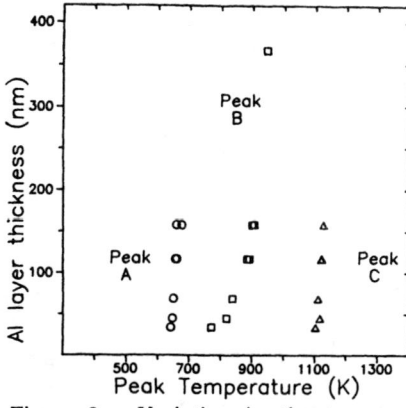

Figure 2 - Variation in the temperatures for peaks A, B, and C as a function of the thickness of the deposited Al layer for 50 K/min scanning calorimetry experiments.

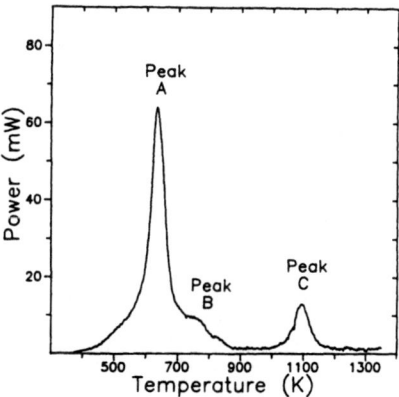

Figure 3 - Scanning calorimetry experiment at 50 K/min using a multilayer film sample of 29 mg of 25 at.% Al overall composition (d_{Al} = 34 nm, d_{Nb} = 109 nm).

Quantitative Interpretation

The heat measured by calorimetry is related linearly to the volume of the intermetallic compound formed, the proportionality given by the heat of formation of the compound. This value is readily obtained by integration of the power signal. For $NbAl_3$ a heat of formation of 170 ± 30 kJ/mol $NbAl_3$ is found in the temperature range of 600 to 1000 K. This value for the heat of formation of $NbAl_3$ agrees fairly well with the value of 200 kJ/mol at 870 K obtained from thermodynamic calculations (free energy curves) of the Nb-Al system by R. Bormann.[8]

The relative areas of peak A and B can be used to estimate the thickness of the $NbAl_3$ layer when coalescence occurs, L_C. Inspection of Figs. 1 and 3 indicate L_C as approximately 20 nm, and this is consistent with the grain size observed at the initial interface by TEM (see Fig. 4 of Ref. 2).

The positions in temperature of the peaks are expected to shift slightly when experiments at heating rates of 20 K/min and 50 K/min are compared. These shifts can be used to determine the effective activation energies Q for the reactions represented by each peak.[9] The effective activation energy calculated for peak A is 1.5 ± 0.2 eV. The effective activation energy for peak B is affected by the presence of peak A, but for samples with Al layer thicknesses between 100 and 200 nm is equal to 1.4 ± 0.3 eV. The activation energy calculated for peak B is consistent with grain boundary diffusion, based upon reports of $NbAl_3$ layer growth kinetics in bulk systems.[10] The effective activation energy for peak A, which includes nucleation events, is not significantly higher than that of peak B (growth only). The nucleation events therefore occur with small Δg^*, the activation barrier for critical nucleus formation.

The velocity of the growing interface during product phase formation can be estimated by direct inspection of the calorimetry data. The area under a peak between an upper and lower temperature limit is proportional to the amount of product phase formed and hence to the distance travelled by a planar reaction front. The time required for the reaction front to move this distance is readily determined from the difference between the temperature limits and the heating rate. The area under peaks A and B in Fig. 3 results from the reaction of 34 nm Al layers with Nb to form 45 nm $NbAl_3$ layers (actually two 22.5 nm $NbAl_3$ layers grow from each Al layer). Inspection of the central region of peak A shows that 50 % of the total area of peaks A and B is found between the temperature limits of 580 and 680 K; that is, the $NbAl_3$ layer is 50 % formed between 580 K and 680 K. For a heating rate of 50

K/min this corresponds to two 12 nm layers of NbAl$_3$ being formed in 120 s. If growth is assumed to initiate in a single plane, then 12 nm/120 s gives an average interface growth velocity (perpendicular to the Nb/Al interface plane) for NbAl$_3$ of 10^{-8} cm/s at an average temperature of 630 K. Since this velocity is much greater than the velocity expected from studies of bulk Nb/Al diffusion couples,[10] 10^{-10} cm/s, grain boundary diffusion, rather than diffusion through the product phase lattice, is the dominant atomic transport mechanism in these fine-grained multilayer films.

DISCUSSION

The most striking feature of the calorimetry curves is the occurrence of peaks A and B, two maxima in the reaction rate, during the formation of a single product phase. The peak A reaction has been previously modeled as the growth of isolated nuclei to form a planar product phase layer,[5] and recently this modeling has been extended to explain the slight shift in peak A position as nucleation at preferred sites in the interface that vary with deposited layer thickness.[11] This shift can be observed in Fig. 2 of this work, and we believe nucleation at preferred sites in the interface is the mechanism for Nb/Al as well.[7] Of general interest in understanding first product phase formation is why the coalescence length, L_C, is so large. For L_C to be large, significant nucleation barriers must prevent nucleation at the interfacial sites between the isolated nuclei. To understand the implications of the large L_C observed we can examine models based upon nucleation and growth kinetics. The equations describing phase transformations that proceed by nucleation and growth are well known in three-dimensional systems and were extended by Cahn to planar systems.[12] The case of an isothermal transformation will be examined, because it is mathematically simpler and conveys the physical understanding required.

L_C can be defined as the extent of growth perpendicular to the Nb/Al interface when the Nb/Al interfacial area is 90 % consumed. For constant nucleation and growth rates the area fraction transformed, Y, is given by

$$Y = 1 - \exp(-\pi I G^2 t^3), \tag{1}$$

where I is the nucleation rate, G is an isotropic growth rate, and t is time elapsed. At Y = 0.90, t = t_C and the above equation can be solved with the relation $L_C = Gt_C$ to yield

$$L_C^3 = (2.3/\pi) \, G/I, \tag{2}$$

which compactly expresses how the coalescence thickness (which is also the grain size) is controlled by the competition between growth and nucleation rates. The nucleation rate can be expressed as[13]

$$I = N \, (kT/h) \, \exp\{-(\Delta g^* + g_m)/kT\}, \tag{3}$$

where N is the nucleation site density, kT/h is an attempt frequency, and g_m is an activation energy for atomic motion at the interface. Since the motion required for nucleation and the motion of the subsequent growth rely on similar atomic mechanisms, they can be expected to have similar activation energies and attempt frequencies giving

$$G = d \, kT/h \, \exp\{-g_m/kT\}, \tag{4}$$

where d is the distance of each atomic motion. Using equations (3) and (4) in equation (2) and neglecting the 2.3/π prefactor gives

$$L_C^3 = (d/N) \, \exp(\Delta g^*/kT). \tag{5}$$

The right side of equation (5) should be considered an upper bound for L_C^3 because equation (4) is an upper bound for G that neglects long range diffusion.

Equation (5) can now be used to calculate values for Δg^* that are consistent with the experimental observations. For nucleation at preferred sites in the Nb/Al interface at T = 600 K, with L_C = 20 nm, d = 0.2 nm, and N = (100 nm)$^{-2}$, Δg^* can be calculated as 0.07 eV, consistent with the similar effective activation energies found for peak A and B. However, for the nucleation at preferred sites to be observed, the competing mechanism of random nucleation at all sites in the Nb/Al interface must be significantly slower. A suitable criterion is that the L_C upper bound calculated by equation (5) for random nucleation be a factor of 10 larger than that observed for nucleation at preferred sites. Then for random nucleation in the Nb/Al interface at T = 600 K, with L_C = 200 nm, N = 1/d2 (all sites are now considered), d = 0.2 nm, and T 600 K, Δg^* is approximately 1 eV.

This value of Δg^* of 1 eV for random nucleation in the Nb/Al interface requires further consideration. The simplest expression for Δg^* is $16\pi\sigma^3/3\Delta g_v^2$, for a isolated spherical nuclei, where σ is the interfacial energy and Δg_v is the change in volumetric free energy. For heterogeneous nucleation at an interface, the calculation of Δg^* is complicated by the loss of Nb/Al interface as well as the creation of Nb/NbAl$_3$ and NbAl$_3$/Al interfaces. The nonwetting of Al when deposited on Nb at room temperature indicates that the Nb/Al interfacial energy is large; hence the formation Nb/NbAl$_3$ and NbAl$_3$/Al interfaces should only moderately increase the energy of the system, especially if coherency of one of the nuclei interfaces is assumed as the lowest energy configuration. Thus the nucleation barrier is assumed to derive from a low thermodynamic driving force for nucleation. Using the spherical nucleus model for simplicity, with σ = 200 mJ/m2 to reflect the change in total interfacial energy of the system, Δg^* equal to 1 eV gives Δg_v as 0.9 kJ/cm^3, well less than the 4.5 kJ/cm^3 expected from the free energy change of the reaction. If a local metastable equilibrium is achieved by interdiffusion at the Nb/Al interface, the free energy at the interface will be lowered and Δg_v will be reduced. While this effect is potentially significant, it can account for only a 40 % reduction from 4.5 KJ/cm3, based upon thermodynamic models for the Nb/Al system developed by R. Bormann.[14] It is our interpretation that a truly nonequilibrium condition exists and that the low driving force is due to interdiffusion of Nb and/or Al at the grain boundaries. Thus nucleation and growth occur before a metastable equilibrium is achieved. The importance of grain boundary diffusion to this reaction is apparent in the high reaction rates subsequent to the nucleation event and it is unreasonable to assume that grain boundary interdiffusion does not precede the nucleation event as well. For grain boundary interdiffusion, the diffusion in the Nb/Al interphase boundary will not be uniform. The Nb and Al grain boundaries that intersect the interphase boundary will act as sources and sinks of Nb and Al flux and hence result in variations in Nb and Al chemical potential. Thus the interdiffusion would result in a lower thermodynamic driving force for nucleation that would also not be identical at all points in the interphase boundary. The chemical potential variations in the Nb/Al interface due to interdiffusion, may be as important as the surface energy effects associated with specific microstructural features (e.g., grain boundary triple points) to the nucleation process.

CONCLUSION

Scanning calorimetry, XRD and TEM have been used to examine the reaction kinetics for NbAl$_3$ formation in thin films. The NbAl$_3$ formation proceeds by a two stage reaction process as shown by two separate peaks in the calorimetry data. The first stage is the nucleation of the product phase at isolated sites in the Nb/Al interface and the subsequent growth to coalescence of these nuclei to form a 20 nm NbAl$_3$ layer. The second stage is the one-dimensional thickening of the NbAl$_3$ layer.

The most interesting conclusion is that the first phase to form upon annealing these Nb/Al diffusion couples shows preferred nucleation at isolated sites. This is unexpected, because it cannot be accounted for if the thermodynamic driving force for the nucleation of this phase is the bulk free energy of formation of this intermetallic compound. The observation of heterogeneous nucleation at preferred sites in the interface is consistent with a low thermodynamic driving force for nucleation equal to 20 % of the expected bulk free energy change. We suggest that the appropriate thermodynamic driving force is defined by the nonequilibrium grain and interphase boundary interdiffusion at the Nb/Al interface.

ACKNOWLEDGEMENTS

K.C. Russell is acknowledged for helpful discussions. Support provided by: DOE BES DE-FG02-85ER45206, CMSE/NSF DMR-8802613, ATT, and IBM Fellowships

+ Contribution by U.S. Government, not subject to copyright.

REFERENCES

1. J.L. Jorda, R. Flukiger, and J. Muller, J. of the Less Common Metals 75, 22(1980).

2. K. Barmak, K.R. Coffey, D.A. Rudman, and S. Foner, J. Appl. Phys. 67, 7313 (1990).

3. R. Bormann, H.U. Krebs, and A.O. Kent, Adv. Cryo. Eng. ICMC 32, 1041 (1986).

4. Y. Im, P. Johnson, L. McKnelly, Jr., and J.W. Morris, J. of the Less Common Metals 139, 87 (1988).

5. K.R. Coffey, L.A. Clevenger, K. Barmak, D.A. Rudman, and C.V. Thompson, Appl. Phys. Lett. 55, 852 (1989).

6. K. Barmak, K.R. Coffey, D.A. Rudman, and S. Foner, J. Appl. Phys. 67, 3780 (1990).

7. K.R. Coffey, Ph.D Thesis, Massachusetts Institute of Technology, 1989.

8. R. Bormann (private communication).

9. P. G. Boswell, J. Thermal Anal. 18, 353 (1980).

10. G. Slama and A. Vignes, J. of the Less Common Metals 29, 189 (1972); 23, 375 (1971); 24, 1 (1971); 29, 189 (1972).

11. E. Ma, C.V. Thompson, and L.A. Clevenger, presented at the 1990 MRS Fall Meeting, Boston, MA, 1990 (unpublished).

12. J.W. Cahn, Acta Met. 4, 449 (1956).

13. J.W. Christian, The Theory of Transformations in Metals and Alloys, 2nd eds. (Pergamon Press, Oxford, 1975), p. 441.

14. K. Barmak, Ph.D. Thesis, Massachusetts Institute of Technology, 1989.

EFFECT OF MICROSTRUCTURE ON PHASE FORMATION IN THE REACTION OF Nb/Al MULTILAYER THIN FILMS

Katayun Barmak[*], Kevin R. Coffey[**], David A. Rudman[*** +], Simon Foner[****]
* IBM, GTD, East Fishkill Facility, Hopewell Jct., NY 12533
** IBM, SSPD, 5600 Cottle Rd., San Jose CA 95193
*** National Institute of Science and Technology, 325 Broadway, Boulder, CO 80303
**** National Magnet Laboratory, MIT, Cambridge, MA 02139

ABSTRACT

We investigated the phase formation sequence in the reaction of multilayer thin films of Nb/Al with overall compositions of 25 and 33 at.% Al. We report novel phenomena which distinguish thin-film reactions unequivocally from those in bulk systems. For sufficiently thin layers composition and stability of product phases are found to deviate significantly from that predicted from the equilibrium phase diagram. We demonstrate that in the Nb/Al system the length scales below which such deviations occur is about 150 nm. We believe that these phenomena occur due to the importance of grain boundary diffusion and hence microstructure in these thin films.

INTRODUCTION

We have previously presented the detailed phase formation sequence in the reaction of Nb/Al multilayer thin films [1]. In summary we showed that the reaction progresses through the following stages:

1. $Nb + Al => Nb + NbAl_3$. This stage was completed by 980 K and all the Al was consumed.

2. $Nb + NbAl_3 => Nb + Nb_{3-x}Al + Nb_{2-y}Al + NbAl_3$. During this stage all the phases of the equilibrium phase diagram (except for Al) were present.

3. Advanced stages of the reaction in which the phases present depended on both the overall composition and the multilayer periodicity.

The kinetics of the first stage are presented elsewhere [2]. In the present paper we will concentrate on Stages 2 and 3 for the overall compositions of 25 and 33 at.% Al with the aim of presenting novel phenomena which distinguish thin-film reactions unequivocally from those in bulk systems. In particular we will show that thin film reactions result in deviations from local equilibrium during the reaction and from the equilibrium phase diagram when all reactants are consumed. We will demonstrate that in the Nb/Al system the length scales below which such deviations from equilibrium occur is about 150 nm.

EXPERIMENT

All the films used in this study were 1 μm thick and deposited in an ultrahigh vacuum system on liquid nitrogen cooled sapphire substrates [1]. Under these deposition conditions, the Al layers were continuous and their grain size was of the order of the Al layer thickness, while the grain size of the Nb was < 30 nm independent of the

Table I. Composition, periodicity and layer thicknesses of Nb/Al multilayer films

Composition at.% Al	Periodicity $d_{Nb} + d_{Al}$	d_{Al} nm	d_{Nb} nm
33	143	45	98
	500	158	342
25	70	16	54
	143	34	109
	330	77	253
	500	117	383

layer thickness. The overall compositions of the films were controlled by varying the individual layer thicknesses and were selected at the stoichiometric composition of the two intermetallic phases A15 (25 at.% Al) and sigma (33 at.% Al) [3]. The A15 and sigma phases have a range of compositions and become stoichiometric only at very high temperatures. In particular the composition 25 at.% Al is in the two phase region of A15 and sigma for the temperatures of this study (< 1100 °C). The multilayer compositions and periodicities are presented in Table I. Calorimetry results of Coffey et al. [2] were used to determine the best annealing conditions for mapping the phase formation sequence. The samples were annealed either isothermally or at a constant heating rate in the calorimeter furnace. Additional long isothermal anneals were carried out in a tube furnace. The details of cross-sectional transmission electron microscopy (XTEM) sample preparation are given elsewhere [4]. The superconducting transition temperature T_c was measured by a four point resistive technique.

RESULTS AND DISCUSSION

As Stage 2 progresses, the Nb and NbAl$_3$ are consumed and the A15 and sigma grow. The presence of all four phases can clearly be seen in Fig. 1. The Nb grains have grown to give one row of grains per remaining layer of Nb. The A15 phase has coalesced into a continuous layer approximately 100 nm in thickness with approximately 100 nm lateral grain size. The sigma phase is a highly twinned phase with grain size 10-30 nm and has grown by penetrating the grain boundaries of the NbAl$_3$ phase. Thus the effective area of contact between the sigma and NbAl$_3$ depends on the grain boundary area of the NbAl$_3$. We believe that this effective grain boundary area will be proportional to the thickness of the NbAl$_3$.

The same four phases are observed for a 25 at.% Al film annealed isothermally for 4 hours at 750 °C. We found the same sequence of phases for both 25 and 33 at.% Al films, whether annealing was carried out isothermally or at a constant heating rate. Thus Stage 2, with all four phase present, is a universal behavior for the system.

Although the phase formation sequence for films of 25 and 33 at. % Al is the same through Stage 2, the composition of the A15 shows a notable difference. Figure 2 shows the superconducting transition temperature T_c versus the maximum-anneal-temperature T_m for films of 143 nm periodicity annealed into Stage 2. Since T_c is sensitively dependent on the composition of the superconducting A15 phase [5], Fig. 2 qualitatively represents the composition of the A15 phase for the two films. (Sigma is not superconducting and Nb has a lower transition temperature than A15). The film of 33 at.% Al consistently shows a lower T_c than the 25 at.% Al film.

This is indeed surprising since the A15 is forming from chemically identical reactant layers (Nb and NbAl$_3$). At this early stage of A15 formation, the reaction is incomplete, i.e., the reactant layers are still present, and hence the overall composition of the film

Fig.1 - XTEM micrograph of a 33 at.% Al-
500 nm periodicity film annealed at 50
°C/min to 930 °C (Stage 2). The 2-
headed arrow marks the periodicity.
Four phases co-exist at this stage.

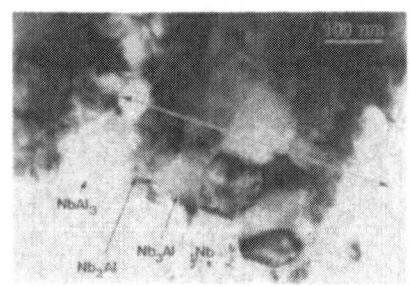

Fig.2 - The superconducting critical
temperature T_c versus maximum-anneal-
temperature T_m for films of 143 nm
periodicity and compositions of 25 and
33 at.% Al. The films were annealed at
50 °C/min. This temperature range
corresponds to Stage 2 of the reaction.

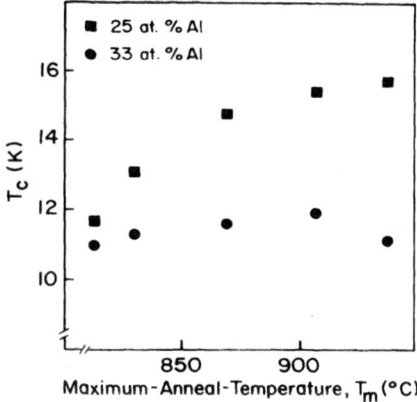

should not yet be "known" to the A15 phase. The existing theories of phase formation
which assume local equilibrium at the interfaces would have predicted identical
composition for the A15 forming from chemically identical reactant phases. Moreover, the
conventional interpretation of interface controlled phase formation is not adequate to
explain Fig. 2, because even though in these models the interface compositions differ
from that predicted by local equilibrium, chemically identical reacting phases would still
lead to identical compositions of the product phase. This is clearly not the case in Fig.
2 where the A15 phase from the very early stages of its formation has a different
composition whether it forms in a film of 25 at.% or 33 at.% overall composition.

We believe a more appropriate explanation of Fig. 2 is that fast grain boundary
diffusion in these fine-grained thin films allows the reacting phases far from the immediate
vicinity of the interface to affect the reaction. The composition of the product phase is
then determined by the ratio of the fluxes of each species to the growing product grains
and not simply the chemical nature of the reacting phases. In other words, this is a
kinetically determined composition and below some characteristic grain boundary diffusion
length a thicker reactant layer which has a larger effective area of reaction will give rise
to a higher flux of that species to the product. Consequently it becomes possible to have
different product compositions even if the reactant layers are chemically identical.

In the present case the Al flux available for the growth of A15 and sigma is proportional to the total surface area of NbAl$_3$ grains and therefore will depend on the thickness of the NbAl$_3$ phase, again assuming that the effective area scales with thickness. Since the NbAl$_3$ in the 33 at.% Al film is thicker than in the 25 at.% film, the growing A15 will be of a different composition in the two films [5].

The films were annealed into Stage 3 and the evolution of phases was studied as a function of overall composition and multilayer periodicity. For films of 33 at.% Al the final phase, independent of layer periodicity, was the sigma phase, as predicted from the equilibrium phase diagram. For films of 25 at.% Al, on the other hand, the multilayer periodicity played a role in the survival or consumption of the sigma phase. The composition 25 at.% Al is in the two phase region of A15 and sigma in the equilibrium phase diagram. Figure 3 is a film of 25 at.% Al-500 nm periodicity annealed at a constant heating rate of 50 °C/min to 1050 °C. It contains both the sigma and A15 phases with approximately 2/3 A15 and 1/3 sigma. Although the exact compositions of the phases are not known, this is in reasonable agreement with the phase diagram. A film of the same composition, i.e., 25 at.% Al, but of 330 nm periodicity, is shown in Fig. 4. This film was annealed at 800 °C for 16 h. Some of the sigma phase remains at the original position of the Al layers, as indicated by the arrows, but far less than seen in Fig. 3 for thicker periodicity film. Finally, for a film of 25 at.% Al-143 nm periodicity, all of the Nb, sigma and NbAl$_3$ are eventually consumed by the A15 phase. Figure 5 shows such a film annealed isothermally at 800 °C for 16 h. The film is fully A15; no other phases are seen in XTEM.

Therefore as the periodicity is changed from 500 to 143 nm for the fixed composition of 25 at.% Al, the multilayers change behavior from one close to that predicted from the equilibrium phase diagram, with both of the equilibrium phases A15 and sigma present in the final film, to one that deviates significantly from equilibrium and becomes single phase A15 with the sigma phase initially forming and then being consumed. The A15 is therefore kinetically stabilized relative to the sigma phase as layer thickness is reduced. This suggests that the layers in the 330 nm periodicity film are of a characteristic length at which the multilayers change from the "bulk" to the "thin-film" regime in this system. This characteristic length is therefore about 150 nm (i.e., (d_{Nb} + d_{Al})/2). This apparently large value is due to the dominance of grain boundary diffusion in these fine grained films.

The fact that 150 nm is a characteristic length for the cross over from "bulk" to the "thin-film" regime is borne out by the measurement of T$_c$ as a function of anneal time at 800 °C for films of 330, 143 and 70 nm periodicity of overall composition 25 at.% Al. The results are presented in Fig. 6. The latter two films reached their final T$_c$ after relatively short annealing times and were essentially indistinguishable from each other. The 330 nm periodicity film had a much slower approach to its final T$_c$ and never reached the maximum value achieved by the other two films because of the remaining sigma phase.

CONCLUSIONS

We investigated the phase formation sequence in the reaction of multilayer thin films of Nb/Al of overall compositions 25 and 33 at.% Al. Both the multilayer composition and periodicity are important in determining the final phases present. We find that the microstructure of thin films is an integral part of the phase formation. Fast grain boundary diffusion allows different fluxes of the reactant species to reach the growing product phases depending on the reactant layer thickness (effective reactant area). Thus chemically identical reactants can give rise to a product with different compositions. We provided experimental evidence that in the Nb/Al system the characteristic diffusion length is about 150 nm. We find that within this length the stability of phases when all reactants are consumed is kinetically determined.

Fig. 3 - XTEM micrograph of a 25 at.% Al-500 nm periodicity film annealed at 50 °C/min to 1050 °C (Stage 3). The arrow marks the periodicity.

Fig. 4 - XTEM micrograph of a 25 at.% Al-330 nm periodicity film annealed isothermally for 16 hours at 800 °C (Stage 3). The 2-headed arrow marks the periodicity. The short arrows indicate the location of the original Al layers now occupied by a small amount of sigma phase.

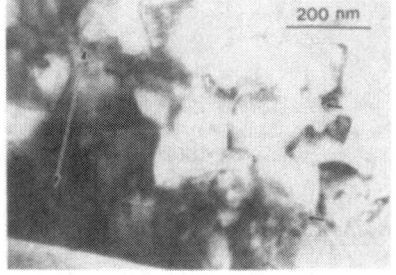

Fig. 5 - XTEM micrograph of a 25 at.% Al-143 nm periodicity film annealed isothermally for 16 hours at 800 °C (Stage 3). The film is fully A15 Nb_3Al. The 2-headed arrow marks the periodicity.

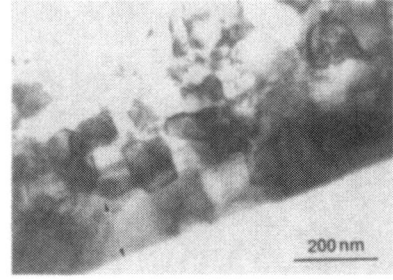

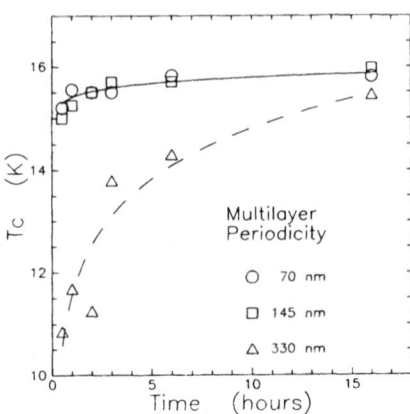

Fig. 6 - The superconducting critical temperature T_c versus annealing time at 800 °C. This corresponds to annealing through Stage 2 to Stage 3.

We wish to thank A. Roshko for critical reading of the manuscript and K.C. Russell for helpful discussions. This work was supported by DOE BES DE-FG02-85ER45206, CMSE/NSF DMR-8802613, AT&T and IBM Fellowships.

REFERENCES

+ Contribution by U.S. Government, not subject to copyright.

1. K. Barmak, K.R. Coffey, D.A. Rudman, S. Foner, J. Appl. Phys. **67**(12), 7313 (1990).
2. K.R. Coffey, K. Barmak, D.A. Rudman, S. Foner, these proceedings and to be published.
3. J.L. Jorda, R. Flükiger, J. Muller, J. Less Common Metals **5**, 227 (1980).
4. K. Barmak, D.A. Rudman, S. Foner, J. Electron Microscopy Technique **16**, 249 (1990).
5. R. Flükiger, <u>Superconductor Materials Science, Metallurgy, Fabrication and Applications</u>, edited by S. Foner and B.B. Schwartz, (Plenum Press, New York, 1981), 511.

Θ CuAl$_2$ PRECIPITATE COARSENING IN Al-2% Cu THIN FILMS

John E. Sanchez, Jr.*, L. T. McKnelly**, J. W. Morris, Jr.***
*Max-Planck Institute Für Metallforschung, Seesstrasse 92, D-7000 Stuttgart 1, Germany
** Vitesse Semiconductor, Camarillo, CA 93010
***Department of Materials Science and Mineral Engineering, University of California, and
Center For Advanced Materials, Lawrence Berkeley Laboratory, Berkeley, CA 94720

ABSTRACT

θ phase CuAl$_2$ precipitate size evolution during coarsening at 310°C in 0.5 μm thick Al-2% (wt) Cu thin films was characterized by transmission electron microscopy. Films were sputter deposited onto oxidized Si substrates by standard techniques. The coarsening process preferred the growth of blocky θ morphologies at Al triple points. Coarsening was via solute Cu diffusion along Al grain boundaries during annealing. The average θ size dependence on annealing time (t) is approximately $(t)^{1/4}$ in general agreement with models for particle coarsening along grain boundaries. Concurrent Al grain growth was shown to initially enhance the θ coarsening rate above $(t)^{1/4}$ behavior. This boundary coarsening process leads to a grain size dependence of the coarsening rate which has been observed in related and other previous work in thin films. These results are shown to be relevant for effects produced during accelerated electromigration testing, such as previous 'curious' θ morphologies at triple points observed by others, the enhanced flux of Cu during testing, and possible mechanisms affecting electromigration failure processes.

INTRODUCTION

Al-Cu alloy thin films (up to 4 wt.%) are widely used as the metallization interconnect material in integrated circuit devices. This is primarily due to the significant increase in electromigration lifetimes, as first reported 20 years ago [1], when Cu is added to Al or Al-Si films. This effect is assumed to be the result of Cu solute segregation to the Al grain boundaries and the resulting decrease in Al boundary diffusion rates. Since it is generally assumed that electromigration failure mechanisms are also determined by boundary diffusion processes, Cu enrichment at boundaries leads to longer interconnect lifetimes. At equilibrium the bulk Cu solubility at low temperatures (less than 200°C) is quite small, less than 0.1%. Cu segregation is not expected to enrich the boundary significantly above this amount. Typically, the alloys in use contain 0.5 % wt. or more, so that after typical anneals [2] most of the Cu exists in the CuAl$_2$ precipitate forms θ (stable and incoherent) and θ´ [3] (metastable and semicoherent). The θ phase has been shown [2] to reside principally at grain boundaries and grain triple points. As pointed out by others [1] during electromigration testing at elevated temperatures these grain boundary θ phases may continually resupply their local boundary region with solute Cu, leading to prolonged lifetime enhancement. Failure eventually occurs when the boundary θ phases completely dissolve in a region, allowing for rapid Al boundary diffusion and failure. Thus the size and relative stability of boundary θ phases are expected to affect electromigration lifetimes. Results of early work [4] in Al-Cu films did not show an obvious θ phase size effect on electromigration lifetimes. However, that study was on films containing greater Cu concentrations (4-19% wt.) than typically used now, i.e., 0.5-4%. In addition the difficulty of plasma etching Cu rich phases [5] with standard fluorine and chlorine based ambients places a restraint on the size of allowable θ phases and amount of Cu used in films for Very Large Scale Integration (VLSI) technologies. Therefore the control and optimization of θ phase distributions of size and morphology are relevant for both electromigration properties and integrated circuit fabrication technologies.

The present work was undertaken to determine the θ phase morphological evolution during relatively low (310°C) temperature anneals. It was previously shown [2] that during low temperature annealing, significantly below the solvus temperature, the θ phase in the Al 2% Cu film coarsens via the classical Ostwald process producing large particles at grain triple points. For the 2% Cu film, 345°C allowed appreciable Cu diffusion with most of the θ volume fraction undissolved. This coarsening process is significant since it will predominate the θ phase evolution process as Al-Cu films undergo other low temperature treatments during circuit fabrication and characterization, such as lithographic resist baking, depositions of passivation dielectric films and accelerated electromigration testing. The Ostwald coarsening results will be extended to briefly describe the effects of electromigration-induced mass flux on the coarsening mechanism and the possible relevance for electromigration failure processes.

EXPERIMENTAL

Al-2±0.02% Cu thin films 0.5 μm thick were magnetron sputter deposited onto 450Å of thermal oxide grown on (100) 100 mm Si substrate wafers. Base pressure of the sputter system was less than $2x10^{-5}$ mTorr. Film composition was verified by high temperature plasma mass spectrometry. Film sputter deposition rate was 1200Å per minute. Substrates were unheated and without applied bias. Small (1 cm x 1 cm) sections from the same wafer were heat treated in (N_2/H_2) in a hot wall tube furnace for up to 40 minutes at 310°C. Disks 3 mm in diameter were cut from wafer sections for TEM sample preparation. Sample preparation and Si substrate removal were by standard TEM foil preparation methods. Final sample thinning was by Ar ion milling. Al grain size and coarsening θ phase sizes, both within grains and at triple points, were measured from several TEM micrographs taken from at least two TEM foils for each film condition. Intercept lengths were measured using a digitizing pad connected to a PC running custom software for data retrieval and analysis, except in two cases noted below where the measurements were done by hand. Grain and precipitate sizes were calculated as the average of two orthogonal length measurements chosen in random directions for each particle and grain.

RESULTS

The as deposited θ particles, about 200Å in diameter, are uniformly distributed along grain boundaries and grain interiors. Cross section TEM results on similar as deposited films [2] show a relatively even dispersion of θ through the film thickness. Annealing for 5 minutes at 310°C showed the initiation of θ phase growth at triple points and at grain interiors, as well as growth of the Al grains, Figure 1. Annealing for 40 minutes at 310°C shows extensive Al growth and much larger θ coarsening phases at triple points, Figure 2. The θ particles remaining in grain interiors have reached a relatively stable size at this anneal time. The results of grain growth and coarsening are summarized in Table I.

The Al grain growth kinetics were determined using the relation [6-8]

$$d/d_0 \cong (kt)^n \qquad (1)$$

where d is the annealed grain diameter, d_0 is the as deposited grain size, k is the thermally activated mobility, t is the anneal time and n is the kinetic exponent. Analysis gives n ≈ 0.5, in agreement with models [6,7] and simulations [6-8] for thin film grain growth. This suggests grain growth is not limited here by θ that can act as pinning obstacles to boundary motion [9].

The data for θ coarsening, Table 1, show that the growth of θ in grain interiors ("in-grain") is saturated after 20 minute anneals at about 3 times the as deposited size, while at triple points the blocky θ continues growth at least until the 40 minute anneal times. The θ coarsening rate behavior at triple points is shown in Figure 3. The growth and coarsening of Al grains, blocky θ and in-grain θ are shown in Figure 4, with each phase size plotted in reduced form, normalized by the appropriate as deposited size. Note that the blocky θ phase shows an initial rapid coarsening rate which decreases to about $t^{1/4}$ with prolonged annealing times, while the in-grain θ saturates growth at about 0.07 μm (20 minutes).

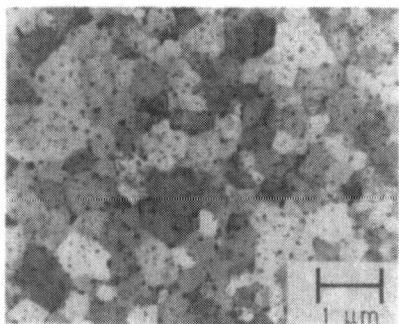 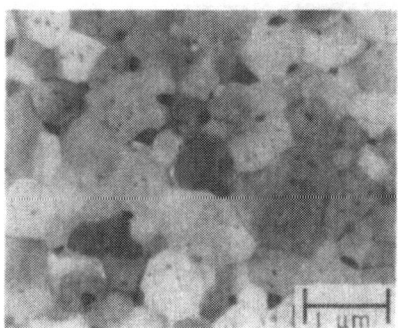

Figure 1. TEM micrograph of Al-2% Cu thin film after 5 minutes at 310°C. Small dark areas in-grain θ, with blocky θ growing at triple points.

Figure 2. TEM of same Al-2%Cu film after 40 minutes, 310°C. Note extensive growth of Al and large θ at triple points.

DISCUSSION

It is evident that the blocky θ phase at triple points coarsens primarily by boundary diffusion, and that this process continues to a larger extent and for longer times than the bulk in-grain θ coarsening process [10]. The in-grain θ growth, limited by bulk diffusion, is kinetically isolated from the boundary coarsening pathways and process. Note in Figure 2 that most of the in-grain θ are at grain centers, that is, there exists a precipitate depleted zone adjacent to most boundaries. Matrix θ which are within ≈ 0.1 μm of the boundary dissolve with the Cu diffusing to the boundary and rapidly diffusing to the coarsening blocky θ. This process will slightly increase the blocky θ growth rate by increasing the θ volume fraction available for coarsening.

The concurrent Al grain growth will similarly enhance the boundary coarsening process when a moving/migrating boundary encounters a matrix θ. The θ phase, now on a rapid diffusion pathway, rapidly dissolves in favor of θ phase located at a triple point. The boundary may continue its growth motion when the particle is completely dissolved. This process also enhances the blocky θ coarsening rate by continually increasing the volume fraction of previously kinetically isolated θ phase available for coarsening. This is consistent with the results, Table 1 and Figure 4, which show an increased coarsening rate during the time of greatest Al grain growth, i.e., the first 10 minutes of annealing. It is this portion of the anneal that the greatest volume of the film will be swept by migrating boundaries which collect a significant amount of θ phase for boundary coarsening. For the longer anneal times, when Al grain growth is less extensive, i.e., grain size ≅ film thickness, the θ growth rate is approximately $t^{1/4}$ as predicted for boundary diffusion controlled coarsening [10,11]. Interestingly, the Al growth kinetics (n ≈ 0.5) do not indicate significant precipitate pinning of the boundary motion. This is due to the probably rapid in-grain θ phase dissolution rate when impinged by a boundary as compared to the relative average boundary velocity.

Table I. Mean intercept Al grain size and coarsening θ phase diameters for 310°C furnace annealing of an Al-2% Cu 0.5 μm thick film. "Blocky θ" are located at Al grain triple points, "in-grain θ"are located in grain interiors.

Film treatment	Al grain size	Blocky θ size	In-grain θ size
as deposited	0.16 μm	0.020 μm	0.020 μm
5 min at 310°C	0.39 μm	0.088 μm	0.045 μm
10 min. at 310°C	0.53 μm	0.159 μm	---
20 min. at 310°C	0.64 μm	0.186 μm	0.066 μm
40 min. at 310°C	0.60 μm *	0.212 μm	0.066 μm

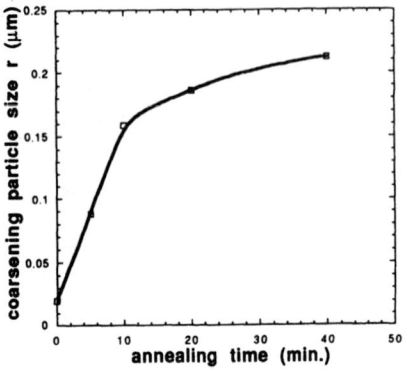
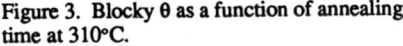
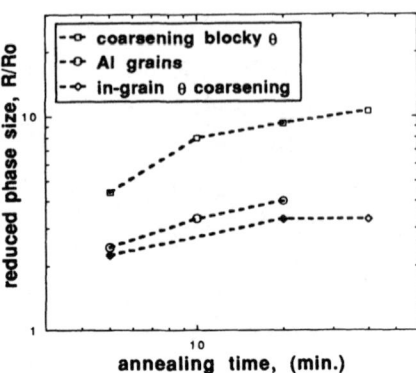

Figure 3. Blocky θ as a function of annealing time at 310°C.

Figure 4. Reduced size (normalized by as-deposited size) during annealing for blocky θ (□), Al (o) and in-grain θ (◊).

There are at least several effects of the Al grain size in these films. It is generally believed that large grain Al film interconnects are desirable as they lead to increased electromigration failure resistance [12]. However, since the average triple point spacing is determined by the average grain size, larger grained Al films will produce fewer but larger and more widely spaced θ particles during annealing and θ coarsening. This would be an especially important affect in, for example, the widely used Al-4% Cu films. For these films annealing at the typical "alloy" temperature of 425°C would rapidly produce large θ phases at triple points by coarsening since θ is stable with respect to dissolution at this temperature and film composition.

As mentioned above, lithographic patterning of films with large Cu rich phases is difficult. Since the projected area of θ in the film largely determines the "etchability" during dry etch patterning, larger grained films with large θ phases would greatly complicate circuit fabrication. Straightforward geometrical analysis shows that for a film made of columnar grains of diameter d, the projected planar area of average cylindrical triple point θ phases (A_θ) depends on the θ phase volume fraction (F_θ) and grain size (d) as

$$A_\theta \approx F_\theta \, d^2 \qquad (2).$$

Here it is assumed that both Al grains and triple point θ are columnar of height equal to film thickness, and that a phase decorates approximately every third triple point. These assumptions correspond to the microstructures seen [2], Figure 2. Thus the measure of coarsened θ phase size relevant for patterning purposes increases at a rate greater than the grain Al size (d). Since large Al grains are desirable from an interconnect reliability standpoint, the obvious strategy for optimizing Al-Cu interconnect microstructures is to reduce the Cu content and thereby reduce the θ volume fraction (F_θ) and size (A_θ). Further, as the Al grain size is increased, the relative density of boundary area in the film (boundary area per film volume) is decreased, reducing the amount of Cu required to segregate at each boundary. These factors may help explain the general trend towards reduced Cu contents in the latest interconnect alloys.

Coarsening During Electromigration Testing

The coarsening process will generally occur in a two phase region of phase stability at conditions (temperature and composition) where solute diffusion is appreciable but where significant portion, at least 50%, of the coarsening phase is stable and undissolved. It is this regime of conditions for Al-Cu alloys that are typical of processes involved in the fabrication of integrated circuits as well as for accelerated electromigration testing. At higher temperatures phase dissolution and subsequent reprecipitation on cooling will determine the precipitate morphology [2]. The mechanism of coarsening will be briefly reviewed in order to apply the process to the conditions during electromigration testing.

The coarsening driving force is the decrease in the total θ-Al interfacial free energy by θ phase coalescence. However the mechanism for solute diffusion results from the size dependent equilibrium solute concentration in the matrix adjacent to the coarsening phases [13]. This is described by the Gibbs-Thompson relation for spherical particles

$$\mu(r) - \mu_0 = 2\,\gamma\Omega/r \tag{3},$$

where $\mu(r)$ is the size dependent chemical potential (free energy per atom) of material in the particle, μ_0 is the 'bulk' chemical potential of the particle material, γ is the interfacial energy between the martix and particle, Ω is the atomic volume, and r is the particle radius. Here γ is assumed to be independent of size and concentration, which is reasonable since θ is incoherent and the solute concentrations are low. By the usual free energy-composition common tangent construction [13], small particles, due to their higher chemical potential described by equation (3), are at local equilibirum with greater solute concentrations in their surrounding matrix than are larger particles. This concentration gradient provides the driving force for solute flux from small to large particles. During coarsening the small particles dissolve at an accelerating rate in order to maintain an increasingly richer solute surrounding while the larger particles grow by incorporating the incoming solute in order to maintain a relatively low solute surrounding region.

It has been shown that Cu electromigrates "faster" [1,14] than Al during accelerated testing, producing Cu enriched regions at the positive terminals and depleted regions at the negative terminals of Al-Cu test conductors. What evidently occurs during testing is that θ precipitates at the negative terminal dissolve to resupply their local region with Cu that has electromigrated down the interconnect. However as they shrink their local equilibrium solute concentration increases inducing even further θ dissolution as described earlier. Similarly at the positive terminal solute flux arriving due to the electromigration driving force is incorporated into existing θ phases as they maintain their lower equilibrium solute concentration in the surrounding matrix. Thus the coarsening process is increased by the induced electromigration solute flux to an extent equal to the length of the interconnect, much greater than possible by the thermal coarsening process alone. In fact, since the local solute concentration is increasingly enriched as the θ phases at the negative region shrink, the total solute flux may be increased by the coarsening mechanism [15]. The fundamental coarsening mechanism can at least qualitatively explain the significant Cu migrations reported previously [1,14,16].

Rapid θ coarsening may also affect electromigration failure processes. In contrast to earlier reports [1,14,16] electromigration voids and failures were routinely found [17] at large θ phases. In that study open circuit failures formed adjacent to plate like θ phases along bamboo boundaries in narrow interconnects. Enhanced coarsening during accelerated testing can be responsible [15] for these θ shapes which may serve as electromigration flux barriers and the sites for flux divergence and failure.

As described above, grain boundaries serve as pathways for rapid θ dissolution and flux. In one TEM in situ study of electromigration in Al-Cu thin films [18] unusual "star" shaped θ phases at triple points were reported, but not explained, after pulsed high current density testing. In that study the electrical current density was sufficient to cause significant heating in the conductor so that the large θ phases partially dissolved during the period of current flow and joule heating. The portion of the θ adjacent to the boundary dissolved fastest, producing sharp erosion cusps there. The film rapidly cooled during the current off cycle, forcing the dissolved boundary Cu to precipitate there in a "finger" of θ phase. Subsequent cycling of current on/off allowed the formation of deeply cusped θ with elongated θ fingers along boundaries, producing the "star" morphology.

The fundamental mechanisms of preciptate shape and size evolution serve to explain observed θ morphologies, to characterize interconnect microstructural effects on device fabrication, and to help understand basic mechanisms that may be responsible for electromigration failure. Further work is required to more fully describe the effects of the combination of second phase coarsening and electromigration solute flux on particle size evolution and interconnect reliability.

CONCLUSIONS

Al grain growth kinetics at 310°C were determined to be approximately $t^{1/2}$, consistent with normal grain growth in thin films, indicating that the matrix θ phases did not appreciably pin migrating boundaries during grain growth. Triple point θ phase coarsening kinetics at 310°C were initially enhanced due to concurrent grain growth, but slowed to approximately $t^{1/4}$ kinetics appropriate for boundary diffusion controlled growth. Electromigration induced solute flux leads to enhanced long range coarsening along the entire length of the interconnect. Finally, enhanced θ coarsening during accelerated electromigration testing may affect failure processes, especially in narrow linewidth bamboo grain structure interconnects.

ACKNOWLEDGEMENT

Experimental work was done at the University of California Dept. of Materials Science and Center For Advanced Materials, Lawrence Berkeley Laboratory. The authors are appreciative to Larry Lamont (MTI, San Jose, CA) for supplying the sputtered films. This work was sponsored by the Director, Office of Energy Research, Office of Basic Energy Sciences, Materials Science Division, U.S. Department of Energy, Contract # DE-AC03-76SF00098.

REFERENCES

1. F. M. D'Heurle, *Met. Trans.* **2**, 683-689 (1971).
2. D.R. Frear, J.E. Sanchez, Jr., A.D. Romig, J.W. Morris, Jr., *Met. Trans.* **21A**, 2449-2458 (1990).
3. S. Mader, S. Herd, *Thin Solid Films*, **10**, 377-389 (1972).
4. B.N. Agarwala, B. Patnaik, R. Schnitzel, *J. Appl. Phys.*, **43 (4)**, 1487-1493 (1972).
5. T. Abraham, *J. Electrochemcal Soc.*, **135 (11)**, 2809-2814 (1987).
6. H.J. Frost, C.V. Thompson, C.L. Howe, J. Whang, *Scripta Met.*, **22**, 65 (1988).
7. W. W. Mullins, *J. Appl. Phys.*, **59**, 1341 (1986).
8. H.J. Frost, C. V. Thompson, *J. Electr. Mater.*, **17**, 447 (1988).
9. C.J. Tweed, B. Ralph, N. Hansen, *Acta Metallurgica*, **32**, 1407-1414 (1984).
10. H.B. Aaron, H. I. Aaronson, *Acta Metallurgica*, **16**, 789-798 (1968).
11. M.V. Speight, *Acta Metallurgica*, **16**, 133-135 (1968).
12. M. J. Attardo, R. Rosenberg, *J. Appl. Phys.* **41 (#6)**, 2381-2386 (1970).
13. D.A. Porter, K.E. Easterling, "Phase Transformations in Metals and Alloys", Van Nostrand, Berkshire, UK, (1981).
14. A.J. Learn, *J. Elect. Mat'ls.* **3 (4)**, 531-552 (1974).
15. J.E. Sanchez, Jr., to be published.
16. B.N. Agarwala, G.Digiacomo, R.R. Joseph, *Thin Solid Films*, **34**, 165-169 (1976).
17. J.E. Sanchez, Jr., L.T. McKnelly, J.W. Morris, Jr., *J. Electr. Mater.*, **19**, 1213-1220 (1990).
18. P. Lobotka, I. Vavra, *phys. stat. sol. (a)*, **63**, 655-661 (1981).

PHASE TRANSFORMATIONS IN Co/Nb AND Co/Zr MULTILAYER FILM STACKS.

J. C. LIN AND R. A. HOFFMAN
Alcoa Laboratories, Alcoa Center, PA 15069

ABSTRACT

Ion-beam mixing and isothermal annealing techniques were applied to induce phase transformations in Co/Nb and Co/Zr multilayer film stacks. Transmission electron microscopy(TEM) and x-ray diffraction(XRD) were used to characterize the microstructure of the films. Interfacial morphology and chemistry of the films were examined by Rutherford backscattering spectrometry(RBS) and cross-section transmission electron microscopy(XTEM). The formation of amorphous phases was found in both systems by either technique. A comparison of the phase transformation mechanisms induced by ion-beam mixing and isothermal annealing is given. The thermodynamic and kinetic factors controlling the phase formation and stability are discussed. For the isothermal annealing, the final stable configuration is predicted by equilibrium phase diagrams.

INTRODUCTION

Amorphization of Co-Nb and Co-Zr alloys has attracted many research activities from both the practical and fundamental points of view. Superior soft magnetic properties of Co/Nb and Co/Zr amorphous films make them suitable for thin-film head applications. On the other hand, since solid-state amorphization from two crystalline phases was first found by Schwarz and Johnson[1] in a Au-La diffusion couple, numerous studies have focused on trying to understand the mechanism of this process. Due to the reasonable growth rates of their amorphous phases at low temperatures[2-3], Co/Nb and Co/Zr couples are good candidates for fundamental mechanism studies. In the present study, two techniques, ion-beam mixing and isothermal annealing were chosen for the amorphization of Co/Nb and Co/Zr multilayered film stacks. A comparison of these two methods from thermodynamic and kinetic points of view will be useful in understanding the amorphization mechanism. In addition to amorphization, higher temperature annealing was also performed to study the whole range of phase transformations of these alloys.

EXPERIMENTAL

Multilayer film stacks were fabricated by sequential e-beam evaporation. Each bilayer thickness of the film stacks was made thin to facilitate mixing; the bilayer thicknesses were 50A for Co/Nb and 100A for Co/Zr. Total stack thickness was about 400A. Individual layer thicknesses were adjusted to give overall compositions of $Co_{.80}Nb_{.20}$ and $Co_{.50}Zr_{.50}$. A 200A SiO_2 layer was deposited on Si wafers prior to the film stack deposition to prevent interaction between film and substrate. The uniformity and the composition of the film stacks were examined by Rutherford backscattering spectrometry(RBS) and energy dispersive spectroscopy(EDS) attached to a transmission electron microscope. The microstructure of the films was investigated by transmission electron microscopy(TEM) using plan-view and cross-section specimens. X-ray diffraction(XRD) was also used to determine the film structure. Ion-beam mixing was performed at room temperature using 180Kev Kr^+ for Co/Nb and 160 Kev Kr^+ for Co/Zr at a dose of 1×10^{16} ions/cm^2. For isothermal annealing, samples were encapsulated in quartz tubes under a vacuum of $\sim 10^{-4}$ Torr and then annealed at various temperatures (400-700 °C) for 1 hour.

RESULTS

Fig. 1 gives the cross-section TEM micrographs of the as-deposited film stacks. Both film stacks show some intermixing at the layer interfaces before any post-treatments. These phenomena were also confirmed by the RBS results given in Fig. 2. After ion irradiation, as indicated in Fig. 2, both film stacks became single layer films with homogeneous composition. Cross-section and plan-view TEM micrographs given in Fig. 3 show featureless images for both films. Electron diffraction patterns (Fig. 3) indicate that amorphous structure was achieved by ion-beam mixing in both cases. It was also noticed that a surface oxide was formed in Co/Zr film after mixing, which contributes the rings to the select area diffraction patterns. Table I lists the results of isothermal annealing of the Co/Nb and Co/Zr film stacks. For the Co/Nb films, a single amorphous phase was found after annealing at 400 °C for 1 hour. A mixture of amorphous and microcrystalline Co_2Nb phases was detected in the sample annealed at 500 °C for 1 hour. After annealing at 600 °C or 700 °C for 1 hour, the film became a mixture of Co_2Nb and Co_7Nb_2 crystalline phases. The plan-view TEM micrographs and diffraction patterns of the samples annealed at 400 °C or 700 °C are given in Fig. 4. For the Co/Zr films, a mixture of amorphous and crystalline hexagonal-ZrO_2 phases was found for the samples annealed at the temperatures below 600 °C. At higher annealing temperatures, a mixture of CoZr, Co_2Zr and ZrO_2 phases forms. The plan-view TEM micrographs and diffraction patterns of the samples annealed at 400 °C or 700 °C are given in Fig. 5.

DISCUSSION

Metastable amorphous structures of $Co_{0.8}Nb_{0.2}$ and $Co_{0.5}Zr_{0.5}$ alloys were sucessfully made either by ion-beam mixing or isothermal annealing of multilayer film stacks. . Although it has been demonstrated that the properties of the amorphous phase does not depend on the fabrication method[4], the mechanisms for the amorphization process could be different for various techniques.

Ion Beam Mixing

In ion-beam mixing, high energy ions bombard the sample generating collision cascades and thermal spikes, which results in mixing of the dissimilar elements in a multilayer film. The whole process seems to have only physical interactions; however, Cheng et al.[5] have pointed out that a negative enthalpy of mixing of elements can enhance this mixing effect. Therefore, an ion-beam mixing process can actually be separated into two procedures: one is driven by

(a) (b)

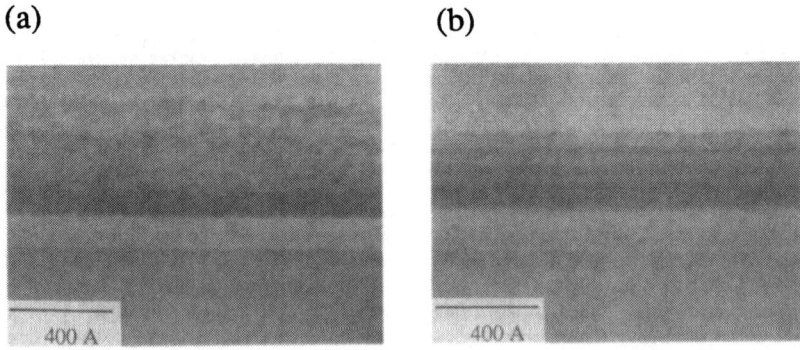

Figure 1. Cross-section TEM micrographs of the as-deposited film stacks (a) Co/Nb (b) Co/Zr.

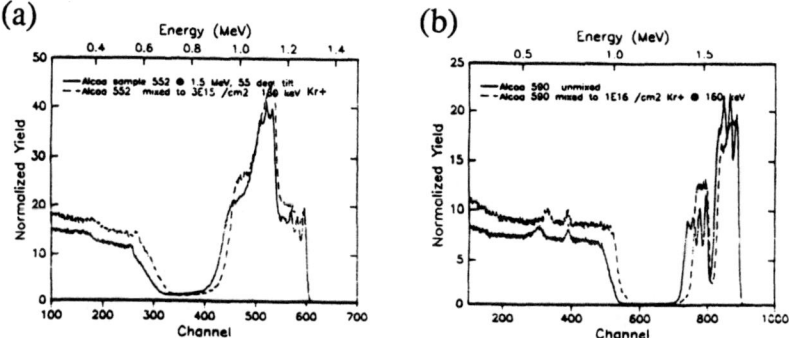

Figure 2. Rutherford backscattering spectra showing intial layered structure and the same sample after mixing (a) Co/Nb (b) Co/Zr.

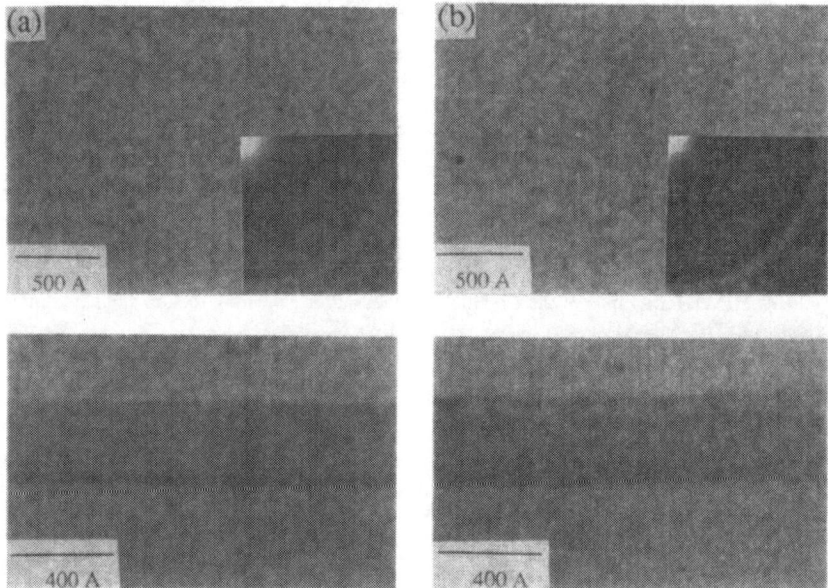

Figure 3. Plan-view and cross-section TEM micrographs of the film stacks after mixing (a) Co/Nb (b) Co/Zr.

physical interactions and the other is driven by chemical interactions. The physical interactions are the collision cascades, which have disordering and mixing effects. The chemical interactions originally come from the chemical potential gradients. The effects (ordering or disordering, mixing or demixing) of chemical interactions are determined by the free energy diagrams of the system. Since the rate of the chemical interactions relies on the mobilities of the atoms, physical interactions will dominate for ion-beam mixing at low temperature (eg. 77K). With increasing mixing temperature, the influence of the chemical interactions will also increase; hence, the structure with lower free energy will be preferred.

Table I. Phase formation after annealing for one hour at various temperatures

	$Co_{.80}Nb_{.20}$	$Co_{.50}Zr_{.50}$
400 °C	Amorphous	Amorphous + ZrO_2
500 °C	Amorphous + Co_2Nb	Amorphous + ZrO_2
600 °C	Co_2Nb + Co_7Nb_2	Co_2Zr + $CoZr$ + ZrO_2
700 °C	Co_2Nb + Co_7Nb_2	Co_2Zr + $CoZr$ + ZrO_2

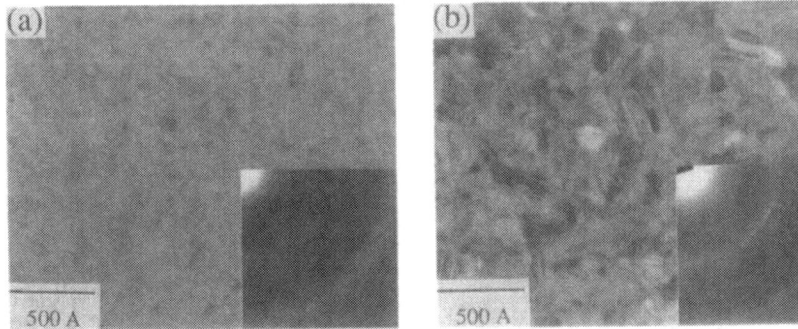

Figure 4. TEM micrographs of the Co/Nb film Stacks after annealing for one hour (a) 400 °C (b) 700 °C.

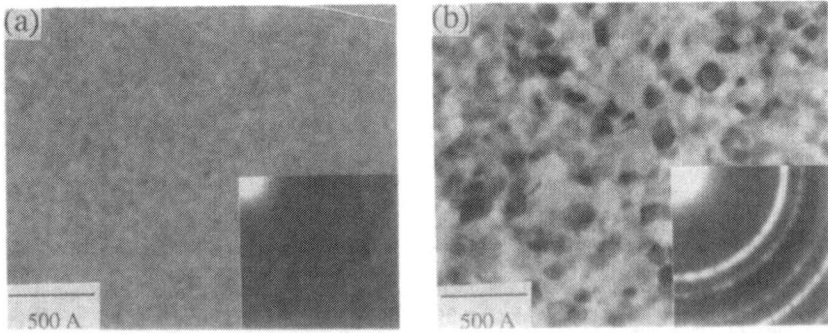

Figure 5. TEM micrographs of the Co/Zr film Stacks after annealing for one hour (a) 400 °C (b) 700 °C.

In order to understand the chemical interactions of Co-Nb and Co-Zr systems, thermodynamic analyses were performed using available data in the literature. Based on equilibrium phase diagrams[6] and lattice stabilities[7], the free energy diagrams of Co-Nb and Co-Zr were calculated and are given in Fig. 6. According to this calculation, an undercooled liquid is the most stable solution phase (not including the compound phases) for $Co_{.50}Zr_{.50}$; on the other hand, for $Co_{.80}Nb_{.20}$, a bcc structure has the lowest free energy. Considering the amorphous phase as an undercooled liquid, it is quite understandable that the $Co_{.50}Zr_{.50}$ multilayer film stack transforms to an amorphous structure to lower the free energy after ion-beam mixing. According to Fig. 6, a bcc phase is the most favorable structure to form after mixing of the $Co_{.80}Nb_{.20}$ film stack. However, an amorphous phase forms instead, and no bcc structure is detected after ion-beam mixing in the present study. This suggests that either the chemical interaction is not strong enough to reach a more stable bcc state or the free energy diagram shown in Fig. 6 is not correct. If the first assumption is right, higher processing temperatures should lead to a bcc structure. However, to our knowledge, forming a supersatuated bcc structure in the Co-Nb system has not been reported in the literature. Actually, it has been verified experimentally[8] that the free energy of the amorphous phase is about 10 KJ/mole lower than that of the undercooled liquid. Futhermore, due to the size difference of the constituents, strain energy would be generated in a supersatuated crystalline solid solution which will raise the free energy of this phase[9]. Considering these two effects, the amorphous phase could be more stable than the bcc phase for $Co_{.80}Nb_{.20}$.

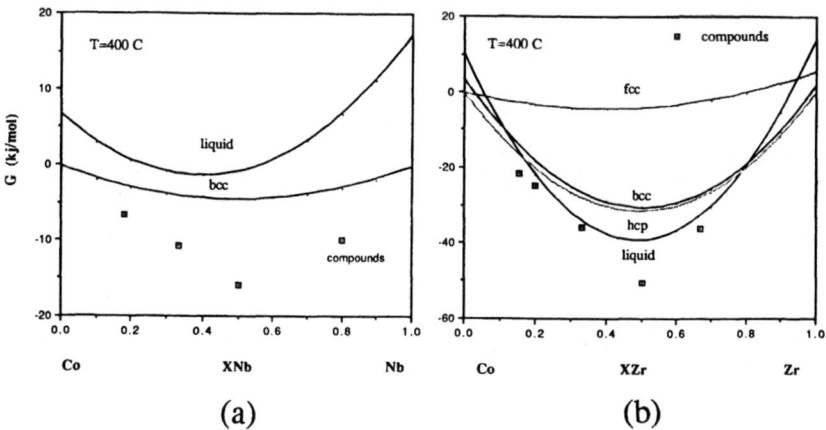

Figure 6. Free energy diagrams at 400 °C (a) Co-Nb (b) Co-Zr.

Isothermal Annealing

For an isothermal annealing process, chemical interactions are the only sources of mixing. Therefore, a higher processing temperature is needed to obtain a reasonable mixing rate. However, for forming a metastable structure like amorphous or supersatuated solid solution, the chosen temperature must be lower than the crystallization temperature of the stable compounds. For $Co_{.80}Nb_{.20}$ films, a single amorphous phase was obtained after annealing at 400 °C for 1 hour, which indicates the crystallization temperature of the compound phase is higher than 400 °C. When annealed at 500 °C, a mixture of amorphous and crystalline Co_2Nb phases was detected, which is in good agreement with the reported crystallization temperature(~500 °C) for this composition[10].It is also interesting to note that from the phase diagram the Co_2Nb phase is the phase that is in equilibrium with the liquid phase at this composition (ie. $Co_{.80}Nb_{.20}$) At higher annealing temperatures (>500 °C), complete crystallization was achieved within an hour. A mixture Co_7Nb_2 and Co_2Nb with a majority of Co_7Nb_2 was identified using electron diffraction patterns. This is a stable configuration for this alloy[11]. For $Co_{.50}Zr_{.50}$ films, annealed at 400 °C or 500 °C for 1 hour, an amorphous phase as well as a crystalline surface oxide ZrO_2 was found. As mentioned early, this oxide was also detected in an ion-mixed sample. The film stack transforms to a mixture of $CoZr$ and Co_2Zr after annealing at 600 °C or 700 °C for 1 hour. This result also agrees with the reported crystallization temperature[12], which is around 535 °C. Based on the above results, both $Co_{.80}Nb_{.20}$ and $Co_{.50}Zr_{.50}$ film stacks transform to an amorphous transition stage, prior to the final stable condition being achieved. As pointed out by several authors [1-3], asymmetrical diffusion is responsible for the solid state amorphization process from diffusion couples. For the Co/Nb and Co/Zr systems, fast diffusion of Co into the refractory metal lattice induces the mixing effect. At the same time, the low mobility of the refractory metal atoms prevents the nucleation and growth of the compound phases. It is very likely that the fast diffusion of Co is interstitial-type, which generates a supersatuated interstitial solution. This solution possesses a very high strain and chemical energy. To relax this energy, a more structurally open and more chemically stable amorphous phase forms. With further annealing, this metastable transition state (amorphous phase) will transform to a thermodynamic equilibrium state(crystalline compounds). From a kinetic point of view, it is possible that the transition-stage amorphous phase also exists even when annealing is performed at temperatures higher than the crystallization temperature. However, the time frame for its existence might be too short to be detected here. Finally, the formation of the $Co_7Nb_2+Co_2Nb$ and $CoZr+Co_2Zr$ films in the present study are also predicted by bulk thermodynamic data. This indicates that the energy differences between thin-film and bulk are not enough to alter the phase equilibrium.

CONCLUSION

Amorphous $Co_{.80}Nb_{.20}$ and $Co_{.50}Zr_{.50}$ films can be fabricated either by ion-beam mixing or isothermal annealing of the multilayer film stacks. These amorphizations can be explained by themodynamic and kinetic considerations. The metastable amorphous phases transform to stable compounds with further annealing as predicted by equilibrium phase diagrams.

ACKNOWLEDGMENTS

The authors thank Professor G. S. Was and Dr. V. H. Rotberg at University of Michigan Ion Beam Laboratory for ion beam mixing and RBS measurements. They also thank N. J. Panseri for film deposition and J. J. Ptasienski for assistance with the TEM micrographs.

REFERENCES

1. R. B. Schwarz and W. L. Johnson, Phys. Rev. Lett. 51, 415 (1983).
2. J. Gfeller, A. Blatter and U. Kambli, J. Less-Comm. Metals 145, 105 (1988).
3. H. U. Krebs and K. Samwer, Europhys. Lett. 2, 141 (1986).
4. R. Bruning, Z. Altounian, J. O. Strom-Olesen and L. Schultz, Mater. Sci Eng. 97, 317 (1988).
5. Y. T. Cheng, M. Van Rossum, M.-A. Nicolet and W. L. Johnson, Appl. Phys. Lett. 45, 185 (1985).
6. Metals Handbook Vol. 8, 8th ed. (American Society for Metals, Ohio, 1973).
7. L. Kaufman and H. Bernstein, Computer Calculation of Phase Diagrams (Academic Press, New York, 1976).
8. F. Sommer, T. Lang and B. Predel, Z. Metallkd. 78, 648 (1987).
9. A. Blatter and M. von Allmen, Mater. Sci. Eng. 97, 93 (1988).
10. N. S. Kazama, H. Fujimori and H. Hirose, IEEE Trans. Magn. 18, 1185 (1982).
11. B. J. Piearcey, R. Jackson and B. B. Argent, J. Inst. Metals 91, 257 (1962-63).
12. E. Hellstern and L. Schultz, Phil. Mag.B 56, 443 (1987).

ATOM PROBE STUDIES OF INTERFACES IN METALLIC MULTILAYERS

ALFRED CEREZO, JONATHON M. HYDE, MARK G. HETHERINGTON[†] AND AMANDA K. PETFORD-LONG
Department of Materials, University of Oxford, Parks Road, Oxford OX1 3PH, U.K.

ABSTRACT

The atom probe field-ion microscope has been used to study the diffusion at interfaces in metallic multilayers deposited directly onto field-ion specimens and to develop models for the solid state reactions occuring at the atomic-scale in multilayer systems. Results are presented for the low temperature annealing of a Co-Ni multilayer. Intermixing over about 2 atomic planes is found even in as-deposited samples, extending to 1nm after heating at 300°C for 1 hour. Using atom probe results from bulk alloys, a Monte Carlo simulation has been developed for the Fe-Cr system, in which a miscibility gap exists, and is being used in an attempt to model the behaviour of interfaces in Fe-Cr multilayers. Preliminary results are presented, showing that interfaces which are initially mixed over 10 atomic planes become sharper by an 'interface spinodal' reaction.

INTRODUCTION

Multilayer films (MLF) are an exciting new class of materials, in which composite films are produced by alternate deposition of two different elements or compounds. The electronic and mechanical effects which arise in the multilayer configuration can produce a material with properties not found in bulk materials. For example, MLF with suitable magnetic properties are being developed for applications as permanent magnet materials and high density magnetic storage media. In all applications for which MLF are being considered, the parameter with the greatest influence on the properties of interest is the quality of the interfaces between adjacent layers [1], and even atomic-scale variations between specimens are likely to result in important differences in final properties.

The ultra-high spatial resolution of the atom probe field-ion microscope [2, 3] provides a powerful method by which information can be gained on the chemistry of interfaces in MLF. In this technique, field evaporation is used to remove atoms from the surface of a specimen in the form of a sharp needle point, with end-radius about 100nm. The evaporated atoms from a region about 2nm in diameter on the specimen surface pass through a selection aperture into a time-of-flight mass spectrometer. The high field sensitivity of field evaporation makes this technique highly surface sensitive, resulting in a composition-depth profile with 2nm lateral resolution, and atomic layer depth resolution. In the more recently developed position-sensitive atom probe (POSAP) [4, 5], both the chemical identity and source position is obtained for atoms field-evaporated from a region 20nm in diameter on the specimen surface. Although the technique has lower mass resolution than the conventional atom probe, it is capable of reconstructing, in three dimensions, the chemical variations originally present in the sample, and so permitting analysis of the interface structure with sub-nanometre (but not atomic) resolution. Both these techniques have been used to study interfaces in metallic multilayers grown on field-ion specimens [6].

ANNEALING STUDIES

The results presented here describe experiments on Co-Ni layered films. In order to produce multilayer specimens suitable for analysis in the atom probe, metal layers were evaporated directly onto pre-prepared Co field-ion specimens. This technique obviously makes it difficult to correlate directly any observed chemical effect with the physical parameters measured from multilayers deposited on planar substrates (such as silicon wafers). However, the specific aim of the present work is to observe the diffusion occuring at metal-metal interfaces

[†] Mark Hetherington died tragically in a car accident during the course of this work.

in a multilayer configuration, and the geometry of the specimen is thus not expected to be problematic. (Fabrication of field-ion specimens from MLF grown on planar substrates has been carried out [7], although it is considerably more difficult than for bulk specimens.) For the diffusion experiments field-ion specimens were made from Co wire using conventional two stage perchloric acid solutions, and then field-evaporated in the atom probe (the VG FIM100 at Oxford) to form atomically clean tips with an approximately hemispherical end-form. The specimen is then withdrawn into a separately pumped preparation chamber (base pressure around $5x10^{-10}$mbar where it is allowed to warm up to room temperature before the metal layers are deposited. Deposition is from a pair of electron-beam assisted evaporation sources (VSW ME10 micro-evaporators) with evaporation rates for Co and Ni of about 0.1nm/min. Typically, about 4nm of Ni and 4nm of Co are deposited on the specimen, as monitored with a quartz oscillator thin-film thickness monitor, with a chamber pressure during deposition of better than $5x10^{-9}$ mbar. The specimen is then annealed on a heating stage within the same chamber, before being returned to the main chamber for atom probe analysis. The layered film can thus be deposited, annealed and analysed without being removed from a UHV environment.

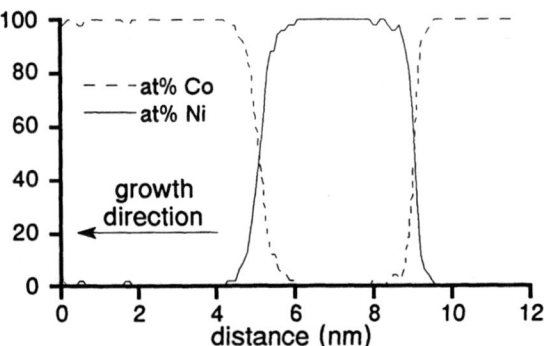

Figure 1. Composition-depth profile through a Co-Ni-Co layered film, annealed at 300°C for 1 hour.

A composition-depth profile through a Co-Ni-Co film annealed at 300°C for 1 hour is shown in figure 1. The depth scale on this profile is an approximate one, obtained from a calibration of the number of ions which would be expected from the field evaporation of a single layer of material. However the scale shows very good agreement with the layer thickness measured by the quartz crystal monitor. Both interfaces show a degree of interdiffusion, shown by the characteristic sigma-shaped curve, with the upper interface (between the two deposited layers) being more diffuse than the lower interface (between deposited layer and substrate). This result is generally observed for these layered films, and may indicate that the deposited layers are not completely flat. It should be stressed, however, that any interfacial roughness must be on a very small scale if it is to influence the composition profile measured in the atom probe, due to the high lateral resolution (2nm) of the technique.

A comparison of the abruptness of the upper interface (Co evaporated on Ni) is shown in figure 2 for an as-deposited film and for films annealed at 200°C and 300°C for 1 hour. In this case the composition-depth profiles are drawn in the form of cumulative plots, which show the sequence of detection of individual ions field-evaporated from the specimen. This type of plot is used extensively for displaying atom probe analyses of interfaces, where composition variations occur over a few atomic planes. The lines show the number of Co or Ni ions detected as a function of the total number of ions, with changes in the slope showing differences in the chemical composition on an atomic scale. For example, in each of the Co plots detection of a Co ion results in the line being extended in both x and y directions (a 45° line), with an increment only along the x-axis if a Ni ion is detected. A 45° line thus indicates 100% of an element, a horizontal line 0%, with a region of composition 50at% having a slope of 27°. Each plot represents the analysis of a region approximately 1.6nm in depth, using the same calibration as for figure 1, although the number of atoms detected per plane varies with the end radius of the specimen and is not the same for each plot. Even in the as-deposited case (figure 2a) a certain degree of interface mixing is apparent, extending over about 0.4nm, which is more than would be expected from the atom probe analysis of an atomically abrupt interface. The extent of the

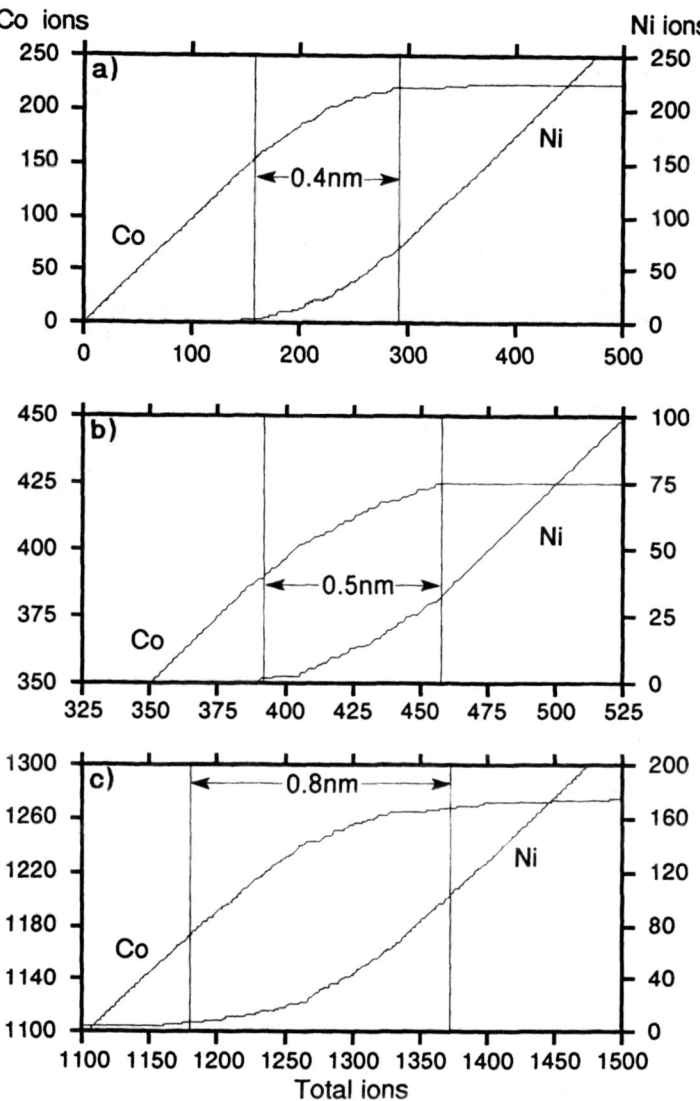

Figure 2. Cumulative plot showing the atom probe analysis of the upper interfaces (Co deposited on Ni) of the Co–Ni–Co layered films: a) as deposited, and annealed for 1 hour at b) 200°C and c) 300°C. The approximate size of the intermixed region is marked in each case.

observed mixed region, from the detection of the first Ni ions to that of the last Co ions, is indicated on each figure. Cumulative plots are not as easily interpretable in terms of composition as a depth profile, but the marked region corresponds approximately to the 5at%–95at% range. Annealing at 200°C for 1 hour slightly increases the extent of this mixing, as seen in figure 2b. After 1 hour at 300°C a mixed region of about 0.8nm width results (figure 2c, which can be compared directly with figure 1). Additional diffusion 'tails' are seen extending on each side of the mixed region out to a total of 1nm on either side of the interface. This increase in extent of the interfacial region is not large in absolute terms, but in many of the applications for which multilayer films are intended, the layer thicknesses are very small (as low as 0.4nm), and a monolayer increase in mixing extent at the interfaces will completely alter the physical properties.

MODELLING OF INTERFACES

The POSAP provides the means by which interfaces in MLF can be characterised not only for chemical abruptness, but also for their atomic scale morphology. Since composition data is acquired with sub–nanometre resolution using this technique, it is possible to distinguish interfaces which are diffuse from those which are abrupt but nonetheless rough on a nanometre scale. The data will not only allow characterisation of the effect of interface morphology on the MLF properties (such as magnetic anisotropy), but will also allow detailed modelling of the solid phase reactions which occur at the interfaces. A Monte Carlo simulation has been developed for atomic-scale diffusion processes in the Fe-Cr system, and this is being used to model the interface reactions which occur on annealing Fe-Cr multilayers. This model has been found to provide a very good fit to POSAP data on the spinodal decomposition in Fe-Cr alloys [8] and so should provide a reliable simulation for the processes occuring in MLFs. During the course of this work, a number of techniques have been developed to parameterise the morphology of the complex interconnected structures observed in POSAP analysis of these alloys. The same morphological parameters can be used to characterise interfaces in MLF, allowing the simulated and experimental data to be compared. It is hoped that further development of the modelling, using POSAP analysis for standardisation of the parameters used in the model, will extend the range of systems that may be simulated.

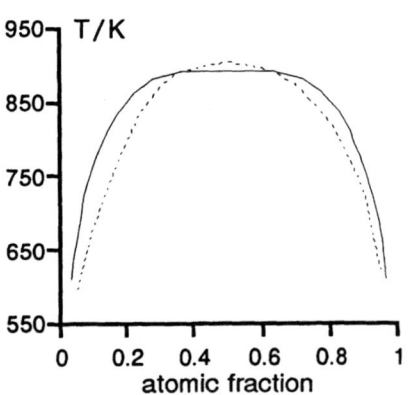

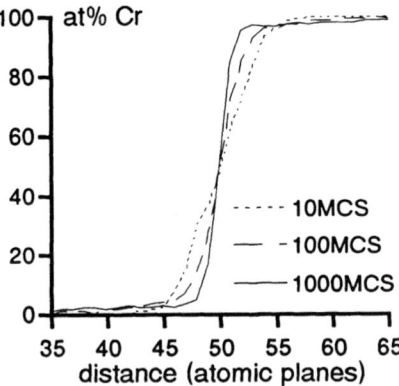

Figure 3. Comparison of the phase diagram for the Monte Carlo model (solid line) with that for the Fe-Cr binary system as calculated using the ThermoCalc database (shown as a dotted line).

Figure 4. Composition profiles across the interface in the Monte Carlo simulation for different ageing times. (MCS = Monte Carlo steps, the average number of nearest neighbour exchanges per atom.)

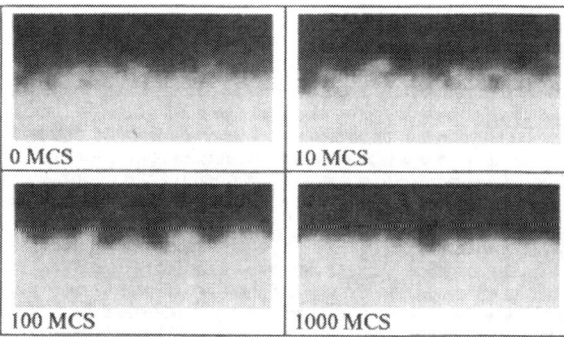

Figure 5. Composition maps drawn through the interface in the Monte Carlo simulation of a layered film being thermally aged. In these plots, white represents 100% Cr and black 100% Fe.

In the multilayer model used here, a crystal of 50x50x100 atoms is used, with an Fe/Cr interface half way along the length of the crystal. Initially, the interface region is linearly graded over a width of 10 atomic layers, and atoms are placed at random within any atomic layer in order to obtain the appropriate composition. The crystal is then aged by random nearest neighbour exchanges, where the probability of two atoms being swapped is related to the change of energy, ΔH, with the energy of the crystal defined only in terms of nearest neighbour bonds. If the energy decreases on swapping a pair of atoms, such an exchange is always accepted. However for $\Delta H > 0$, the probability for the exchange (changing the existing lattice configuration X_i, to the new configuration X_{i+1}) is given by

$$P(X_i \rightarrow X_{i+1}) = \exp(-\Delta H/kT)$$

An absolute temperature scale for the model is obtained by comparing the phase diagram for the simulation with the phase diagram for the Fe-Cr system as calculated with the ThermoCalc database [9]. The good agreement between our model and the calculated phase diagram can be seen in figure 3.

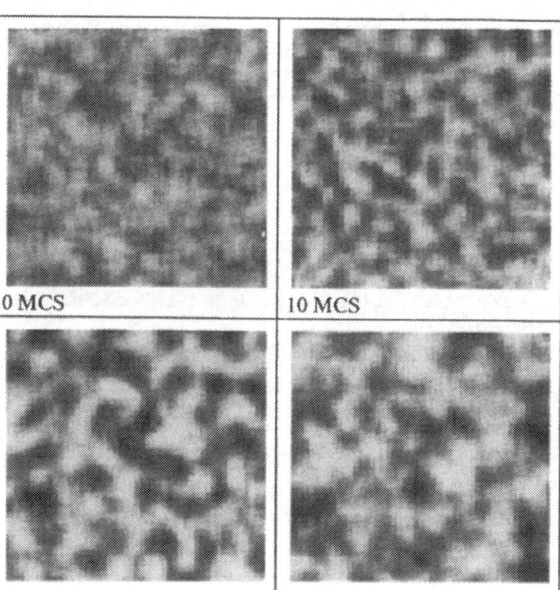

Figure 6. Composition maps drawn parallel to the interface in the Monte Carlo simulation of a layered film: a) initially, and after ageing for b) 10 MCS, c) 100MCS and d) 1000MCS. In these plots, white represents 100% Cr and black 100% Fe.

Composition profiles across the interface in the Monte Carlo simulation are shown in figure 4, for a crystal annealed in the model at a temperature of 600K. The crystal is aged up to 1000 Monte Carlo steps (MCS), namely each atom site in the lattice has been selected for an exchange 1000 times on average. It is immediately apparent that as the crystal is aged, the interface becomes more abrupt, although at longer ageing times an extended diffusion 'tail' is seen on either side of the interface. These composition profiles are calculated from the whole of the crystal, and each point thus represents a composition measurement integrated over the 250 atoms in a single plane. A more complete picture of the process is obtained if we look at the morphology of the interface, as seen in figures 5 and 6. The composition map through the interface, figure 5, shows that a structure develops from the initially random distribution of species which bears a marked similarity to that observed in a spinodally decomposing material, despite the fact that the region which is undergoing 'decomposition' is only 2nm thick.

The composition maps drawn in cross–section through the interface (figure 6) show that some structure is formed in this plane as the material is aged, with the interface becoming sharper as the material and, at late stages of ageing, flatter. Thus while the interface is chemically abrupt after only a short anneal, it is only when the interface becomes flat that the composition profile becomes sharper. We have used some of the techniques developed for the characterisation of interface morphology in spinodally decomposed materials [8] to determine a scaling parameter (loosely speaking, a fractal dimension) for the interfaces in the Monte Carlo simulation. It was found that the fractal dimension decreased with ageing, from 2.8 for the unannealed case, to 2.4 after 1000 MCS, where a value of 2 would be expected for a flat, perfectly smooth interface. This indicates that the use of characterisation techniques we have already developed can provide a means of quantifying the morphology of interfaces in multilayer films.

The sharpening of interfaces by this 'interface spinodal' reaction will obviously have a profound effect on the physical properties of multilayers with layers composed of immiscible elements such as iron and chromium. A number of other systems may be expected to show similar behaviour, such as Co–Cu and Co–Au, and a smoothing of interfaces has in fact been observed after annealing of Co–Au multilayers [10].

ACKNOWLEDGEMENTS

We would like to thank Professor Sir Peter Hirsch FRS for the provision of laboratory space. AC is grateful to to The Royal Society and Wolfson College, Oxford for support during the course of this work. AKPL is grateful to the Glasstone Benefaction and to Corpus Christi College, Oxford. The Oxford Atom Probe facility is funded by the Science and Engineering Research Council.

REFERENCES

1. Synthetic modulated structure materials, edited by L.L. Chang and B.C. Giessen (Academic Press, New York 1986).
2. E.W. Müller, J.A. Panitz and S.B. McLane, Rev. Sci. Instrum. 39, 83 (1968).
3. M.K. Miller and G.D.W. Smith, Atom Probe Microanalysis: Principles and Applications in Materials Science (MRS, Pittsburgh, 1990).
4. A. Cerezo, T.J. Godfrey and G.D.W. Smith, Rev. Sci. Instrum. 59, 862 (1988).
5. A. Cerezo, T.J. Godfrey, C.R.M. Grovenor, M.G. Hetherington, J.M. Hyde, J.A. Liddle, R.A.D. Mackenzie and G.D.W. Smith, EMSA Bulletin 20, 77 (1990).
6. A. Cerezo, M.G. Hetherington and A.K. Petford-Long, J. de Phys. 50–C8, 349 (1989).
7. A.J. Melmed, J. de Phys. 50–C8, 547 (1989).
8. A. Cerezo, J.M. Hyde, M.G. Hetherington and G.D.W. Smith, to be published in Scripta. Metall. et Mater. 25, (1991).
9. B. Sundman, B. Janson and J. Andersson, Calphad 9 (2), 13 (1985).
10. F.J.A. den Broeder, D. Kuiper, A.P. van der Mosselaer and W. Hoving, Phys. Rev. Lett. 60, 2771 (1988).

IN SITU HVEM OF CRYSTALLIZATION OF AMORPHOUS TiNi THIN FILMS

WARREN J. MOBERLY*, J. D. BUSCH**, A.D. JOHNSON** AND M. H. BERKSON***
* Dept. of Materials Science & Engineering, Stevens Institute of Technology, Hoboken, NJ
** TiNi Alloy Company, Oakland, CA
*** Dept. of Materials Science & Engineering, University of California at Berkeley, CA.

ABSTRACT

TiNi alloys have been extensively studied for their shape memory properties, arising from a martensitic transformation between the cubic B2 crystal structure and the monoclinic B19' structure. However, only recently has the application of TiNi alloys as thin film actuators been considered. Well-controlled thicknesses of TiNi films have been deposited via d.c. magnetron ion sputtering, with as-deposited films exhibiting an amorphous structure. These ductile metallic glass films may be bent into various macroscopic shapes and then heated to crystallize the B2 structure, thereby determining its parent (memory) shape. *In situ* high voltage electron microscopy (HVEM) heating and cooling experiments are utilized to observe the crystallization of the B2 structure and the martensitic transformation from B2 to B19'.

The crystallization process is observed to occur over a range of temperatures, from ~500°C to ~600°C. Isothermal annealing determines the kinetics of the crystallization process to be nucleation limited. The nucleation of B2 crystallites in thicker regions of TEM specimens, which occurs prior to nucleation in thinner regions, indicates that nucleation does not occur preferentially at surfaces, but rather homogeneously. After crystallization is completed, the existence of random orientations of B2 grains provides films that will exhibit shape memory in any desired orientation. Video-recording of the crystallization processes has been acquired.

INTRODUCTION

Thin film materials are quickly emerging in all aspects of technology; yet certain issues have to be considered to determine if the thin film will behave similarly to corresponding bulk materials. Will the thin film, as processed, have the same structure (and microstructure) as the bulk material, and will the thin film provide the same properties as the bulk material? If thin films can be processed to have desirable properties, then often the materials cost-savings in using a thin film can compensate any added costs arising during thin-film processing. As an example, TiNi alloys, which are well-known for exhibiting a shape memory effect, have been limited in their bulk use primarily because of their prohibitive manufacturing costs. Whenever an application is considered for the TiNi shape memory alloy, a less expensive solution is often chosen. When utilized in thin film form, on the other hand, the cost of the material is minimal, the cost of processing is in line with other thin film materials, and TiNi alloys may often provide the best solution; for example, a thin-film actuator used as a microvalve [1].

Numerous investigators [2, 3, 4] have recently shown that thin films of TiNi alloys can exhibit shape memory properties. In at least one case, [2] the thin film has shown excellent reliability for cyclic shape changes, with millions of cycles being achieved. However, this reliability of shape memory properties depends on the consistent processing of the thin films, which in turn requires an accurate assessment of the microstructure(s) at various processing steps. The most common method for deposition of TiNi thin films is ion sputtering, which typically results in the deposition of an amorphous thin film. Although such as-deposited amorphous films are utilized as corrosion-resistant and wear-resistant coatings, it is necessary to crystallize these films in order to subsequently realize the shape memory effect associated with a martensitic phase transformation. This research considers the crystallization of amorphous TiNi films, and how the shape memory effect is thereby influenced by this crystallization.

Mat. Res. Soc. Symp. Proc. Vol. 230. ᶜ 1992 Materials Research Society

EXPERIMENTAL PROCEDURES

Single targets of polycrystalline TiNi were utilized to ion sputter deposit the thin films of TiNi. The d.c. magnetron sputtering was conducted in a diffusion pumped vacuum system, with a pressure below 10^{-6} Torr. Chemical analysis of the as-deposited films have previously [2] determined oxygen contamination to be less than 0.1 atomic per cent. Thin films of TiNi may be deposited on various substrates, with single crystal Si or glass slides being judicious choices. Si wafers with either an oxide coating and/or a "soap" coating are quite useful if it is desirable to "peel" the thin film off the substrate after deposition, thereby providing free-standing films. As shall be discussed, all films ion-sputter-deposited in this manner exhibit an amorphous structure after deposition. (Amorphous films also result from depositions on substrates which are heated to temperatures as high as 425°C.) Various thicknesses of thin films are easily obtained, with films in excess of five to ten microns being quite durable as free standing films. Films thinner than one micron in thickness exhibited good mechanical integrity, when provided by support at the edges. Films of various thicknesses up to thirty microns were sputter deposited for this crystallization study.

In situ high voltage electron microscopy was utilized to study the crystallization of films of various thicknesses ranging from less than ten nanometers to more than one micron in thickness. The preparation of TEM samples, using a Fishione electrochemical jet polisher and an electrolyte consisting of three parts methanol and one part nitric acid, provided thin foils that were roughly wedge-shaped. Such samples, as are common for TEM metal foils, provides for the observation of crystallization in regions of various thicknesses in the same sample, during a single *in situ* experiment. In addition, diffraction contrast provides insight into strain developing between the crystallite and the surrounding amorphous matrix, which in turn affects the shape of the crystallite [5]. In order to observe the crystallization in film regions as thick as 1.5 microns, microscopy experiments were performed on the 1500 keV high voltage electron microscope at the National Center for Electron Microscopy, Lawrence Berkeley Laboratories. A Gatan double-tilt heating holder, which had been calibrated to within +/- 20° C using high melting temperature waxes, was used for heating experiments. Subsequent experiments to observe the martensitic transformation of films which had been crystallized in the microscope required using a liquid-nitrogen-cooled Gatan cold stage.

The crystallization of "bulk" TiNi thin films, thicker than seven microns, were considered as *ex situ* annealing experiments. Such *ex situ* annealing was performed by either heating the substrate in the deposition chamber or by vacuum annealing free-standing films. Films which had been annealed *ex situ* were subsequently thinned to electron transparency for TEM analysis and to compare to *in situ* crystallization experiments. In addition, possible artifacts arising from the bombardment of TiNi by high voltage electrons had to be addressed. Crystalline thin films, prepared from bulk TiNi alloys, have been observed to be amorphized due to exposure of a high dosage of high energy electrons [6, 7]. (As a historical note, the identical microscope used for the present analysis was utilized to observe the first amorphization of a metal, a polycrystalline TiNi thin film, by high energy electrons! [6]) During this study, amorphization of the thin films after crystallization was also observed. However, if the electron beam was not condensed amorphization of crystalline regions was avoidable. Electron-induced amorphization was avoided even for experiments where samples were exposed to the 1500 keV electrons for greater than four hours.

The crystallization of thicker films was assessed using Differential Scanning Calorimetry. The films of thicknesses comparable to that investigated during *in situ* crystallization experiments could not be studied using DSC. Two primary reasons are: the requirement of a finite, accurately measured mass of material for providing a sufficient total heat of crystallization for detection by DSC; and the consummate oxidation of thin films during the heating in the calorimeter. These thicker films of TiNi were also investigated, both in the as-deposited amorphous condition and after the crystallization anneal by X-Ray Diffraction.

RESULTS & DISCUSSION

A preliminary study of the as-sputter-deposited thin films was conducted to determine if the films were truly amorphous or consisted of very fine crystallites. High resolution electron microscopy observed no fine crystalline regions [8]. Electron diffraction results, presented in reference [9], exhibited amorphous rings identical to those observed in other studies of amorphous TiNi films which have been prepared both by sputter deposition [10] and by mechanical alloying [11].

All amorphous TiNi films crystallized in this study, whether annealed *in situ* in the TEM or annealed *ex situ* followed by TEM and/or XRD analysis, exhibited the B2 structure upon crystallization. Figure 1a exhibits an electron diffraction, acquired at a temperature of ~400° C of a single crystallite in a surrounding amorphous matrix. Both the amorphous rings and the reflections indicative of the [111] B2 zone axis are exhibited. Prior studies of the crystallization of TiNi thin films have suggested the possible formation of the disordered BCC structure [12]. However, theoretical calculations of the free energy of formations of 50::50 compounds of TiNi having a B2 ordered cubic structure, a disordered BCC structure, and an amorphous phase predict that it is energetically unfavorable for an amorphous phase to transform to a disordered BCC structure prior to forming the equilibrium B2 phase. Medina calculations in reference [13], determine the energy of formation of a 50::50 amorphous TiNi alloy to be ~ 2 kcal/mol greater than the energy of the B2 structure. The theoretical free energy of formation of disordered BCC TiNi, if it were to form, is ~ 5 kcal/mol greater than the energy of the amorphous phase, as based on CALPHD calculations [14]. No other study of bulk TiNi has provided conclusive results of the possible disordering of the B2 structure at any temperature. In addition, with all the various modes by which TiNi may be amorphized: ion beam radiation, electron beam radiation, mechanical alloying, interdiffusion of Ti and Ni, sputter deposition, etc., no disordering of the BCC structure has been observed prior to amorphization.

DSC scans acquired while annealing TiNi films thicker than seven microns indicate the amorphous to crystalline transformation occurs at 490° C (+/- 20° C) and is exothermic. The energy released is 1.1 kcal/mol of TiNi [2], which is approximately half of the theoretically calculated value [13]. The energy released is comparable to that observed during crystallization of other binary intermetallics, for example, 1.5 kcal/mol released when Cu_3Ti_2 crystallizes [15]. Upon cooling in the calorimeter, the transformation from B2 austenite to martensite may also be detected. For the films prepared in this study, the martensitic transformation temperature (M_S), is ~60° C [2]. Numerous chemical and physical changes in the microstructure of the thin films can result in a change, typically a decrease, in the transformation temperature. The primary means by which the M_S temperature is decreased is via an increase in the relative Ni concentration in the matrix. The suppression of the martensitic transformation often results in the realization of the intermediate "R" phase. Oxygen, fine grain size, thin film effects [16], and precipitates [1] have all been determined to depress M_S.

Thicker films (greater than seven microns thick) which were annealed in the DSC have an M_S temperature above room temperature. However, all thin films (less than one micron thick) which were crystallized in the high voltage electron microscope exhibited an M_S temperature below room temperature. In most grains, electron diffraction acquired at room temperature exhibited the R phase. Figure 1b was acquired at room temperature on the same crystallite (in a surrounding amorphous matrix) as for Figure 1a.

a)
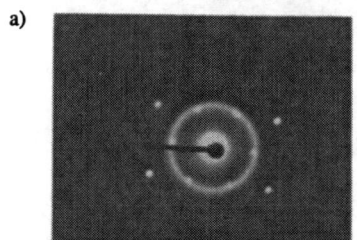
b)

Figure 1:   a) [111] B2 zone axis electron diffraction pattern acquired from a submicron size crystal, which is nucleating in a surrounding amorphous matrix. The diffuse rings are indicative of the amorphous matrix which still exists above and below the crystallized grain. Diffraction pattern was acquired at > 400° C, where nucleation and growth have been halted.
b) [111] diffraction pattern acquired from the same region as 1a, except after cooling to room temperature. The extra 1/3 <110> reflections are indicative of the R phase, which has begun to form from the B2 structure. The rings are indicative of the surrounding amorphous matrix.

The imaging of the *in situ* crystallization provided many results which are described briefly in the following paragraphs (and in more detail in reference [9]). To begin with, *in situ* crystallization affords the ability to observe the initial crystallization by slowly raising the temperature. In addition, the crystallization process may be abruptly terminated by turning off the heater. The stability of the heating stage enables the acquisition of images, as well as video recordings, at the temperature of crystallization. Figure 2 is a bright field image acquired at a temperature of ~540° C of a region of TiNi film having a uniform thickness of ~ 300 nm. These crystallites exhibit a "spherulitic" [17] microstructure, where the overall grain has a spherical shape, yet on a nanoscale the interface exhibits dendritic growth. Other amorphous thin films of compound materials (CuZr [17], CuTi [18], and NiSi$_2$ [19]) also exhibit a similar microstructure upon crystallization. Whereas other crystallizing films often exhibit nucleation of new grains at the interface of pre-existing grains, new grains of TiNi homogeneously nucleate. Figures 3a and 3b are acquired of the same region at 540° C, with greater than ten minutes time lag between them. Grains continue to spherically grow, with new grains nucleating in the remaining amorphous matrix, until the grain boundaries impinge.

Figure 2: Bright field TEM image of a partially crystallized region of TiNi thin film of uniform thickness, acquired at 540° C. The growth is dendritic at the nanometer scale, yet "spherulitic" for each overall grain.

Figure 3: a) BF TEM image of grains growing in an amorphous matrix at ~ 540° C.

b) Image acquired of same area as Figure 3a after ten more minutes of annealing at 540° C.

2 μ

3a

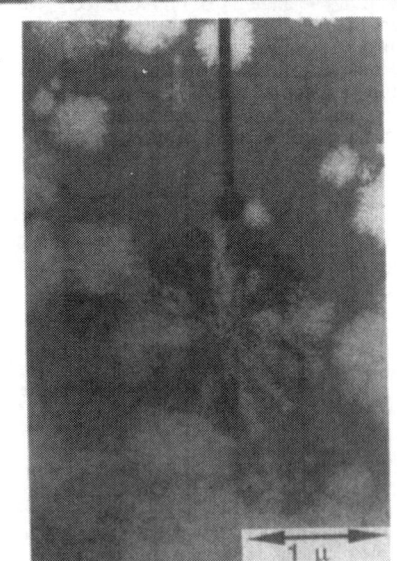

1 μ

3b

1 μ

Whereas the crystallization temperature observed by DSC, for films greater than seven microns in thickness, is quite reproducible at 490° C, +/- 10° C, the temperature of crystallization observed in the HVEM exhibited considerable variability. For films thicker than one micron, nucleation began once the film was heated to ~540° C. However, regions of TiNi films which were less than 100 nm thick would not crystallize until the temperature was greater than 580° C. (The complete crystallization of thicker films would occur in less than 15 minutes at 490° C. However, no nucleation had even begun in films less than 100 nm thick after annealing *in situ* for greater than one hour at 550° C.) As plotted in reference [9], nucleation temperature (for a fixed volume fraction of nuclei) increases as the film thickness decreases. Another interpretation of this data is that for a given temperature, nucleation will occur first and fastest in thicker regions of a TEM thin foil [2, 9]. This can be explained as a consequence of the crystallization process in TiNi involving homogeneous nucleation [9].

Once critical size (> 10 nm) nuclei form, they quickly grow until their size is greater than the thickness of the film. Two-dimensional growth then continues at a constant, though slower, rate. Since nucleation is quite sluggish, however, the grains can undergo extensive growth in the two-dimensional foil before impinging other grains. Figure 4 plots the radial two-dimensional growth of a crystallite while sequentially raising the temperature. From such an experiment an activation energy for the rate of growth in the two-dimensional regime is 4 eV, +/- 1 eV. (A more thorough analysis of three dimensional growth at earlier times is being conducted [9].) Whereas further nucleation is essentially terminated by lowering the temperature, once grains are nucleated, they can continue to grow at measurable rates in the TEM even for temperatures as low as 450° C.

The nucleation component of the crystallization process is more temperature dependent than the growth regime. Thus once a nuclei forms, it will quickly consume the thickness of the film. Even though nucleation occurs more readily in thicker regions, growth is still the dominant stage. As a consequence, grains grow to a larger size in the thicker regions. TEM analysis of TiNi films (greater than 7 microns in thickness) crystallized *ex situ* [1, 2] determine the grain size to be 1 - 2 microns, independent of the film thickness. However, regions of TiNi films greater than one micron in thickness, which were annealed in the TEM, often exhibited dramatic grain growth, with some grains growing in excess of 10 microns. Films of less than one micron thickness exhibited grain sizes that were correspondingly smaller. The grain size was always greater than the local film thickness, with regions of film having more than one grain through the film's thickness being rare. Figure 5 plots the grain sizes measured for various thicknesses. Such excessive grain growth is attributed to the difficulty of nucleation. It appears that after the onset of two dimensional growth, strain builds up in the amorphous matrix as a crystallite grows [5, 20] and hinders further nucleation in the surrounding matrix.

Possible artifacts affecting the *in situ* crystallization experiments due to the high energy electrons of the TEM and/or nonuniform heating have been addressed. Firstly, converging the electron beam to provide a higher local current density in thin areas did not lead to crystallization occurring there prior to thicker regions not exposed to the electron beam. *In situ* TEM heating experiments of other materials (especially thermally conductive metals) have not resulted in nonuniform heating at the micron scale. Generally, the electron beam results in additional

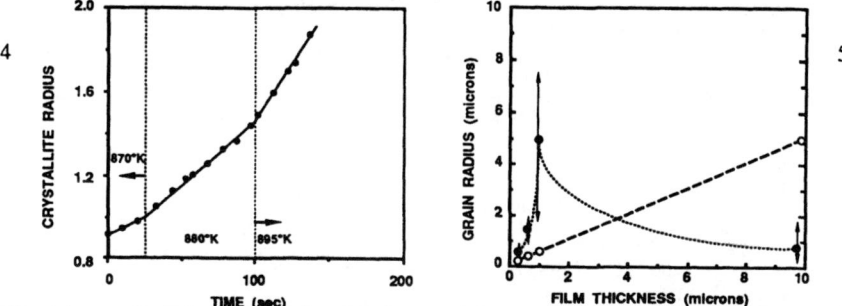

Figure 4:   Radial two-dimensional grain growth increases as the temperature is increased. Data acquired by video-recording TEM images.   Figure 5:   For thin films crystallized *in situ* the grain size is significantly larger than the film thickness. Dashed line represents typical case where grains nucleate at surfaces and grow to a grain size ~ 1/2 the film thickness.

heating in thin regions of a TEM foil. If the thinner regions of the TiNi TEM foil were hotter, they would be expected to nucleate first. If nucleation were slower in thinner regions as a result of these regions being cooler, then the larger grain size in the thick regions would imply that the growth rate is more temperature dependent than the nucleation rate. Annealing at a higher temperature in the thin regions in order to increase the nucleation rate would more dramatically increase the growth rate and produce even larger grains in the thin regions of the foil. Conversely, if thicker regions of a TEM foil were hotter, nuclei would be expected to form more readily in the thick regions, producing a finer grain size than would occur in the thin regions of the foil.

CONCLUSIONS & FUTURE WORK

The presence of larger grains in thicker regions of a TEM foil, as well as nucleation occurring first in the thick regions indicate that nucleation occurs homogeneously and is the rate limiting step of the crystallization process. The resulting random orientations of grains is partially responsible for insuring these thin films will have shape memory properties comparable to bulk TiNi parts. For films that are thinner than 1 - 2 microns, the limited nucleation enables the growth of grains to a size much larger than the thickness of the film. The shape memory properties of such films have not yet been addressed. However, using bend contour analysis to interpret the strain effects associated with the two-dimensional growth regime may provide insight into why such large grains can form in a thin film.

ACKNOWLEDGEMENTS

The authors wish to acknowledge the National Center for Electron Microscopy at Lawrence Berkeley Labs, Berkeley, CA where the majority of this research was conducted, and Uli Dahmen for many enlightening conversations. Additionally, Robert Sinclair is acknowledged for the use of the electron microscopy facility at Stanford University. This work was partially funded by TiNi Alloy Co.

REFERENCES

1.   J. D. Busch, M. H. Berkson and A. D. Johnson, presented at the 1991 MRS Spring Meeting, Anaheim, CA (unpublished).
2.   Johnson, A.D., "Vacuum-deposited TiNi shape memory film: characterization and applications in microdevices." J. Micromech. Microeng., $\underline{V1}$, p34-41, (1991).
3.   C. Hu-Simpson, L. Chang, and D. S. Grummon, presented at the 1990 MRS Fall Meeting, Boston, MA (unpublished).
4.   A. P. Jardine and L. D. Wasielesky, presented at the 1990 MRS Fall Meeting, Boston, MA (unpublished).
5.   W. J. Moberly, to be published in Ultramicroscopy.
6.   G. Thomas, H. Mori, H. Fujita and R. Sinclair, Scripta Met. $\underline{V16}$, p589-592, (1982).
7.   A. R. Pelton, in Proc. 7th Int. Conf. on HVEM, edited by R. M. Fisher, R. Gronsky and K. H. Westmaccott, p. 245 (1983).
8.   C. Hetherington and W. Moberly (unpublished results).
9.   W. J. Moberly, J. D. Busch and A. D. Johnson, to be published in J. of Matls Sci.
10.   P. Moine, A. R. Pelton and R. Sinclair, J. Non-Cryst. Sol. $\underline{V101}$, p213-222 (1988).
11.   B. S. Murty, M. M. Rao and S. Ranganathan, Scripta Met. $\underline{V24}$, p1819-1824 (1990).
12.   J. D. Busch, A. D. Johnson, C. H. Lee and D. A. Stevenson, J. Appl. Phys. $\underline{V68}$, p6224-6228 (1990).
13.   B. M. Clemens, Phys. Rev. B $\underline{V33}$, p.7615-7624 (1986).
14.   W. L. Johnson, Prog. in Matls. Sci. $\underline{V30}$, p81-134 (1986).
15.   A. F. Marshall, Y. S. Lee and D. A. Stevenson, Acta Met. $\underline{V31}$, p1225-1231, (1983).
16.   G. Michal, Ph.D. Thesis, Dept. Matls. Sci. and Eng., Stanford University (1979).
17.   A.F. Marshall, R. G. Walmsley and D. A. Stevenson, Matls. Sci. and Engin. $\underline{V63}$, p215-227 (1984).
18.   A.F. Marshall, Y.S. Lee and D.A. Stevenson, J. Non-Cryst. Sol. $\underline{V64}$, p399-419(1984).
19.   J. L. Batstone and D. A. Smith in Proc. EMSA, p524-525 (1990).
20.   I. E. Bolotov, V. Y. Kolosov and A. V. Kozhyn, Phys. Stat. Sol. A $\underline{V72}$, p645-654 (1982).

PHASE TRANSFORMATIONS IN SPUTTERED Ni-Ti FILM: EFFECTS OF HEAT TREATMENT AND PRECIPITATES

J.D. BUSCH*, MITCHELL H. BERKSON**, AND A.D. JOHNSON*
*   TiNi Alloy Company, 1144 65th Street, Unit A, Oakland CA 94608
**   Department of Materials Science, University of California at Berkeley, CA 94720

ABSTRACT

This research investigated the influence of heat treatment and precipitation on the phase transformation temperatures of sputter-deposited nickel-titanium films. Films 5 to 10 microns thick were subjected to isochronal and isothermal heat treatments in vacuum. Four-point resistance measurements were made and the T_p, T_R, M_s, and M_f temperatures identified. The correlation between transformation temperatures and the film's microstructure was studied using transmission electron microscopy. Existence of Ti_2Ni and $Ti_{11}Ni_{14}$ precipitates was seen to adversely influence the phase transformation temperatures. It was concluded that precipitate-free film is preferable for mechanical actuator applications.

INTRODUCTION

Actuation of mechanisms in the size range 10 microns to 1 mm poses operational requirements that are not well-matched by electrostatic, electromagnetic, or piezoelectric technologies. In contrast, shape-memory alloy film, deposited by processes compatible with microelectronic manufacture, offers an opportunity for high density energy conversion using TTL voltages. One of the applications now being developed is a shape-memory alloy microvalve [1,2].

The nickel-titanium shape-memory alloy (SMA) is a unique material which can be plastically deformed by as much as 8% at one temperature, yet will return to its original shape when heated to a slightly higher temperature. The mechanism behind this phenomenon is a diffusionless phase transformation from the low temperature monoclinic B19' crystal structure to the parent BCC B2 structure, respectively referred to as martensite and austenite.

The ideal shape-memory film for mechanical actuation would be arbitrarily thin, reveal no fatigue after millions of cycles, exhibit large strain recovery, and have a phase transformation near 100°C. The purpose of this work is to determine the effects of heat treatment on the most sensitive of these shape-memory properties: transformation temperature. Most micro-actuator applications which utilize thin film shape-memory alloy will require transformation temperatures well above room temperature.

EXPERIMENTAL PROCEDURE

Sputtered Ni-Ti film used in this work was produced by TiNi Alloy Company. Films 5 to 10 microns thick were sputtered on [100] silicon in an inert argon atmosphere by a d.c. magnetron source. The films were deposited in an amorphous state by

choice, which was verified by selected area diffraction patterns
(SADP) and x-ray diffraction (XRD). After sputtering, the
amorphous films were peeled from their 75 mm diameter silicon
substrates and cut into smaller specimens. All the films used in
this research were produced in the same sputtering run so as to
minimize variations in composition or contamination levels.

Two separate sets of film samples were subjected to
isothermal and isochronal heat treatments in vacuum. A previous
study using differential scanning calorimetry showed that
amorphous Ni-Ti film crystallizes at approximately 500°C [2]. To
ensure that all film samples were crystalline, the sequences were
chosen as follows:

Isothermal: 540°C for   30 min.   2 hrs.   8 hrs.   16 hrs.
Isochronal: 30 min. at   540°C   600°C   700°C   800°C

The samples were separately encapsulated in quartz ampoules
and evacuated to approximately 1×10^{-3} torr. The sealed
ampoules were individually placed in a preheated furnace, held at
temperature for the prescribed time, then removed and allowed to
cool in air. Each sample was then cut into two pieces, half for
resistance measurements and half for transmission electron
microscopy.

The phase transformation temperatures for each sample were
determined by plotting resistance versus temperature data. A
collinear four point probe apparatus was used in which voltage
was measured between the two inner contacts while constant
current was passed through the outer two. The samples were
uniformly heated and cooled between 110°C and -120°C at rates of
approximately 8°C per minute.

The microstructural characteristics of each sample were
examined using Philips 301 and 400 transmission electron
microscopes (TEM) at 100kV equipped with cold stages. The
samples were prepared by ion milling and showed no evidence of
ion mill damage.

Quantitative composition measurements in localized areas
were made by energy dispersive x-ray spectroscopy (EDS) in the
Philips 400 analytical TEM.

RESULTS

As mentioned above, the as-deposited films were amorphous
and thus could not exhibit the B2 to B19' phase transformation
associated with a shape-memory effect. DSC results indicate that
film crystallization occurs near 500°C. Supporting this
observation, the TEM micrographs showed that every annealed film
sample was indeed crystalline and exhibited a typical grain size
of 1 to 2 microns. In addition, all samples having a martensite
transformation above -50°C demonstrated a definite mechanical
shape-memory effect.

Results of the resistance versus temperature measurements
for each sample are presented in Figures 1 and 2. To interpret
these curves, it should be noted that, on cooling, the Ni-Ti
alloy first undergoes a second order transition from its high
temperature austenite phase to an incommensurate phase [3]. This
is followed by a transformation to the R-phase, and then a
martensite transformation which is characterized by distinct
start and finish temperatures. Depending on composition and
annealing conditions, the R-phase and martensite transformations
can overlap or be entirely separated [4]. On heating, the alloy

will transform back into the austenite state, characterized by the austenite start and finish temperatures which are not always distinguishable in resistance versus temperature plots.

The nomenclature used is:

T_p: Commencement of incommensurate ("premartensitic") phase
T_R: Appearance of R-phase
M_s: Beginning of martensite transformation
M_f: Completion of martensite transformation
A_s: Beginning of austenite transformation
A_f: Completion of austenite transformation

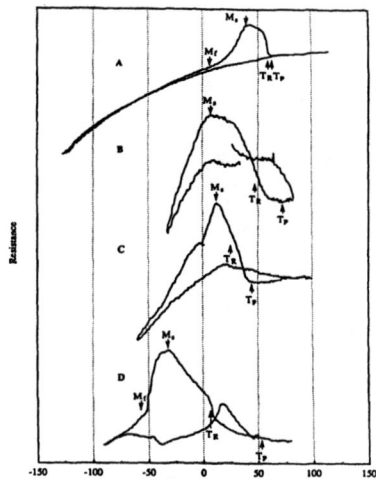

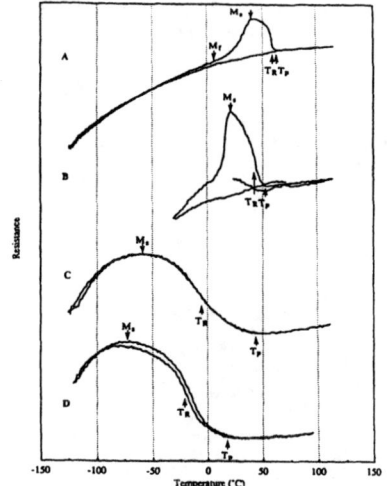

Figure 1: 540°C Isothermal Anneal for (A) 30 minutes, (B) 2 hours, (C) 8 hours, and (D) 16 hours

Figure 2: 30 minute Isochronal Anneals at (A) 540°C, (B) 600°C, (C) 700°C, and (D) 800°C

Figure 1 shows the resulting phase transformation temperatures for the first series of 540°C isothermal anneals. As is evident from these data, longer annealing times result in lower transformation temperatures on cooling and a more pronounced incommensurate phase.

Figure 2 presents results for the series of 30 minute isochronal heat treatments. With an increase in annealing temperature, there follows a drop in the transformation temperatures and a widening of the incommensurate phase.

Each specimen was examined by TEM to determine the specific effects of heat treatment leading to the observed decreases in transformation temperatures. Generally, a correlation was seen between precipitate density and decreased transformation temperatures.

Two types of precipitates were observed, those forming at the grain boundaries and those forming within grain interiors.

No grain boundary precipitation occurred in the 540°C anneals for 30 minutes and 2 hours, a slight degree of precipitation was noted in the 540°C anneals for 8 and 16 hours, still more was observed in the 30 minute 600°C anneal, and much larger quantities of boundary precipitates were seen in the 30 minute anneals at 700°C and 800°C. Interior precipitates were seen in conjunction with grain boundary precipitation, and only in the 700°C and 800°C heat treatments.

Microstructural characterization of the film annealed at 700°C for 30 minutes is presented below for purposes of identifying the two major precipitates observed. Figure 3 presents a magnified image of a single grain from this film in which grain boundary precipitates, interior precipitates, and a precipitate-free zone are clearly seen. Figure 4 presents a dark field image and SADP of the grain boundary precipitate. The SADP's of these precipitates are consistent with the structure's identification as FCC with a_o = 11.3A [5].

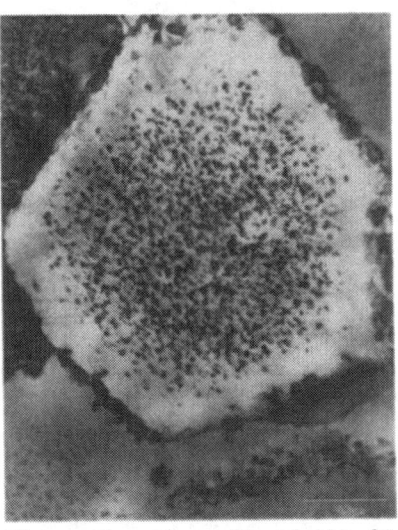

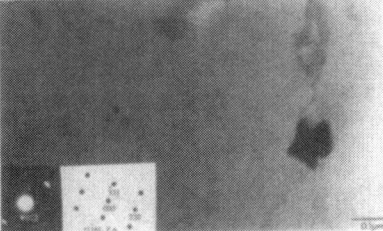

Figure 4: Bright field and SADP of a Grain Boundary Precipitate in 30 minute 700°C annealed film indicative of Ti_2Ni.

Figure 3: Close-Up of the 30 minute 700°C film sample showing precipitates in both the grain boundaries and interior.

Both Ti_2Ni and $(Ti_2Ni)_{1-x}O_x$ ($0<x<1/3$) have the Fe_3W_3C type FCC structure with lattice spacings of a_o = 11.319A and a_o = 11.328A [6], respectively; hence they are indistinguishable by SADP within the accuracy of the technique. To determine whether oxygen was present in the grain boundary precipitates, EDS was performed using a windowless detector. Nearly the same, small amount of oxygen was found in the grain boundary precipitates as was found in the adjacent region with no precipitates. This supports the conclusion that the oxygen detected was on the film surface and that the precipitates at the grain boundaries were Ti_2Ni.

Figure 5 shows a dark field and SADP for a precipitate within the grain interior. The large interior precipitates with circular cross-sections were identified as $Ti_{11}Ni_{14}$, a rhombohedral unit cell with a_o = 6.61A and α = 113.65 [6]. This precipitate was studied previously by Nishida et al. [7] who determined the precipitate's variants, habit plane, unit cell, and relationship to the matrix.

95

DISCUSSION

All 540°C heat treatments exhibited a complete austenite to martensite phase transformation above -100°C. However, progressively longer annealing times were accompanied by both a drop in the transformation temperatures and an increase in grain boundary precipitation. The coincidence of these events can be explained by a change in the film's effective composition. Increasing nickel content in the Ni-Ti system by several atomic percent is known to cause a substantial decrease in the transformation temperatures [8]. Growth of the Ti$_2$Ni precipitates within grain boundaries is expected to decrease M$_s$ due to the corresponding nickel enrichment within the grains. Since no other precipitates or microstructural aberrations were

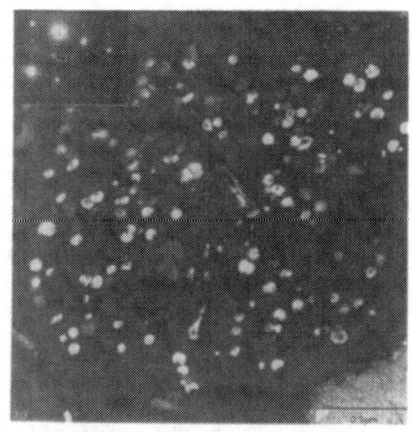

Figure 5: Dark Field and SADP of an Interior Precipitate from the 30 minute 700°C annealed film indicative of Ti$_{11}$Ni$_{14}$.

observed in these samples, it could be concluded that Ti$_2$Ni growth was the primary cause of decreased transformation temperatures.

Resistance plots of the isochronal heat treatments revealed two distinctly different phase transformations. First, the 540°C and 600°C annealed films, which were free from interior precipitation, exhibited complete martensitic transformations at relatively high temperatures. The slight decrease in transformation temperatures observed in the 600°C sample can be attributed to grain boundary precipitation as described above. In contrast, films annealed at the two higher temperatures showed substantial interior precipitation as well as grain boundary precipitation. These samples did not exhibit a complete transformation. Appearance of Ti$_{11}$Ni$_{14}$ precipitates caused a broadening of the incommensurate phase and R-phase transitions, a significant drop in the M$_s$ temperature, and prevented the completion of the martensite transformation.

A drop in the M$_s$ temperature cannot be explained by a composition change induced through growth of Ti$_{11}$Ni$_{14}$ precipitates. This argument works in favor of raising the transformation temperature because these precipitates will remove nickel from the matrix, not add it. Recent research by Xie et al. states that coarse Ti$_{11}$Ni$_{14}$ (which they refer to as Ti$_3$Ni$_4$) precipitates favor nucleation of martensite but constrain it from further development [9]. It may be inferred that the relatively smaller decreases in T$_p$ and T$_R$ for the 700°C and 800°C annealed films are due primarily to the growth of Ti$_2$Ni grain boundary precipitates, but that the much larger shift in M$_s$ and complete absence of M$_f$, is due to a constraint imposed by the Ti$_{11}$Ni$_{14}$ precipitates. This effect is by far more dominant than any effects due to variation in the matrix composition.

Upon comparing the effects of both types of precipitation, it was clear that the existence of grain boundary precipitates alone did not have as strong an effect on transformation temperatures as did the existence of interior precipitates.

CONCLUSIONS

The results of this study show that annealing temperatures well above the crystallization point of 500°C, or annealing times beyond 30 minutes, can cause a definite decrease in the transformation temperatures of Ni-Ti sputtered films. Depression of the martensitic transformation, as well as a broadening of the incommensurate and R-phase, was observed to be coincident with increased precipitation. Grain boundary precipitates appeared to induce moderate decreases in the transformation temperatures whereas precipitates within the grain interior, observed only in combination with grain boundary precipitates, had a much more pronounced effect.

Future research should address the relationship between precipitate density and the mechanical properties of sputtered Ni-Ti films.

Figure 6: Applications for thin film shape-memory alloy include millimeter and micron size valves, high aspect ratio foil actuators, and silicon-based micromachines.

ACKNOWLEDGEMENTS

This research was supported by a NASA Phase II SBIR contract, number NAS2-13113.

REFERENCES

1. A.D. Johnson, presented at the 1990 American Vacuum Society Conference in Toronto, Canada, 1990 (not yet published).
2. J.D. Busch, A.D. Johnson, C.H. Lee, and D.A. Stevenson, J. Appl. Phys. 68(12), 6224-6228 (1990).
3. C.M. Wayman in Proceedings of the Conference on Phase Transformations '87, edited by G.W. Lorimer, pp. 16-22.
4. S. Wu and C. Wayman, Acta metall., 37, 2805 (1989).
5. M. Nishida, C.M. Wayman, T. Honma, Met. Trans. A, 17A, 1505 (1986).
6. M.H. Berkson, Masters thesis, Materials Science Department, University of California at Berkeley, 1990.
7. M. Nishida, C.M. Wayman, Mat. Sci. and Eng., 93, 191 (1987).
8. H. Funakubo (editor), Shape Memory Alloys, OPA, Amsterdam, 1987, p. 99.
9. C.Y. Xie, L.C. Zhao, T.C. Lei, Scripta metall., 24, 1753 (1990).

STRUCTURAL AND ELECTRICAL PROPERTIES OF TITANIUM-NICKEL FILMS DEPOSITED ONTO SILICON SUBSTRATES

KATHLEEN R. COLLEN,* ARTHUR B. ELLIS,* J. D. BUSCH ** and A. D. JOHNSON**
*University of Wisconsin-Madison, Department of Chemistry, Madison, WI 53706
**TiNi Alloy Co., Oakland, CA 94608

ABSTRACT

Thin films of the shape memory alloy NiTi have been sputter-deposited onto p-type silicon substrates. Films that are initially amorphous may be crystallized by vacuum annealing. The crystalline films exhibit the B2->B19' phase change associated with the shape memory effect while remaining in contact with the silicon substrate. Transition temperatures were determined by resistance measurements and x-ray diffraction. The NiTi - Si contacts are diodes, as evidenced by their current-voltage characteristics; however, the effect of the phase change on the barrier height could not be determined.

INTRODUCTION

The intermetallic nickel-titanium alloy is the most common material exhibiting the shape memory effect [1]. A thermoelastic martensitic phase transformation provides the mechanism for the shape memory. In NiTi, the high temperature phase (austenite) has the CsCl, or B2, structure. An intermediate rhombohedral structure, the R - phase, has been identified. The low temperature phase (martensite) is monoclinic, and may exist in any one of twenty-four geometrically-related variants. Deforming martensitic NiTi interconverts these variants to whichever ones most reduce the internal energy. Upon heating, the NiTi must recover the high temperature structure: The only low energy pathway to the CsCl structure exactly retraces the atomic movements responsible for the deformation, and this causes the shape memory.

There has been growing interest in the fabrication of free-standing shape memory alloy films [2], and also in films that remain bound to a substrate [3]. The films prepared to date can be resistively heated to produce a mechanical response. They have been incorporated into the control element in microvalves and may prove useful as microactuators.

To our knowledge, semiconductor-metal structures have not previously been fabricated from NiTi films. Diodes constructed from such structures could in principle exhibit current-voltage properties reflecting the phase of the NiTi overlayer. In this paper we lay the groundwork for such a study by demonstrating that crystalline NiTi films can be deposited onto p-Si, yielding structures that exhibit rectifying behavior.

EXPERIMENTAL

NiTi films were prepared by standard d.c. magnetron sputtering from a composite NiTi target. Polished (100) p-type silicon substrates were chemically etched with a 10% solution of HF and rinsed with methanol prior to deposition. No intermediate or adhesion layers were deposited between the silicon and the NiTi. The films were annealed in sealed glass ampules, at 540°C under vacuum (10^{-3} torr), to initiate crystallization. Higher temperature annealing was not used because it is reported to cause precipitation within the metal film and reaction between the film and the substrate, degrading the properties of the film [3,4]. It was also possible to crystallize films during deposition at temperatures less than 500°C [5].

The films were structurally and electrically characterized. Four- probe resistance measurements in a temperature-controlled chamber were made on the NiTi side of the sample. X-ray diffraction (XRD) powder patterns were taken on a Nicolet R3m/V Polycrystalline X-ray Diffraction System (Cu Kα radiation), using a heated sample holder.

Auger electron spectroscopy measurements were performed on a Perkin Elmer Model 660 Scanning Auger Multiprobe; Ar+ sputter etching provided a depth profile of the films.

Electrical contacts for the current-voltage measurements were made by a spring-loaded pressure contact (NiTi side) and In/Ga eutectic (Si side).

CHEMICAL AND STRUCTURAL CHARACTERIZATION

The composition of the NiTi films was determined by Auger electron spectroscopy. All films show a surface titanium oxide, due to the high reactivity of titanium even with extremely low oxygen exposure. After annealing, this oxide layer becomes thicker, and carbon is also seen on the surface. Under the oxide layer, the films were composed entirely of nickel and titanium.

As-deposited films do not adhere well to silicon. The XRD patterns of as-deposited films were amorphous, showing only a very broad feature between $2\theta = 40°$-$45°$ and the silicon peak just under $2\theta = 80°$. Since the shape memory effect is the result of a crystallographic phase change, resistance vs. temperature plots of the amorphous films were found to lack features characteristic of the phase change (vide infra).

After the annealing process, adhesion is dramatically improved, presumably due to reaction at the NiTi-Si interface. Once the films are annealed, the characteristic XRD pattern for NiTi is seen, along with the diffraction peak for silicon. The film resistance curves show the typical features of the phase transition (Fig. 1). As the sample is cooled from the austenite into the R-phase, the resistance rises sharply. When the film transforms from the R-phase into the B19' martensite, the resistance falls. Subsequent heating of the sample causes a gradual rise in resistance back to the austenite resistance. In some samples a small peak is seen in the heating curve as the NiTi transforms from the martensite.

The variable temperature XRD patterns also demonstrate the phase change (Fig. 2). In the austenite, a single large peak at $2\theta = 42.3°$ appears. Cooling into the R-phase results in a splitting of this peak. As the martensite phase is reached, the most intense peak shifts to $2\theta = 42.5°$, and its intensity shrinks by about half. A number of other B19' reflections appear at $2\theta = 39.2°$, $41.2°$, $43.8°$, and $44.9°$, indexed by Michal and Sinclair as the (020), ($1\bar{1}1$), (002), and (111) peaks, respectively [6].

ANALYSIS OF THE RESISTANCE VS. TEMPERATURE PLOTS

Fourteen film samples having thicknesses between 0.5 and 8 microns were deposited onto silicon. Eight of them adhered during the annealing process, three detached before annealing, and three more detached during the anneal. Film thickness did not have a pronounced effect on adhesion. The three films that were detached before annealing were annealed as stand-alone foils for comparison to the substrate-bound films.

Resistance vs. Temperature, NiTi Film on Si

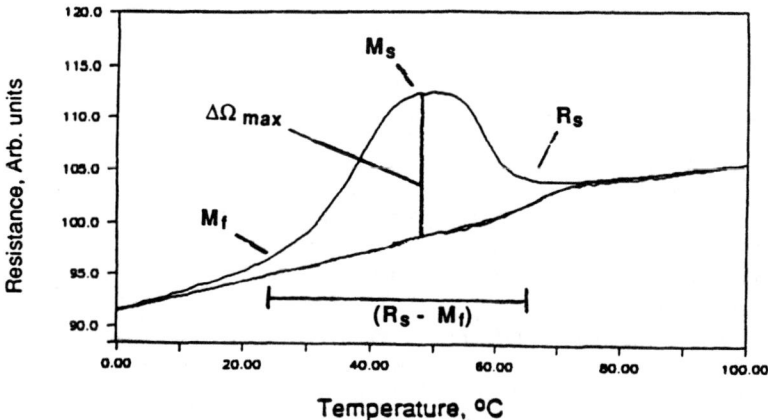

Fig. 1. A typical plot of film resistance vs. temperature. The transformation temperatures for the sample are labeled as follows: beginning of R-phase growth, R_S; beginning of martensite growth, M_S; and the end of the martensite transformation, M_f.

Variable Temperature XRD Powder Patterns of NiTi on Si

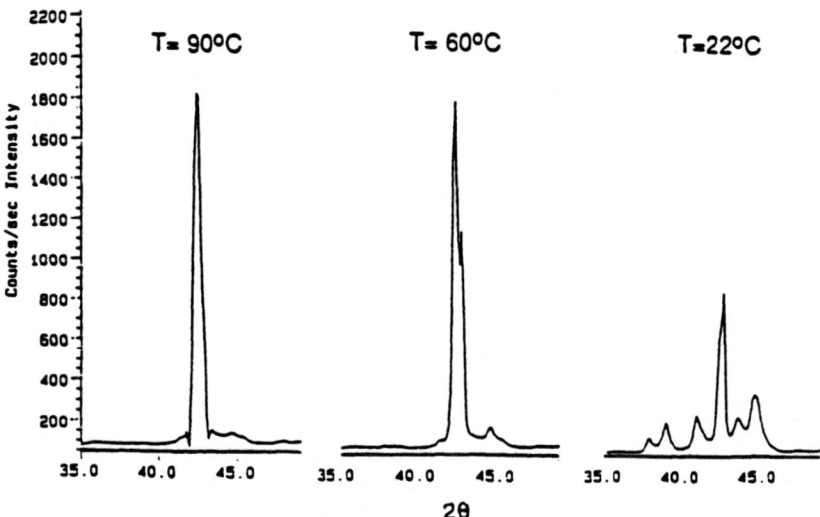

Fig. 2. The XRD powder patterns of NiTi on Si as it is cooled through the phase transformation. At T= 90°C, 60°C, and 22°C, the NiTi is in the austenite phase, R-phase, and martensite phase, respectively.

The temperatures at which transformations into the R and martensite phases began, R_S and M_S, and the temperature at which the martensite transformation finished, M_f, were tabulated for each film, as was the total transformation width, $(R_S - M_f)$. The maximum difference in resistance at a particular temperature between the heating and cooling curves, $\Delta\Omega_{max}$, was also determined for each sample. The parameters are shown on the resistance vs. temperature plot in Fig. 1.

The onset of the phase change during cooling occurred between 40 and 65° C in most of the samples. The width of the transformation on cooling, $(R_S - M_f)$, was generally 30 to 50° C, although a few samples had much larger widths, between 80 and 100° C. Large values of $(R_S - M_f)$ were primarily due to a lower M_f temperature, since the M_S temperature fell between 10 and 30 degrees below the R_S temperature regardless of the total width of the transformation. Values of R_S, M_S and $(R_S - M_f)$ could not be correlated with the film thickness.

The maximum change in resistance at a particular temperature, $\Delta\Omega_{max}$, was a function of film thickness. Thicker films showed a larger absolute change in resistance. For samples taken from the same sputtering run, the change in the resistance is a linear function of thickness ($r^2 = 0.97$). The largest values of $\Delta\Omega_{max}$ were found in samples that did not adhere to the substrate. Regardless of whether they became detached before or during annealing, they had values of $\Delta\Omega_{max}$ that were, on average, 40% larger than those of the adherent samples.

Films under one micron in thickness generally had weak resistance signatures, showing only small changes in resistance as they were cooled.

CHARACTERIZATION OF NiTi-Si SCHOTTKY BARRIERS

The current-voltage (I-V) characteristics of the NiTi - Si junctions were measured. Like many metal - semiconductor contacts, they were capable of rectifying current. In the forward bias direction the film side acted as the cathode.

In order to measure the barrier height of the diodes, semilog plots of the I-V data were made (Fig. 3). However, the data were extremely curved, and it was not possible to extrapolate a value of the saturation current and therefore to calculate a value of the barrier height. Low temperature measurements extended the linear region, but it remained too small for accurate extrapolation.

Curvature of I-V data is often caused by a series resistance due to the semiconductor substrate. An alternative to the semilog I-V curve, introduced by Norde [7], allows the series resistance to be separated from the diode response. The following function is plotted vs. voltage:

$$F(V) = (V/2) - (k_B T/q) \ln (J/A^* T^2) \tag{1}$$

where J is the current density, A^* is the Richardson constant (32 A/cm^{2}K^2) for p-type Si, T is the temperature, and k_B and q have their usual values. The barrier height is calculated from the minimum in the F(V) curve, so long as the minimum occurs at a voltage greater than $3k_B T/q$. For all samples tested, there was either no minimum in the Norde plot, or the minimum was below $3k_B T/q$. This indicates that the series resistance probably is not the cause of the curvature.

Although the NiTi - Si junction acts as a diode, its properties are far from ideal. At high voltages (>1V) the semilog I-V curves are linear, and the ideality constant, n,

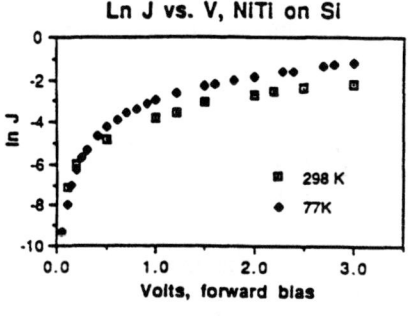 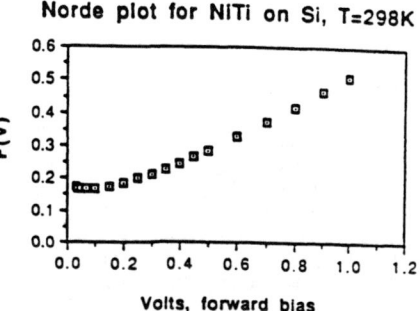

Fig. 3. Two methods of analyzing current-voltage data in the forward bias direction are shown. On the left, ln (J) is plotted against the applied voltage. On the right, a sample Norde plot (see text, equation 1) is presented.

may be calculated from the slope of the line in this region using the I-V relationship for thermionic emission:

$$J = J_0 \, [\exp(qV/nk_BT)] \qquad (2)$$

This equation applies for voltages greater than $3k_BT/q$. Values of n ranged from 5 to 50 for the NiTi - Si contacts, all well above the ideal value of 1.02. This may indicate the presence of a large interfacial layer between the NiTi and Si, possibly an oxide or silicide, or that current transport mechanisms in addition to thermionic emission are in effect.

CONCLUSIONS

NiTi films can be deposited directly onto silicon. The NiTi is amorphous upon deposition and shows no phase change. After annealing, the martensitic phase change that creates the shape memory effect is observed in the films, and temperature cycling does not cause the film to become detached from the silicon substrate.

The width of the phase change is generally 30 to 50 degrees, and the R_s temperatures, indicating the beginning of the phase change while cooling, ranged from 45 to 65 degrees. Extremely wide transitions, larger than 80ºC, were due to low M_f temperatures. Film thickness did not appear to affect any of the transition temperatures, but the value of $\Delta\Omega_{max}$ was a function of thickness.

The NiTi - silicon junctions are capable of rectifying; however, a barrier height could not be extracted from the current-voltage curves. Lower temperature annealing methods to limit the growth of any interfacial layers are currently being explored to improve the quality of the diodes. When barrier heights for the diodes may be measured, the effect of the phase transition on the diode properties will be determined.

ACKNOWLEDGEMENTS

The authors would like to acknowledge Prof. R. West for the use of the high temperature XRD sample stage, Prof. T. Kuech for helpful discussions about Schottky barriers, and Dr. N. Tran for help in obtaining the Auger data. We are grateful to the

Office of Naval Research and the National Institutes of Health for their financial support.

REFERENCES

1. K. R. Collen et al., "Principles of the Shape Memory Effect in Nickel-Titanium" in Chemistry of Advanced Materials, ed. by C.N.R. Rao (Blackwell, in press).
2. J. A. Walker et al., Sensors and Actuators A, 243 (1989).
3. A.P. Jardine et al., unpublished results.
4. M. Berkson, J. D. Busch and A.D. Johnson, these proceedings
5. K R. Collen, A. B. Ellis, J. D. Busch, and A.D. Johnson, in preparation.
6. G.M. Michal and R. Sinclair, Acta Cryst. B37, 1803 (1981).
7. H. Norde, J. Appl. Phys. 50, 5053 (1979).

CORRELATION OF STRESS RELAXATION AND MICROSTRUCTURE CHANGE IN POLYCRYSTALLINE THIN FILMS ON SUBSTRATES:
Au on Si <100> at RT

A.C. Vermeulen, R. Delhez and E.J. Mittemeijer
Delft University of Technology, Laboratory of Materials Science,
Rotterdamseweg 137, 2628 AL Delft, The Netherlands.

ABSTRACT

Stress relaxation in polycrystalline layers can be explained by processes, in which the microstructure plays a dominant role. The microstructure itself may also be subjected to changes. With X-ray diffraction information about both the stress and the microstructure can be obtained without destroying the specimen and without disturbing the stress relaxation process.

In this paper a model system is studied: Au on Si<100>. The specimens showed a simultaneous decrease of macrostress and dislocation density with time at room temperature. This could be interpreted on the basis of a model founded on thermally activated dislocation motion. It followed that the grain size is an important parameter for the change of the dislocation density.

INTRODUCTION

Stress levels in polycrystalline layers tend to decrease [e.g. 1,4], even at room temperature [e.g. 2]. A description of relaxation as a function of time [1,2] results from the recognition that stress induced dislocation glide is hindered by obstacles, which are passed by thermal activation [5]. Hence the dislocation density will influence the relaxation.

In this paper X-ray diffraction has been applied because it does not disturb the relaxation process. Moreover it allows to determine macrostress and dislocation density simultaneously and non-destructively. Au layers on Si<100> have been investigated at room temperature. A strong correlation between stress relaxation and change of dislocation density has been found.

EXPERIMENTAL

Gold layers of about 1 μm thickness were made by physical vapour deposition (deposition rate ~1 nm/s) in high vacuum (pressure <10^{-3} Pa) on Si <100> substrates covered with a sputter deposited 50 nm SiO_2 buffer layer to prevent that Au grows epitaxially and diffuses into the Si. During deposition the temperature of an unheated substrate raised from room temperature to 338 K (specimen A) and for a heated substrate from 473 K to 483 K (specimen B). After 210 days aging at room temperature specimen A was annealed at 483 K for 1.75 h (new specimen code: A') in the same vacuum chamber as used for the layer deposition. Specimens A' and B were cooled to room temperature in about 4.5 h.

X-ray stress measurements were performed with a Siemens D500 diffractometer using a diffracted beam monochromator. CoKα {420} diffraction lines were recorded with steps of 0.05 °2θ using a 1° divergence and a 0.1°2θ receiving slit. Counting times were such that ~10^4 counts were obtained at the top. Top positions were determined by fitting a parabola to intensities larger than 70% of the maximum intensity, after corrections for dead time, linear background, Lorentz-polarization factor, absorption and the Kα_2-component. The macrostresses were determined according to the sin$^2\psi$ method [7].

According to Williamson and Smallman [3] the dislocation density ρ can be derived from the integral breadth of the strain broadened profile β_{strain} ($=2\,\xi\tan\theta$, where ξ is microstrain):

$$\rho = \frac{k_{hkl}}{F_{int}} \left[\frac{\beta_{strain}}{2\,b\,\tan\theta} \right]^2 \tag{1}$$

where F_{int} is a factor which takes into account interaction between dislocations (here we have taken F=1; cf. DISCUSSION), k_{hkl} is a reflection-dependent constant ($k_{420}=11.8$ for a screw dislocation in fcc material with Burgers vector along <110>), b is the length of the Burgers vector and θ is half the diffraction angle. The integral breadth β of the structurally broadened profile (at $\psi=0$) was obtained after correction for instrumental broadening using Voigt functions [6]. The broadening due to small domain size was neglected in this analysis (see RESULTS) and β_{strain} in Eq.(1) was set equal to β.

Grain sizes were determined by transmission electron microscopy taking the square root of the average grain area as obtained from the number of grains on the photograph.

RESULTS

The state of stress has been tested by performing two stress measurements in perpendicular directions. The stresses were equal, confirming the rotational symmetric plane state of stress ($\sigma_1=\sigma_2=\sigma$, $\sigma_3=0$). Fig.1 shows stress σ versus aging time t at room temperature for specimens A, A' and B.

The dislocation density calculated from the integral breadth β is overestimated because size broadening (see below Eq.(1)) is neglected. The error $\Delta\rho$ in the dislocation density can be estimated from the TEM grain size (see Table I) using the Scherrer formula ($\Delta\beta_{strain}=\beta_{size}=\lambda/(D\cos\theta)$). This gives a $\Delta\rho$ of $5\ 10^{12}\ m^{-2}$ and $21\ 10^{12}\ m^{-2}$ for specimens A' and B respectively. So neglecting size broadening seems justified, also because in the further (quantitative) analysis only changes of dislocation densities are considered. Fig.2 shows dislocation density ρ versus aging time t at room temperature for specimens A' and B (β_{size} is too large for specimen A). Although the dislocation densities of specimens A' and B differ by a factor 2, their decreases with time correlate with the decreases of the stresses.

Table I
Experimental details and results for specimens A, A' and B.

specimen code		A	A'	B
substrate temperature during deposition (K)		338	338	483
annealing (K)		-	483	-
<111> fibre texture		weak	weak	strong
grain size from TEM	(nm)	75	475	235
slope dσ/dρ (measured)	(MPa/10^{12} m^{-2})	-	0.915	0.205
(calculated)	(MPa/10^{12} m^{-2})	0.62	3.91	1.94
initial stress σ_i (fitted)	(MPa)	-	175.6	192.2
(calculated)		65	300	300
constant P (fitted)	(MPa)	-	36.2	24.2
constant C (fitted)	(10^{-6} s^{-1})	-	1.01	6.83

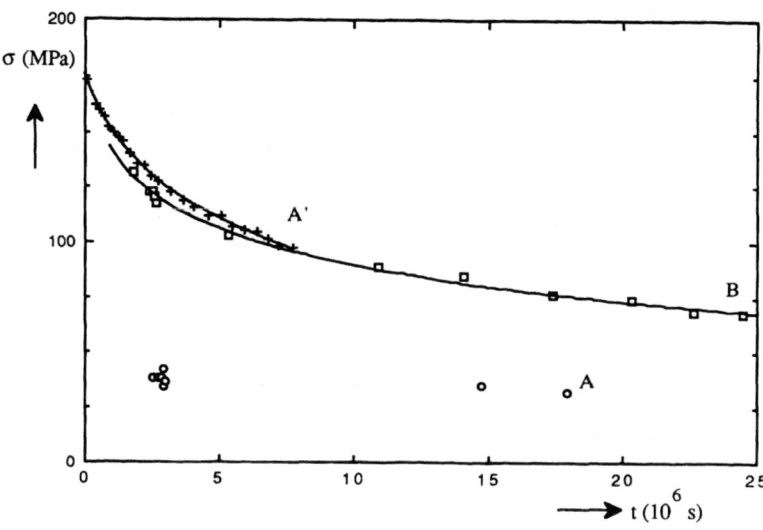

Fig. 1

Stresses σ in polycrystalline Au layers on Si<100> of specimens A, A' and B followed during aging at room temperature for 7, 3 and 9 months respectively. The accuracy of the (X-ray) stress measurements is ~1 MPa. The full lines are fitted according to Eq.(6).

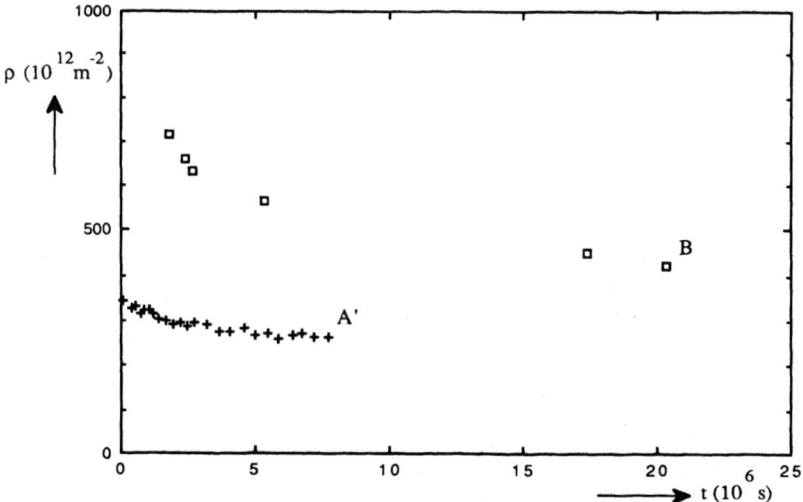

Fig.2

Dislocation densities ρ of specimens A' and B versus aging time t at room temperature. ρ is calculated according to Eq.(1) from the integral breadth of the CoKα {420} X-ray diffraction line.

A MODEL FOR CHANGES IN STRESS AND DISLOCATION DENSITY

The model used here for the correlation between stress relaxation and change of dislocation density in thin polycrystalline layers is based on the idea that stress is relaxed by dislocation glide and that the dislocation density decreases by disappearance of dislocations in grain boundaries.

Plastic deformation at low temperatures ($T < 0.3\ T_m$) is dominantly realized by glide of dislocations [5]. This motion is hindered by obstacles.

Firstly a single slip system j in a single crystallite is considered. The number of successful jumps per unit time of a dislocation segment of length l over obstacles under the action of both a shear stress τ_j and thermal vibration is given by

$$\dot{n}_j = v_d\, e^{-\Delta F(1-\tau_j/\tau_0)/k_B T} \tag{2}$$

where $v_d = v\, b/l$ is the vibration frequency of the dislocation segment, v is the atomic vibration frequency, b is the length of the Burgers vector, ΔF is the Helmholtz free energy required to overcome the obstacle and τ_0 is the flow shear stress at 0 K, k_B is Boltzmann's constant and T is the absolute temperature. Now the plastic-shear rate can be written as

$$\dot{\gamma}_{pl,j} = \rho_{m,j}\, b\, L_j\, \dot{n}_j \tag{3}$$

where $\rho_{m,j}$ and L_j denote the mobile dislocation density and mean obstacle distance for the case considered; Eq.(3) ignores the time of glide between the obstacles.

For an uniaxial state of stress ($\sigma_1 = \sigma_2 = 0$, $\sigma_3 = \sigma$) or a rotational symmetric plane state of stress ($\sigma_1 = \sigma_2 = \sigma$, $\sigma_3 = 0$) in the layer, an (applied) stress σ is related to a shear stress τ by

$$\sigma = m\,\tau \tag{4}$$

where m is a factor that accounts for the geometrical relation between σ and τ (cf. [8]). Relaxation of stress occurs if part of the shear stress, initially accommodated elastically, is accommodated plastically, while the total shear strain is constant. Therefore $\dot{\gamma}_{el,j} = -\dot{\gamma}_{pl,j}$. The contribution to stress relaxation of shear in slip system j is:

$$\dot{\sigma}_j = m_j\, \dot{\tau}_j = m_j\, G\, \dot{\gamma}_{el,j} = -\, m_j\, G\, \dot{\gamma}_{pl,j} \tag{5}$$

where G is the shear modulus. Eq.(5) pertains to stress relaxation in a single crystal where only a single slip system is active. To arrive at an expression for the whole polycrystalline layer, allowance for all possible slip systems should be made and averaging over all grains of the layer should be performed. It is supposed that this can be done by introducing effective values for the parameters with a subscript j (i.e. dropping the subscript j).

Assuming that only stresses are time dependent, integration of the generalized Eq.(5), after combination with the generalized Eqs.(2-4), yields an equation of the form:

$$\sigma = \sigma_i - P\, \ln(1 + C\, t) \tag{6}$$

where P and C are constants and σ_i is the initial stress (at t=0). Eq.(6) is frequently used [e.g.1,2].

The decrease of the dislocation density can be treated on the same basis as the stress relaxation. Gliding dislocation segments will eventually be absorbed by grain boundaries. In a crystal half of the mobile dislocation segments within a distance L_j from a grain boundary can be annihilated (the other half will glide towards the interior of the grain). For a spherical grain it follows for $f_{a,j}$, the fraction of the mobile dislocations that is in a position to be annihilated:

$$f_{a,j} = \frac{1}{2} \frac{3}{D/2} L_j \tag{7}$$

where D is the crystallite size. The rate of change of the density of dislocations (immobile and mobile) obeys:

$$\dot{\rho}_j = - f_{a,j} \, \rho_{m,j} \, \dot{n}_j \tag{8}$$

Combination of Eqs.(3), (5), (7) and (8) yields:

$$\frac{d\sigma}{d\rho_j} = \frac{1}{3} b \, G \, D \, m_j \tag{9}$$

This equation holds for a single slip system in a single crystallite. To arrive at an expression for the whole polycrystalline layer a procedure as that below Eq.(5) should be performed. Then one arrives at:

$$\frac{d\sigma}{d\rho} = \frac{1}{3} b \, G \, \bar{D} \, \bar{m} \tag{10}$$

where $\bar{D}$ is the average grain size and $\bar{m}$ corresponds to the so-called Taylor factor (3.06 for polycrystalline fcc materials with randomly oriented grains [8]).

DISCUSSION

From the difference in thermal shrink $\varepsilon = \Delta\alpha\Delta T$ ($\Delta\alpha = 11.9 \, 10^{-6}/K$, $\Delta T = T-T_{room}$) between Au layer and Si substrate stress values expected from $\sigma = E \, \varepsilon/(1-v)$ are 65 MPa for the layer deposited at 338 K and 300 MPa for the ones deposited or annealed at 483 K (Young's modulus $E_{Au} = 80$ GPa and Poisson's ratio $v_{Au} = 0.42$). The determined (initial) macrostresses are significantly lower than the values calculated. Apparently relaxation occurred also during cooling from the deposition or annealing temperature to room temperature.

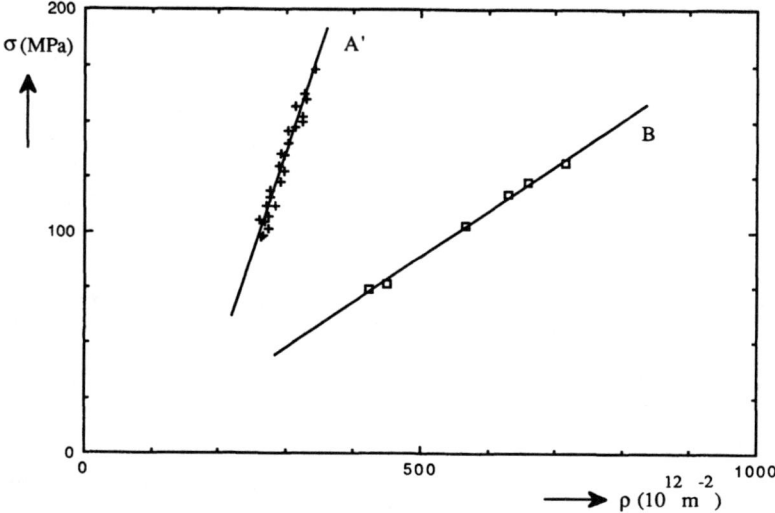

Fig.3
Correlation between stress σ and dislocation density ρ for specimens A' and B.

For the stress relaxation at room temperature it was found possible to fit Eq.(6) to the stress data (see Table I and full lines in Fig.1). This provides support for the first part of the model proposed: stress relaxation by dislocation glide.

During fitting of Eq.(6) to the results of specimen B it turned out that σ_i and C of Eq.(6) are correlated. The value of P, however, did vary little when varying σ_i and C. Comparison of the values obtained for P (=k_B T $\bar{m}$ $\tau_O/\Delta F$) of the specimens A' and B shows a significant difference. This may be due to specimen dependence of the Taylor factor $\bar{m}$, which could be caused by the difference in texture of the specimens (see Table I).

Fig.3 shows that the observed slopes $d\sigma/d\rho$ for specimens A' and B are constant indeed as predicted by Eq.(10). This provides support for the second part of the model proposed: absorption of gliding dislocations by grain boundaries. Further the dependence of $d\sigma/d\rho$ on the average grain size $\bar{D}$ is qualitatively confirmed by the experimental results (see Table I for estimates of $\bar{D}$). However, the measured slopes are smaller then the calculated ones. Even the ratio of the measured values of $d\sigma/d\rho$ for specimens A' and B differs from the ratio of the calculated values: 4.5 versus 2. If $\bar{m}$ is specimen dependent in the way suggested above, implying $\bar{m}_{A'}/\bar{m}_B=P_{A'}/P_B$, the calculated ratio becomes about 3. The remaining difference could be caused by differences in dislocation arrangement (e.g. pile-ups) in both samples, which leads to a difference in the dislocation interaction factor F_{int} (cf. Eq.(1)) for specimens A' and B.

CONCLUSIONS

Macrostress relaxation and decrease of dislocation density occur at room temperature in polycrystalline Au layers under stress on Si substrates.

The relaxation of macrostress can be described by adopting dislocation glide as stress relaxation mechanism.

The correlation between stress relaxation and decrease of dislocation density can be described by adopting, additionally, that gliding dislocations annihilate at grain boundaries. Until now the influence of the change of dislocation density has not been taken into account in the description of stress relaxation in thin metallic layers.

ACKNOWLEDGMENT

These investigations have been supported financially by the Foundation for Fundamental Research on Matter (FOM), The Netherlands.

REFERENCES

1. P.A. Flinn, D.S. Gardner and W.D. Nix, IEEE Trans. ED-34, 689 (1987).
2. M.A. Korhonen, C.A. Paszkiet, R.D. Black and Che-Yu Li, Scripta Met. 24, 2297 (1990).
3. G.K. Williamson and R.E. Smallman, Phil. Mag. 1, 34 (1956).
4. M. Hershkovitz, I.A. Blech and Y. Komem, Thin Solid Films 130, 87 (1985).
5. M.F. Ashby and H.J. Frost, in Constitutive Equations in Plasticity, edited by A.S. Argon (MIT Press, Cambridge, MA, 1975) pp.117-147.
6. R.Delhez, Th.H. de Keijser and E.J. Mittemeijer, Fres. Z. Anal. Chem. 312, 1 (1982).
7. H.P. Klug and L.E. Alexander, X-ray Diffraction Procedures, 2nd ed. (John Wiley & Sons, New York, 1974), pp.755-790.
8. R.W.K Honeycombe, The Plastic Deformation of Metals, (Edward Arnold Publishers, London, 1968), pp.224-226.

THE MICROSTRUCTURAL EVOLUTION OF NANOMETER RUTHENIUM FILMS IN Ru/C MULTILAYERS WITH THERMAL TREATMENTS

Tai D. NGUYEN[1,2] Ronald GRONSKY[2], and Jeffrey B. KORTRIGHT[1]

[1]Center for X-Ray Optics, Accelerator and Fusion Research Division, Lawrence Berkeley Laboratory, University of California, Berkeley, CA. 94720.
[2]National Center for Electron Microscopy, Materials Sciences Division, Lawrence Berkeley Laboratory, and Department of Materials Science and Mineral Engineering, University of California, Berkeley, CA. 94720.

ABSTRACT

The evolution of nanometer Ru films sandwiched between various C layer thicknesses with thermal treatments was studied by plan-view and cross-sectional Transmission Electron Microscopy. Plan-view observation provides information on the Ru grain size, while cross-sectional studies allow examination of the multilayer morphology. After annealing at 800°C for 30 minutes, the grain size in the 2 and 4 nm Ru layers show little difference from each other, while that in the 1 nm Ru layers depends strongly on the thickness of the C layers in the multilayers. It increases with decreasing C layer thickness. Agglomeration of the Ru layers is observed in 1nm Ru / 1nm C multilayers after annealing at 600°C for 30 minutes. The evolution of the microstructures and layered structure stability of the Ru/C system is compared to that of W/C and Ru/B$_4$C systems.

INTRODUCTION

Multilayer structures containing alternating layers of high and low atomic number materials provide highest reflectivity in the region of the electromagnetic spectrum from extreme ultra-violet to x-ray wavelengths.[1] Multilayers having carbon as the low Z material yield practical reflectivities at normal incidence at wavelengths between 4.5 and 12.5 nm. As the operating wavelength decreases, the period or layer thicknesses inside the multilayers also decreases, and imperfections in the layered microstructure have increasing effects on the multilayer reflectance and stability. As-deposited nanometer-period multilayers are in a metastable configuration. The microstructures depend on, among other factors, the evolution of the metal layers with thermal treatments and its reactions with the specific materials which make up the multilayers. Annealing studies therefore are useful in understanding the microstructural evolution of the multilayers toward equilibrium, the relative stability of multilayers composed of different materials, and to simulate long-term or elevated temperature applications of x-ray multilayers.

The microstructural characteristics of the layers of a metal/carbon multilayer depend on the reactions between the metal and carbon constituents which can be predicted from the appropriate phase diagram. The Ru/C system was studied because of its potential high performance compared to other metal/carbon systems at the soft x-ray wavelengths. The Ru-C phase diagram is of simple eutectic type, with Ru and C having very low mutual solubilities at low temperature. Previous studies show that the amorphous Ru layers in a 2 nm period multilayer agglomerate, while thicker crystalline layers in longer period multilayers do not, under the same annealing conditions.[2] The thin amorphous Ru layers are in a high energy non-equilibrium state, which provides a strong driving force for crystallization through diffusive rearrangements of the atoms. Annealing of W/C multilayers of different periods, however, does not result in agglomeration of the metal layers.[3] The W layers in the 2 nm period W/C multilayers remain amorphous, while formation of a carbide phase was instead found in longer periods, in contrast to the formation of elemental Ru crystallites in the Ru/C system. In this paper, we investigate the effects of thermal treatments upon the microstructural evolution of Ru films sandwiched between C layers, and of agglomeration of the thin Ru films after annealing. Existing models for thin film agglomeration, and their relevance to these nanometer thick films are discussed. The phases present in, and the layered structure stability of, the Ru layers in Ru/C multilayers are compared to those of the W layers in W/C multilayers, and of the Ru layers in Ru/B$_4$C multilayers.

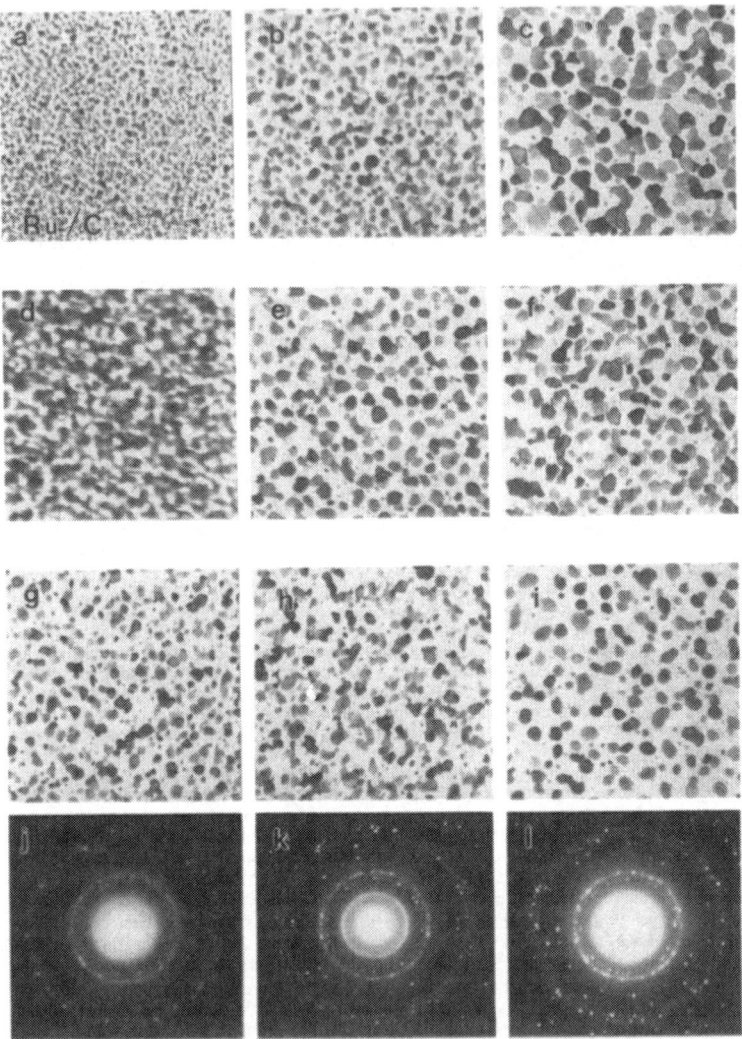

Fig. 1 -- Plan-view TEM images of a) 1/2, b) 1/1, c) 1/0.5, d) 2/4, e) 2/2, f) 2/1, g) 4/8, h) 4/4, i) 4 nm Ru / 2 nm carbon multilayers after annealing at 800°C for 30 minutes. The electron diffraction patterns in j)-l) correpond to the bright field images in g)-i), respectively.

EXPERIMENTAL TECHNIQUES

Samples were prepared by magnetron sputtering at floating temperature at the Center for X-Ray Optics, LBL. The apparatus and procedures of deposition have been described elsewhere.[4] Samples of nominally 1, 2, and 4 nm Ru layer thicknesses were prepared as multilayers sandwiched between various thicknesses of C films. Three thicknesses of C were deposited for each Ru thickness, such that the Ru-to-C layer thickness ratios were 0.5, 1, and 2. The nanometer thicknesses of the Ru and C layers in the Ru/C multilayer samples containing the 1 nm thick Ru layers hence were 1/2, 1/1, and 1/0.5. Similarly, the layer thicknesses of the samples containing the 2 and 4 nm thick Ru layers were 2/4, 2/2, 2/1, and 4/8, 4/4, and 4/2, respectively. Different periods of W/C and Ru/B$_4$C multilayers, with the high Z material layer composed of approximately 40% of the multilayer periods, were also prepared for the studies. A B$_4$C alloy target was used in sputtering of the B$_4$C films.

Cross-sectional and plan-view Transmission Electron Microscopy (TEM) were used to characterize the films. Cross-sectional samples from multilayers deposited on standard Si (111) wafers were prepared by mechanical thinning, followed by ion beam milling in a cold stage.[5] Plan-view samples were prepared so that the multilayers spanned the holes of the TEM copper grids.[6] Annealing of the samples was performed in a vacuum furnace at 10^{-6} torr. The cross-sectional samples were studied in a JEOL JEM 200CX electron microscope equipped with high resolution goniometer, operating at 200 kV. The plan-view samples were studied in a Philips 301 operating at 100 kV.

RESULTS AND DISCUSSIOIN

Plan-view TEM observation

Plan-view TEM observation of the as-prepared samples reveals that the structures are predominantly amorphous for the 1 and 2 nm Ru films, and show signs of micro-crystallites in the 4 nm Ru films, for all thicknesses of C films prepared. The plan-view TEM samples annealed at 400°C for 2 minutes and for 30 minutes show no significant differences in the microstructure from the as-prepared samples. The electron diffraction patterns from all three Ru film thicknesses show very diffuse rings signifying a predominantly amorphous structure. Further annealing at 600°C for 30 minutes results in elemental Ru micro-crystallites with grain size of the order of a few nanometers. Annealing at higher temperature or longer time was required to induce further grain growth or coarsening of these grains.

The plan-view TEM images in Figure 1 exhibit the microstructures of Ru crystallites for the three thicknesses of Ru films sandwiched between different thicknesses of C films after annealing at 800°C for 30 minutes. The bright field images in Figures 1a)-i) display the images of samples with the same Ru layer thickness in each row; 1 nm, 2 nm, and 4 nm thick Ru layers are in the first, second, and third row, respectively. Each column in Figure 1 has the same Ru-to-C layer thickness ratio of 0.5, 1, and 2, in the right, center, and left column, respectively. The electron diffraction patterns of the 4 nm Ru layer thickness samples in Figures 1g)-i) are shown as example in Figures 1k)-m). The polycrystalline rings in these diffraction patterns originate from elemental Ru crystallites.

Several trends in Ru grain size are apparent from Figure 1. First, the grain size distribution in samples with 2 and 4 nm thick Ru layers in Figures 1d)-i) shows little difference from each other. The Ru grain size in these samples is of the same order as the Ru film thickness, consistent with the grain size observed from cross-sectional TEM samples. Second, and in constrast to the 2 and 4 Ru layers, the grain size in the 1 nm Ru film increases with decreasing C film thickness. These thinnest as-deposited Ru layers studied are the most highly disordered and hence have the strongest driving force for crystallization. Higher surface-to-volume ratios of structures with thinner C layers may also enhance the kinetics of crystallization and grain growth of the Ru crystallites. Evidently, the thinner C layers increase the mobility of the Ru atoms in the annealing-induced crystallization and grain growth, which results in larger final grain distribution. Indeed it is possible that the thinnest C layers in Figure 1c) may not be continuous in the as-deposited sample. The microstructure of the 1nm Ru / 2nm C multilayer in Figure 1a) shows a highly dispersed composite of approximately one-nanometer diameter grains of Ru in a C matrix. Third, comparison of Figures 1c), f), and i) of the samples with the same Ru-to-C layer thickness ratio also indicates that the grain size increases with decreasing layer thickness of Ru and/or C. Thus thinner as-deposited Ru layers result in larger Ru grain sizes, contrary to the expectation that grain size is proportional to the film thickness. Evidently, the larger annealed grain size in the thinner layers results either from increased kinetics of crystallization and grain growth with thinner C layers, or from the higher degree of metastability in the thinner Ru layers.

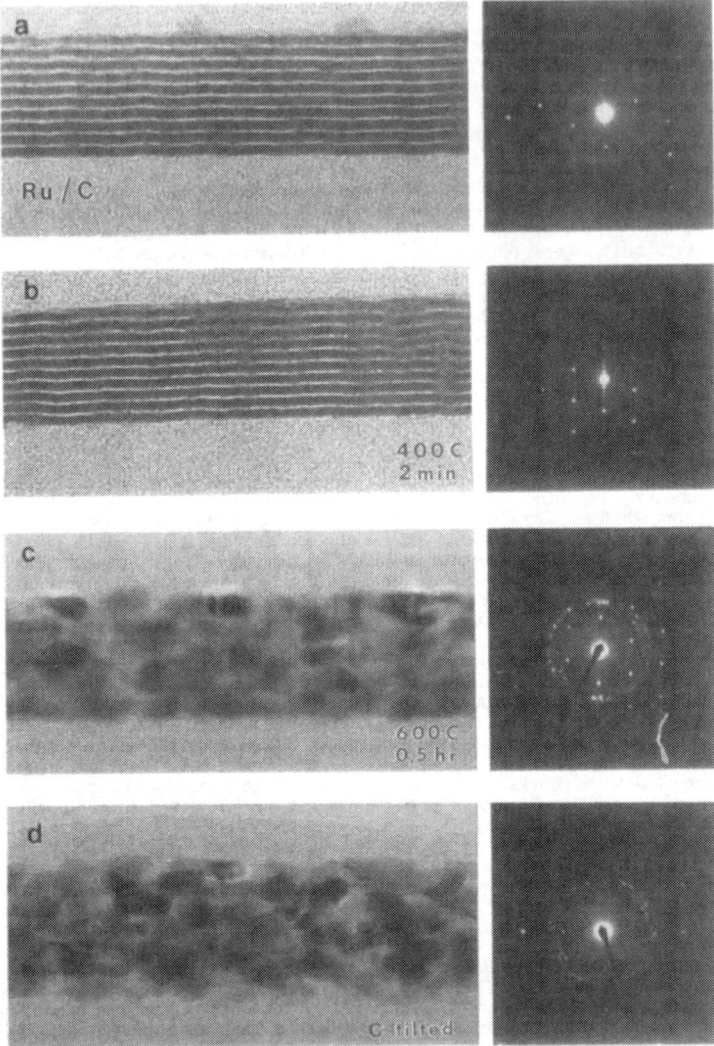

Fig. 2 -- TEM images of as-prepared and annealed 1 nm Ru / 1 nm carbon multilayers:
a) as-prepared, b) 400°C for 2 minutes, c) 600°C for 30 minutes, showing agglomeration
in the Ru films, and d) sample in c) tilted perpendicular to the film surface to show
the morphology and the crystallites and the interface.

<u>Cross-sectional TEM samples</u>

Our previous study indicates that the thicker Ru layers are stable while the thinner layers agglomerate upon annealing.[2] In this study, samples of 1nm Ru / 1nm C treated under various annealing conditions were examined in cross-section to study the microstructures and morphology of the films. Figure 2 demonstrates the evolution of microstructure of this multilayer with different thermal treatments. Shown in each frame of Figure 2a)-d) are the cross-sectional HRTEM image of ten bilayers of Ru and C films deposited on a thick C buffer layer prepared on a Si substrate (not shown), and its corresponding electron diffraction pattern. The amorphous layer on top of the multilayers is the epoxy adhesive used in the cross-sectional TEM specimen preparation process. The small-lattice-spacing diffraction spots in the diffraction patterns of Figures 2a)-c) arise from the epitaxial Si substrate.

The layered structure of the as-prepared sample is shown in Figure 2a). Both the bright field image and the diffraction pattern indicate that the structure is predominantly amorphous. After a thermal treatment of 400°C for 2 minutes in vacuum, the layered structure is still stable and displays an amorphous structure within the layers, as seen in Figure 2b). Annealing at 600°C for 30 minutes, however, results in destruction of the layered structure by agglomeration of the Ru films, shown in Figure 2c). The Ru films have agglomerated and crystallized into almost spherical crystallites with diameters on the order of a few nanometers, similar to that observed in the plan-view samples. The sizes of these crystallites are much larger than the initial bilayer thickness of the multilayer, which suggests that during agglomeration, Ru from adjacent layers may coalesce. The rings in the diffraction pattern indicate a preferred orientation with the <101> direction of the elemental Ru normal to the film surface, similar to results reported earlier.[2]

The interface between the agglomerated multilayer and the C buffer layer appears quite uniform and continuous in Figure 2c). After the sample was ion-milled to a thickness transparent to the electron beam, the multilayer has a trapezoid configuration in the wedged-shape TEM specimen where the multilayer / carbon interface is the base of the trapezoid. Imaging with the electron beam parallel to the base of the trapezoid thus produces an image from a thick region of many Ru grains, and results in an image of a smooth and uniform interface. A single layer of Ru grains at one corner of the base of the trapezoide sample can be viewed by tilting the TEM specimen in the direction perpendicular to the film surface. The interface then appears like that of an agglomerated structure, as seen in Figure 2d). The morphology of single Ru grains is seen to be almost spherical crystallites that do not appear to be connected to each other. Similar morphologies occur at the surface of the aggomerated structure, which is the interface between the multilayer and the epoxy layer in this Figure.

<u>Existing models for agglomeration</u>

Mechanisms for agglomeration of thin films have been studied by various groups,[7-10] although they were developed for films that are substantially thicker than those in this study. One model is analogous to the Rayleigh instability in an infinitely long cylinder with isotropic surface energy, in which perturbation of the straight cylinder walls drives the structure to a spherical shape in order to reduce the surface area. When applied to thin films, linear stability analysis predicts that all small perturbations decay and a flat film should remain stable.[7] Studies of systems with large perturbations (amplitudes of the same order as the film thickness) require nonlinear stability analysis.[8] Energy calculations show that a film can actually be ruptured if the amplitudes of the perturbations are large enough, and that perturbations with amplitudes smaller than the film thickness will decay.

Another model of agglomeration is based upon the mechanism of grain boundary grooving in polycrystalline thin films,[8-10] which predicts that agglomeration in polycrystalline films results from the deepening of the grooves at grain junctions, driven by the equilibrium interfacial energies during grain growth. In general, agglomeration in polycrystalline films from grain boundary grooving is observed when the crystalline grain size to film thickness ratio reaches a critical value that is greater than 1.

The existing models are not necessarily appropriate to explain for the agglomeration mechanisms of the one-nanometer metal films in this study. The initial Ru films are effectively only 3 to 4 atomic layers thick and are amorphous before agglomeration. The multilayer structure here is far from the equilibrium state. The high free energy of the amorphous Ru layers relative to crystalline Ru provides an additional driving force for crystallization and/or agglomeration which is not considered in these models. In addition, it is difficult to differentiate the bulk from the interfacial properties at this thickness. Simple calculation of the surface areas, assuming uniform microstructure and equal interfacial energies for amorphous and crystalline structures, indicates that agglomeration occurs more readily in thinner films, and that a spherical morphology is preferred to a flat film when the grain size is larger than the film thickness. This is consistent with the values in the reported models,[8-10] and with the measured values from the agglomerated structure in Figure 2. It is not clear whether crystalization or agglomeration occurs first for these 1 nm thick layers since they are so thin. Grain boundary grooving of the polycrystalline layered films that have crystallized from the initial amorphous state is possible; the equilibrium dihedral angles at the grain junctions however cannot be observed in the images since the crystalline Ru grains appear to be separated by a thick C grain boundary region.

<u>Comparison with compound forming systems</u>

The evolution of the Ru layers in Ru/C multilayers with annealing is different from that of the W layers in W/C multilayers and that of the Ru layers in Ru/B_4C multilayers. Previous studies found that annealing of W/C multilayers of 2 nm period at 500°C for 4 hours does not result in agglomeration of the W layers.[3] Diffusion of the C atoms into the W films forms a phase that stabilizes in an amorphous state and lowers the energy of the system, and hence the structure remains layered. A carbide (W_2C) is formed in thicker period W/C multilayers after annealing under the same conditions.[2]

Preliminary studies of the microstructure and layered structure stability in Ru films between boron carbide layers reveal similar evolution to that of the W/C system. Examination of 1.1 nm Ru films between 1.7 nm B_4C layers indicates that the films remain layered and continuous upon annealing at 600°C for 30 minutes. Both the as-prepared and annealed Ru/B_4C multilayers show predominantly amorphous structure at this period. Longer period Ru/B_4C multilayers also remain layered after annealing, and crystallization or recrystallization of the Ru layers to form a boride phase (RuB_2) is observed from the electron diffraction patterns of plan-view samples. Compound formation in this system, similar to that of W/C, may have stabilized the layered structure of the thin metal films upon annealing. Agglomeration of the Ru films in the Ru/C system, in contrast to the stable layered structure in the W/C and the Ru/B_4C systems, appears to be related to the immiscible region in its phase diagram, which promotes phase separation between the Ru and the C constituents.

CONCLUSIONS

The microstructural evolution of thin Ru layers sandwiched between C layers was studied using cross-sectional and plan-view TEM. The Ru grain size distribution in the annealed multilayers is a function of the thicknesses of both the metal and the carbon layers, as observed in plan-view samples. The grain size dependence is more apparent in the thinner Ru and C layers, in which high surface-to-volume ratio provides a stronger driving force for crystallization and grain growth of the Ru. Agglomeration of the 1 nm Ru layers between 1 nm C layers is observed after annealing at 600°C for 30 minutes by cross-sectional TEM. Annealing of W layers between C layers and of Ru layers between B_4C layers of comparable thicknesses does not result in agglomeration in the thinner films, and leads to formation of a crystalline compound in the thicker films.

ACKNOWLEDGEMENTS

One of the authors (T.D.N.) would like to thank Z. Weng for performing TEM of the plan-view samples. This work was supported by the Director, Office of Energy Research, Office of Basic Sciences, Materials Sciences Division, of the U.S. Department of Energy under Contract No. DE-AC03-76SF00098 and by the Air Force Office of Scientific Research, of the U.S. Department of Defense under Contract No. F49620-87-K-0001.

REFERENCES

1. E Spiller, AIP Conf. Proc. 75 (New York, 1981) 124; T.W. Barbee, Jr., Ibid., 131.
2. T.D. Nguyen, R. Gronsky, and J.B. Kortright, Mat. Res. Soc. Proc. 187 (1990) 95.
3. T.D. Nguyen, R. Gronsky, and J.B. Kortright, Mat. Res. Soc. Proc. 139 (1989) 357.
4. J.B. Kortright and J. Denlinger, Mat. Res. Soc. Proc. 103 (1988) 95.
5. T.D. Nguyen, R. Gronsky, and J.B. Kortright, accepted for publication in J. Elec. Micros. Tech.
6. T.D. Nguyen, R. Gronsky, and J.B. Kortright, Elect. Micros. Soc. Ame. Proc. (1990) 680.
7. W.W. Mullins, J. Appl. Phys. 28 (1957) 333.
8. D.J. Srolovitz and S.A. Safran, J. Appl. Phys. 60, 1 (1986) 247.
9. K.T. Miller, F.F. Lange, and D.B. Marshall, J. Mater. Res. 5, 1 (1990) 151.
10. T. Nolan, R. Bevers, and R. Sinclair (Mat. Res. Soc. Meeting, Boston, 1990).

ON CONCENTRATION-DEPENDENT SOLID STATE DIFFUSION

Yang-Tse Cheng

General Motors Research Laboratories, Warren, Michigan 48090-9055

ABSTRACT

Using a master equation approach, we derive a general expression for the diffusion coefficient as a function of concentration-dependent jump rates. When this approach is applied to diffusion in a binary solid, Darken's equation for intrinsic diffusion coefficients is derived together with an expression for self diffusion coefficients which satisfies the semi-empirical Ugaste relationship. This analysis suggests that the Darken term and the self diffusion coefficients are in general related.

I. INTRODUCTION

Solid state diffusion is an area which has been studied extensively over the past several decades.[1-16] The standard macroscopic theory of solid state diffusion is the linear response theory which relates fluxes linearly to the corresponding driving forces.[1-5,10] In the absence of external fields and a temperature gradient, the atomic flux, J_i, of the species i is linearly proportional to the gradient of the chemical potential μ_j:

$$J_i = -\sum_j L_{ij} \frac{\partial \mu_j}{\partial x}, \qquad (1)$$

where the L_{ij}'s are phenomenological coefficients which are symmetric (Onsager's theorem): $L_{ij} = L_{ji}$. It is believed[10] that the phenomenological coefficients L_{ij} are kinetic quantities, which in general are functions of thermodynamic variables but are explicitly independent of the driving forces, such as $\partial \mu_j / \partial x$. When applied to diffusion in a binary alloy $A_{1-c}B_c$, the linear response theories lead to the expression of the intrinsic diffusion coefficients D_A and D_B,[1-5,8,10,13,16]

$$D_B = D_B^*(1 + \frac{\partial \ln \gamma_B}{\partial \ln c}), \qquad (2)$$

in which D_B^* is the self diffusion coefficient of B in a homogeneous alloy $A_{1-c}B_c$ and γ_B is B's activity coefficient. Eq. (2) is the familiar Darken equation for intrinsic diffusion coefficients.[13] The Darken term, $(1 + \partial \ln \gamma_B / \partial \ln c)$, contains the dependence on the activity coefficient γ_B. In the regular solution approximation,[14,16] $(1 + \partial \ln \gamma_B / \partial \ln c) = (1 - 2\Delta H_{mix}/k_B T)$, where ΔH_{mix} is the heat of mixing for alloy $A_{1-c}B_c$, k_B is Boltzmann's constant, and T is the absolute temperture. Thus, the Darken term depends on the heat of mixing ΔH_{mix}. In contrast, D_B^*, which is an explicit function of temperature T, concentration c, and phenomenological constants L_{ij}, does not explicitly depend on γ_B or ΔH_{mix} in linear response theories.[10]

However, a semi-empirical relationship exists between self diffusion coefficients and the heat of mixing which was first established by Ugaste and co-workers:[17,18]

$$\ln D_B^* - (c_A \ln D_B^{*A} + c_B \ln D_B^{*B}) = -\frac{\Delta H_{mix}}{kT}, \tag{3}$$

where D_B^{*A}, D_B^{*B}, and D_B^* are the self diffusion coefficients for B in pure A, pure B, and $A_{c_A} B_{c_B}$, respectively. Equation (3) states that the self diffusion coefficients depend on the heat of mixing ΔH_{mix}. Thus, self diffusion coefficients and the Darken term, $(1 + \partial \ln \gamma_B / \partial \ln c) = (1 - 2\Delta H_{mix}/k_B T)$, are not unrelated.

In this paper, we use a master equation approach to derive a general expression for the diffusion coefficient as a function of concentration-dependent jump rates. An expression for intrinsic diffusion coefficients in the form of eq. (2) with self diffusion coefficients agreeing with eq. (3) are derived for a special case of solid state diffusion. This analysis together with Ugaste's relationship (eq. (3)) implies that the Darken term and self diffusion coefficients are indeed related.

II. Master Equation to Diffusion Equation

Consider planar diffusion caused by near-neighbor jumps in the x-direction. A master equation for the number of atoms per unit area in the ith plane N_i is given by:[12,15]

$$\dot{N}_i = N_{i+1}\Gamma_{i+1,i} + N_{i-1}\Gamma_{i-1,i} - N_i(\Gamma_{i,i+1} + \Gamma_{i,i-1}), \tag{4}$$

where $\Gamma_{i,k}$ represents the transition frequency from site i to k (see: Fig. 1). For given initial and boundary conditions and a set of $\Gamma_{i,i\pm1}$'s, master equation (4) specifies the time evolution of N_i's. For an arbitrary set of $\Gamma_{i,i\pm1}$'s, equation (4) cannot always be

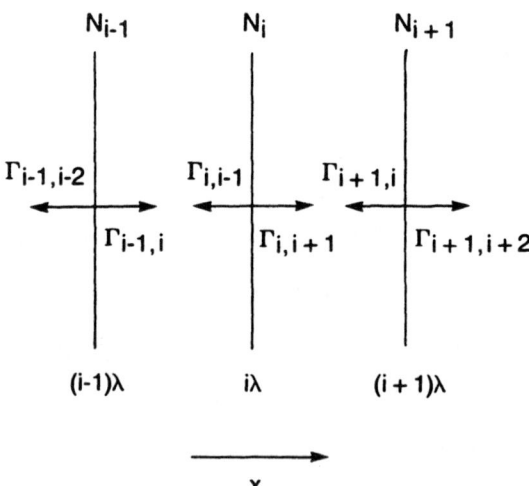

Figure 1 Adjacent lattice planes showing the relevant jump rates.

transformed into a diffusion equation. However, we find that a diffusion equation can be obtained from equation (4) if we assume that $\Gamma_{i,i\pm1}$ is given by

$$\Gamma_{i,i\pm1} = \Gamma f^*(i) g^*(i \pm 1),\tag{5}$$

where Γ has the dimension of a jump frequency and is independent of position and concentration. In other words, $\Gamma_{i,i\pm1}$ is a product of two functions $f^*(i)$ and $g^*(j)$, where f^* depends on the site the atom jumps from (i) and g^* depends on the site the atom jumps to $(j = i \pm 1)$. Consider that N_i, f^*, and g^* take on the values of differentiable functions $N(x,t)$, $f(x)$, and $g(x \pm \lambda)$ at time t and position $x = i\lambda$, i.e., $f^*(i) = f(i\lambda) = f(x)$ and $g^*(i \pm 1) = g(i\lambda \pm \lambda) = g(x \pm \lambda)$, where λ is a constant jump distance. We obtain, after Taylor expansion of eqs. (4) and (5), the standard diffusion equation (Fick's second law)

$$\frac{\partial c}{\partial t} = \frac{\partial}{\partial x}(D\frac{\partial c}{\partial x}),\tag{6}$$

in which the diffusion coefficient is given by

$$D = \lambda^2 \Gamma f g[1 + \frac{\partial \ln(f/g)}{\partial \ln c}],\tag{7}$$

where $c = N/\lambda$ is the concentration.

In general, D is concentration-dependent since f and g may depend on the local concentration. For non-constant jump rates, the diffusion coefficient is not simply the product of jump frequency and the square of the jump distance, i.e., $D \neq \Gamma fg\lambda^2$. Instead, it is modified by the term $[1 + \frac{\partial \ln(f/g)}{\partial \ln c}]$, which may be called, after comparing with eq. (2), a generalized Darken's term. The present derivation of the diffusion equation and diffusion coefficient does not rely on Fick's first law. It also does not require a linear response relationship between fluxes and driving forces.

III. DIFFUSION IN BINARY SILIDS

We now apply eq. (7) to diffusion in a binary solid $A_{1-c}B_c$ describable by the regular solution approximation.[14,16] In this approximation, the chemical activity coefficient for species B is given by $\gamma_B = \exp[\frac{z\epsilon}{kT}(1 - c)^2]$ and the chemical potential is $\mu_B = \frac{z}{2}\epsilon_{BB} + kT\ln(\gamma_B c)$, where z is the coordination number, $\epsilon = \epsilon_{AB} - \frac{\epsilon_{AA}+\epsilon_{BB}}{2}$ is the exchange energy, and ϵ_{ij} is the energy between neighbors i and j. For simplicity, we assume $\epsilon_{AA} = \epsilon_{BB}$ in the following discussion.

We consider diffusion via a vacancy mechanism. As in the usual treatment of vacancy diffusion, vacancies are assumed to be in thermal equilibrium.[3,4,5,10] Using that $\epsilon_{AA} = \epsilon_{BB}$, the chemical potential of vacancies can be shown to be independent of position. Since in general g is determined by the equilibrium properties of vacancies in vacancy mechanism, g is therefore some function of the chemical potential of vacancies at a given temperature and pressure. This implies that g is also independent of position. The general expression for solid state diffusion via a vacancy mechanism where vacancies are in thermal equilibrium is thus given by

$$D = \lambda^2 \Gamma f[1 + \frac{\partial \ln f}{\partial \ln c}].\tag{7'}$$

An inspection of eqs. (2), (3), and (7') suggests that all three equations can be satisfied by letting f be γ_A and γ_B for species A and B, respectively. Indeed by substituting γ_B for f in eq. (7'), we obtain an expression for the intrinsic diffusion coefficient of B:

$$D_B = D_B^*(1 + \frac{\partial \ln \gamma_B}{\partial \ln c}) \text{ with } D_B^* = \lambda^2 \Gamma \gamma_B. \tag{8}$$

Writing explicitly for the regular solution approximation, this diffusion coefficient becomes:

$$D_B = D_B^*[1 - 2\delta c(1 - c)] = D_B^*(1 - 2\frac{\Delta H_{mix}}{kT})$$

with

$$D_B^* = \lambda^2 \Gamma \exp[\delta(1 - c)^2] = \lambda^2 \Gamma \exp[\frac{\Delta H_{mix}}{kT}(\frac{1 - c}{c})], \tag{8'}$$

where $\delta = z\epsilon/kT$ and $\Delta H_{mix} = z\epsilon c(1 - c)$ is the heat of mixing in the regular solution approximation for an alloy $A_{1-c}B_c$. The equations for D_A and D_A^* are obtained by replacing B by A and c by $c_A = 1 - c$ in eqs. (8) and (8'). Simple algebra shows that $D_A^* = \lambda^2 \Gamma \gamma_A$ and $D_B^* = \lambda^2 \Gamma \gamma_B$ satisfy Ugaste's relationship for self diffusion coefficients (eq. 3). Thus, Darken's equation (eq. (2)) and the self diffusion coefficients, D_A^* and D_B^*, can be derived using the present master equation approach to solid state diffusion together with the regular solution approximation and the assumption that $f = \gamma_A$ and $f = \gamma_B$ for species A and B, respectively.

The choice of $f = \gamma_B$ may be justified by the following considerations. At a given concentration the difference in free energy of a B atom in a regular and an ideal solution is given by the difference in their respective chemical potentials, i.e., $\Delta \mu_B = \mu_B - \mu_B^{id} = kT \ln \gamma_B$. We now consider the influence of $\Delta \mu_B$ on jump rates by using a frequently made assumption that the chemical potential influences the jump rate through an Arrhenius temperature dependent term.[8,19,20] Under this assumption, the difference in the forward and backward jump rates is usually expressed in terms of the respective chemical potential difference,[8,19,20] i.e., a chemical driving force. Unlike previous approaches, a linearization of this driving force is not needed in the master equation approach. Instead, the chemical potential at each site relative to a reference state, in this case the ideal solid solution state, is needed in order to estimate the composition dependent jump rate. For a regular solution, the difference $\Delta \mu_B$ in chemical potential of a B atom leads to a term $\exp(\Delta \mu_B/kT) = \gamma_B$ which modifies the jump rate for this B atom jumping from its initial site to a vacancy site. The concentration dependent part of the jump rate is now given by γ_B. From the meaning of f in eq. (5), we obtain $f = \gamma_B$ for a B atom jump to a neighboring vacant site. Although this picture of atomic jump process is oversimplified, it does provide an alternative derivation of Darken's equation (eq. (2)) with self diffusion coefficients consistent with Ugaste's semi-empirical relationship (eq. (3)).

Some of the consequences of eqs. (8) and (8') have been discussed in ref. (21). Eq. (8') shows that both D_B^* and the Darken term contain the heat of mixing ΔH_{mix}. Consequently, a large and negative ΔH_{mix} may not necessarily lead to an enhancement in the magnitude of the intrinsic diffusion coefficient as would be predicted by the Darken term, $(1 + \partial \ln \gamma_B/\partial \ln c) = (1 - 2\Delta H_{mix}/k_B T)$, alone. This is true because γ_B in D_B^* decreases when ΔH_{mix} becomes negative and large in magnitude as seen clearly from eq. (8'). A similar observation has recently been made by Atzmon using Monte Carlo simulations.[22]

IV. SUMMARY

A general expression for the diffusion coefficients is obtained from a master equation approach to concentration-dependent solid state diffusion when the concentration-dependent jump rate is a product of two functions, one depending on the site that an atom jumps from and the other depending on the site the atom jumps to. When the approach is applied to diffusion in a binary solid, Darken's equation for intrinsic diffusion coefficients is derived together with an expression for self diffusion coefficients which satisfies the semi-empirical Ugaste relationship. This analysis suggests that the Darken term and the self diffusion coefficients are in general related.

ACKNOWLEDGMENTS

I would like to thank M. Atzmon, S. Browne, B. K. Cho, G. Eesley, G. B. Fisher, J. G. Gay, W. L. Johnson, V. Laxmanan, J. V. Mantese, W. J. Meng, J. R. Smith, K. C. Taylor, D. Turnbull, and M. W. Verbrugge for helpful discussions and providing valuable suggestions for the manuscript.

REFERENCES

1. D. Lazarus, in *Solid State Physics* Vol. 10, edited by F. Seitz and D. Turnbull (Academic Press, New York, 1960), p. 71.

2. S. R. de Groot and P. Mazur, *Non-equilibrium Thermodynamics* (Dover, New York, 1983).

3. P. G. Shewmon, *Diffusion in Solids* (McGraw-Hill, New York, 1963).

4. R. E. Howard and A. D. Lidiard, Rep. Prog. Phys. **27**, 161 (1964).

5. J. R. Manning, *Diffusion Kinetics for Atoms in Crystals* (Nostrand, New Jersey, 1968).

6. J. Crank, *The Mathematics of Diffusion*, second edition (Oxford University, London, 1975).

7. G. E. Murch and A. S. Nowick, eds., *Diffusion in Crystalline Solids* (Academic Press, Orlando, 1984).

8. K. N. Tu, Ann. Rev. Mater. Sci **15**, 147 (1985).

9. J. W. Haus and K. W. Kehr, Phys. Rep. **150**, 263 (1987).

10. A. R. Allnatt and A. D. Lidiard, Rep. Prog. Phys. **50**, 373 (1987).

11. D. Gupta, A. D. Romig, and M. A. Dayananda, eds., *Diffusion Processes in High Technology Materials* (Trans Tech, Aedermannsdorf, 1988).

12. R. Ghez, *A Primer of Diffusion Problems* (Wiley, New York, 1988).

13. L. S. Darken, Trans Am. Inst. Min. Metall. Engrs. **175**, 184 (1948).

14. L. E. Reichl, *A Modern Course in Statistical Physics* (University of Texas, Austin, 1980).

15. R. Ghez and W. E. Langlois, Am. J. Phys. **54**, 646 (1986).

16. P. Haasen, *Physical Metallurgy* (Cambridge University, Cambridge, 1978).

17. Yu. E. Ugaste, Fiz. Metal. Metalloved. **31**, 57 (1971).

18. I. B. Borovskiy, I. D. Marchukova, and Yu. E. Ugaste, Fiz. Metal. Metalloved. **29**, 86 (1970).

19. J. L. Bocquet, G. Brébec, and Y. Limoge, in *Physical Metallurgy I*, edited by R. W. Cahn and P. Haasen (North-Holland, Amsterdam, 1983), p. 385.

20. A. D. Le Claire, Phil. Mag. **3**, 921 (1958).

21. Y.-T. Cheng, GM Research Publication, GMR-7080 (1990).

22. M. Atzmon, Phys. Rev. Lett. **65**, 2889 (1990).

THE MODIFIED GIBBS-WULFF CONSTRUCTION AND CRITICAL NUCLEUS
MORPHOLOGY AT AN INTERFACE

J. K. LEE, J. H. CHOY AND Y. CHOI
Department of Metallurgical and Materials Engineering, Michigan Technological
University, Houghton, MI 49931

ABSTRACT

The conditions for the equilibrium shape of a second-phase particle
nucleating either at a planar interface or at a spherically curved interface
are reviewed under the assumption that the interfacial structure of the
nucleation site has a torque of an appropriate strength. The results are
then incorporated into the modified Gibbs-Wulff construction in order to
demonstrate a graphical representation of both the equilibrium shape and the
free energy barrier for heterogeneous nucleation.

INTRODUCTION

Heterogeneous nucleation is the process through which the smallest
stable particle of a new phase is formed with the aid of a defect such as a
substrate. In general, the study of heterogeneous nucleation has assumed a
uniform flat substrate even if surface roughening is present to an appreciable extent; the basis of this approximation is that the radius of an embryo
is usually much smaller than the radius of the curvature of the substrate
surface [1-3]. However, the possibility of having comparable radii was
recognized as early as in 1939 by Volmer [4], and Turnbull [5] analyzed the
thermodynamic stability of embryos in cylinders and conical cavities in
order to explain the variation of nucleation rate in liquid to solid transformations upon subsequent cooling. In solid-state phase transformations,
nucleation of crystals of one phase often takes place at the interphase
boundaries of previously formed precipitates of another phase or even of
precipitates of the same phase, as sympathetic nucleation [6]. Under these
circumstances, the precipitates serving as substrata can have radii in the
nanometer range. In this work, we review the equilibrium morphology of a
second phase particle nucleating at an interface, with emphasis on both
spherically curved substrate and spherical interphase boundary case. The
modified Gibbs-Wulff construction, developed originally for a planar interface [7-10], is extended to the present case in order to demonstrate a
graphical representation of both the equilibrium shape and the free energy
change for formation of the critical nucleus.

In the classical theory of nucleation, the rate, J^*, is usually
expressed in terms of a time-dependent equation [2,3]:

$$J^* = Z\beta^*N \exp(-\Delta G^*/kT) \exp(-\tau/t) \tag{1}$$

where Z = Zeldovich non-equilibrium factor, β^* = frequency factor or the
rate at which single atoms join the critical nucleus, N = number of
nucleation sites per unit volume, τ = incubation time, k = Boltzmann's
constant, and T = the absolute temperature. Clearly, the most important
term in the rate theory is the free energy change for formation of the
critical nucleus, ΔG^*. For mathematical simplicity, the analysis will make
use of an isotropic surface free energy, $\sigma_{\alpha\beta}$, between the matrix phase, α,
and second phase particle, β, but it will be later shown that an
anisotropic $\sigma_{\alpha\beta}$ case is equally acceptable for the modified Gibbs-Wulff
construction. In analyzing an equilibrium particle shape on a planar
substrate, it has been customary to assume that the substrate surface
structure has a torque of an appropriate strength to counterbalance the

normal component of $\sigma_{\alpha\beta}$ at the triple junction of the α-β-substrate; otherwise the interfacial elements would undergo rearrangements until the
surface tension reaches at equilibrium [7-10]. Likewise, a spherical interfacial structure is assumed to have a torque of an appropriate strength. The
curvature effect on a surface free energy has been debated in literature for
a long time, but it will not be addressed. Neither the elastic strain due to
a difference in lattice parameters nor the effect of external forces such as
gravity will be considered in the analysis. We note that detailed derivations
for equilibrium conditions are presented elsewhere [11,12].

THE MODIFIED GIBBS-WULFF CONSTRUCTION

In the absence of coherency strain, the original Gibbs-Wulff construction
[1-3] gives the equilibrium shape for a crystal of a second phase in a homogeneous matrix. Winterbottom [7] was the first to demonstrate the construction to find the equilibrium shape of a liquid particle upon a planar substrate with the aid of the concept of negative interface free energy. Later,
Cahn and Hoffman [8] introduced a vector function for interface free energy,
and derived the equilibrium shape of a second-phase particle at a planar
grain boundary. Alternatively, Lee and Aaronson [9] extended the method of
Winterbottom to construct equilibrium shapes at a planar interphase boundary.
Here, we will utilize the method of Winterbottom, often termed as the
modified Gibbs-Wulff construction or the Winterbottom construction [10].

Planar Substrate

The equilibrium shape of an embryo on a planar substrate has been
extensively addressed in the literature [1-5]. A discussion, however, is
re-introduced here for the comparison purpose with the case of spherically
curved substrata. The conditions for an equilibrium shape are obtained by
minimizing the total surface free energy of a second-phase particle at a
constant volume. An alternate way is to minimize the total free energy
change associated with the formation of a second-phase particle. For the
context of nucleation, we follow the latter approach. Consider the total
free energy change, ΔG, for the formation of a β embryo from the α phase on
a planar substrate (Fig. 1a):

$$\Delta G = \Delta G_v \pi r^3 (2 - 3\cos\theta + \cos^3\theta)/3 + 2\sigma_{\alpha\beta}\pi r^2 (1 - \cos\theta) - \Delta\sigma\pi r^2 \sin^2\theta \quad (2)$$

where ΔG_v is the free energy change per unit volume from α to β phase, θ is

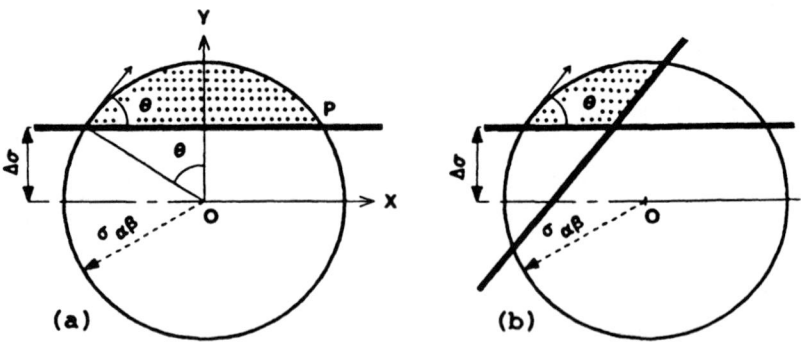

Fig. 1 Planar Substrate

the contact angle, and r is the radius of the nucleus [3]. $\Delta\sigma = \sigma_{\alpha s} - \sigma_{\beta s}$, where $\sigma_{\alpha s}$ and $\sigma_{\beta s}$ are the interfacial free energies between the matrix and substrate, and between the new phase and substrate, respectively. Taking $\partial\Delta G/\partial\theta = 0$ and $\partial\Delta G/\partial r = 0$ and solving for θ and r yields:

$$\sigma_{\alpha\beta}\cos\theta = \Delta\sigma, \qquad (0 \le \theta \le \pi) \tag{3}$$

$$r^* = -2\sigma_{\alpha\beta}/\Delta G_v \tag{4}$$

Eq. (3) is the well-known Young's equation, which is the equilibrium condition for a new phase on a substrate. Note that Young's equation implicitly assumes that a torque associated with the substrate surface structure balances the $\sigma_{\alpha\beta}\sin\theta$ at the junction of the α-β-substrate [3,8]. Eq. (4) is a Gibbs-Thomson expression in which r^* is the radius of the critical nucleus [2,3].

The modified Gibbs-Wulff construction begins with a sphere with its radius equal to $\sigma_{\alpha\beta}$. Let the center of the sphere be the origin (Fig. 1a). Now draw a plane at a distance equal to $\Delta y = \Delta\sigma = \sigma_{\alpha s} - \sigma_{\beta s}$. The equilibrium shape is then the spherical cap in the region of $y > \Delta\sigma$. Since the radius is given by $r^* = -2\sigma_{\alpha\beta}/\Delta G_v$, the critical nucleus becomes the equilibrium shape scaled by a factor of $-2/\Delta G_v$, and its free energy of formation, ΔG^*, is equal to:

$$\Delta G^* = 4V_W/\Delta G_v^2 \tag{5}$$

where V_W is the Wulff volume [13]. For either a planar substrate or a conical cavity case, V_W is exactly equal to the volume of the equilibrium shape in the Wulff space wherein $\sigma_{\alpha\beta}$ replaces r^* as a coordinate. Thus, for the present case, we have:

$$V_W = \pi\sigma_{\alpha\beta}^3(2 - 3\cos\theta + \cos^3\theta)/3 \tag{6}$$

where $\cos\theta = \Delta\sigma/\sigma_{\alpha\beta}$. If $\Delta\sigma > \sigma_{\alpha\beta}$, the location of the substrate plane is out of the $\sigma_{\alpha\beta}$ sphere and consequently $V_W = 0$, indicating complete wetting. On the other hand, if $\Delta\sigma < -\sigma_{\alpha\beta}$, the common Wulff space in the region of $y > \Delta\sigma$ becomes the entire $\sigma_{\alpha\beta}$ sphere. Thus $V_W = 4\pi\sigma_{\alpha\beta}^3/3$, and the situation is equivalent to homogeneous nucleation due to the unwettability of the substrate.

As the Wulff volume, V_W, decreases, the rate of nucleation, J^*, increases. Thus, as shown in Fig. 1b, β embryos sitting on an edge between two intersecting planar substrata render a nucleation rate higher than those on one planar substrate.

Conical Cavity

As an extended case of a planar substrate, a conical cavity provides an interesting site for heterogeneous nucleation [5,12]. Consider a β embryo in a right conical cavity with its apex angle equal to 2ϕ (Fig. 2b). Assume that the depth of the cavity is so large that the β embryo does not touch the cavity opening, and neglect the effect due to the sharp curvature at the cavity bottom. The work to form such an embryo is then equal to:

$$\Delta G = \Delta G_v \pi r^3 [2 - 3\sin(\theta+\phi) + \sin^3(\theta+\phi) - \cos^3(\theta+\phi)\cot\phi)]/3$$
$$+ 2\sigma_{\alpha\beta}\pi r^2 [1 - \sin(\theta+\phi)] - \Delta\sigma\pi r^2\cos^2(\theta+\phi)/\sin\phi \tag{6}$$

Taking $\partial\Delta G/\partial r = 0$ and $\partial\Delta G/\partial\theta = 0$, and solving the results for θ and r^* provide the extremum condition for ΔG, yielding the results of Eqs. (3) and (4).

By substituting Eqs. (3) and (4) into Eq. (6), the Wulff volume, V_W, can be expressed as:

$$V_W = 0, \qquad\qquad \text{if } 0 \le \theta \le \pi/2-\phi \tag{7}$$

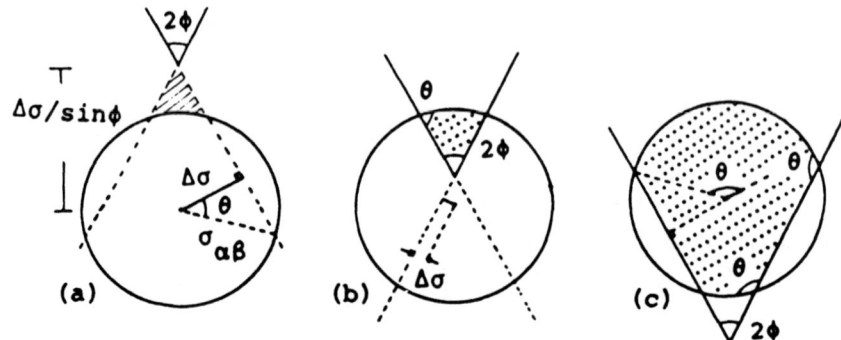

Fig. 2 Conical Cavity

$$= (\pi\sigma_{\alpha\beta}^3/3)\,[2 - 3\sin(\theta+\phi) + \sin^3(\theta+\phi) - \cos^3(\theta+\phi)\cot\phi)]$$

$$\text{if } \pi/2-\phi \le \theta \le \pi/2+\phi, \qquad (8)$$

$$= (\pi\sigma_{\alpha\beta}^3/3)\,[4-3\{\sin(\theta+\phi)+\sin(\theta-\phi)\} + \{\sin^3(\theta+\phi)+\sin^3(\theta-\phi)\}$$

$$- \cot\phi\{\cos^3(\theta+\phi)-\cos^3(\theta-\phi)\}], \text{ if } \pi/2+\phi \le \theta \le \pi \qquad (9)$$

For each of Eqs. (7-9), the Gibbs-Wulff construction is shown in Fig. 2. In the construction, the equilibrium shape is given by the Wulff space shared together by the sphere of a radius equal to $\sigma_{\alpha\beta}$ and a cone whose vertex is located at a distance equal to $\Delta\sigma/\sin\phi$ from the spherical center. Thus as shown in Fig. 2a, if $0 \le \theta \le \pi/2-\phi$, the cone is entirely located outside the sphere, and the Wulff volume, V_w, vanishes and consequently there is no nucleation barrier, indicating complete wetting. Fig. 2b represents the second case of $\pi/2-\phi \le \theta \le \pi/2+\phi$, and the shared volume between the sphere and cone is equal to $\bar{V}_w$. The third case of $\pi/2+\phi \le \theta \le \pi$ is displayed in Fig. 2c, where now the equilibrium shape is shown not to utilize the tip portion of the cone due to a significant decrease in $\Delta\sigma$.

Fig. 2a also displays a β embryo in a conical cavity at a temperature, T, above the $\beta \to \alpha$ transition temperature, T_o [5,12]. Consider the up-side-down cone, i.e., the one drown with broken lines. The conical Wulff space on the top of the $\sigma_{\alpha\beta}$ sphere (shaded region) represents the shape of the β embryo, and its Wulff volume is given by:

$$V_w = (\pi\sigma_{\alpha\beta}^3/3)\,[2 - 3\sin(\theta+\phi) + \sin^3(\theta+\phi) - \cos^3(\theta+\phi)\cot\phi)] \qquad (10)$$

If $0 \le \theta \le \pi/2-\phi$, V_w of Eq. (10) is always negative, indicating that the β embryo can be thermodynamically stable at $T > T_o$. Consequently, for a given θ, the stability of such a β embryo increases with decrease in ϕ.

Concave Spherical Substrate

Consider a β particle on a concave substrate with the two principal radii of curvature equal to a, as shown in Fig. 3. If we divide the contact angle into θ and ϕ by the plane of the α-β-substrate triple junction, the free energy change for the formation of a β particle becomes:

$$\Delta G = \Delta G_v \pi\,[r^3(2 - 3\cos\theta + \cos^3\theta) + a^3(2 - 3\cos\phi + \cos^3\phi)]/3$$

$$+ 2\sigma_{\alpha\beta}\pi r^2(1 - \cos\theta) - 2\Delta\sigma\pi a^2(1 - \cos\phi) \qquad (11)$$

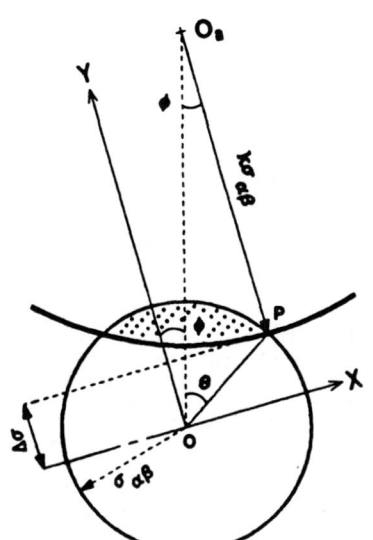

Fig. 3 Concave Substrate

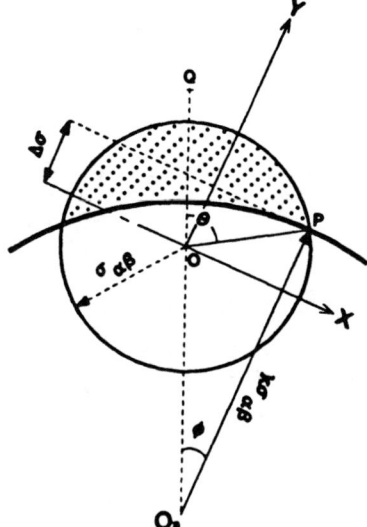

Fig. 4. Convex Substrate

In addition, we have now a geometrical constraint, $r \sin\theta = a \sin\phi$. In order to solve for θ, ϕ, and r^* for given $\sigma_{\alpha\beta}$, $\Delta\sigma$, ΔG_v, and a, we use Lagrange's method. Define a new function, $F = \Delta G + \lambda (r \sin\theta - a \sin\phi)$, where λ is the Lagrangian multiplier. Taking $\partial F/\partial r = 0$, $\partial F/\partial \theta = 0$, and $\partial F/\partial \phi = 0$ yields the Gibbs-Thomson equation and a relationship for the contact angle:

$$\sigma_{\alpha\beta}\cos(\theta+\phi) = \Delta\sigma \qquad (0 \leq \theta+\phi \leq \pi) \qquad (12)$$

When the radius of a concave substrate, a, approaches infinity, the geometrical constraint requires ϕ to vanish. Thus the limiting behavior of Eq. (12) predicts Eq. (3), the result of a planar substrate.

As in the planar substrate case, we draw a plane at a distance equal to $\Delta\sigma$. Let the junction between the plane and the $\sigma_{\alpha\beta}$ sphere be P, as shown in Fig. 3. Locate the center, O_a, of a sphere with radius equal to $k\sigma_{\alpha\beta}$ (= $a\sigma_{\alpha\beta}/r^*$) on the straight line passing through P, away from the origin O but parallel to the y-axis. Then the spherical zone of radius $\sigma_{\alpha\beta}$ contained within the sphere of $k\sigma_{\alpha\beta}$ is the equilibrium shape desired. A proof that this is so is easily made with the two angles, $\theta = <POO_a$, and $\phi = <PO_aO = <O_aOY$. Since $\sigma_{\alpha\beta} \cos <POY = \sigma_{\alpha\beta} \cos (<POO_a + <O_aOY) = \Delta\sigma$, Eq. (12) is satisfied in the construction.

An _effective_ Wulff volume, V_W, may be defined as:

$$V_W = \pi\sigma_{\alpha\beta}^3[(2 - 3\cos\theta + \cos^3\theta) + k^3(2 - 3\cos\phi + \cos^3\phi)]/3$$
$$+ \pi\sigma_{\alpha\beta}^3[\cos\theta\sin^2\theta - 2\nu k^2(1-\cos\phi) - k^3(2-3\cos\phi+\cos^3\phi)] \qquad (13)$$

where

$$\cos\theta = (k\nu + 1)/(k^2 + 2\nu k + 1)^{1/2} \qquad (14)$$

$$\cos\phi = (k + \nu)/(k^2 + 2\nu k + 1)^{1/2} \qquad (15)$$

and $\nu = \Delta\sigma/\sigma_{\alpha\beta}$. In this case, the effective Wulff volume, which is related to ΔG^* via Eq. (5), is <u>not identical</u> to the geometrical volume in the Wulff space; in Eq. (13), the <u>first term</u> is equivalent to the geometrical volume, but the finite curvature of the substrate yields an additional term, which vanishes as k approaches infinity.

Convex Spherical Substrate

For a convex substrate case (Fig. 4), we define the contact angle to be $\theta-\phi$. The free energy change for the formation of a β particle becomes:

$$\Delta G = \Delta G_v \pi [r^3(2 - 3\cos\theta + \cos^3\theta) - a^3(2 - 3\cos\phi + \cos^3\phi)]/3$$
$$+ 2\sigma_{\alpha\beta}\pi r^2(1 - \cos\theta) - 2\Delta\sigma\pi a^2(1 - \cos\phi) \tag{16}$$

The same geometrical constraint, $r \sin\theta = a \sin\phi$, is still required. Note that the only difference between the present and the concave substrate case is the sign in front of the term involving a^3. Hence the procedure of the Lagrange method yields Eq. (4) and a relationship for the contact angle:

$$\sigma_{\alpha\beta}\cos(\theta-\phi) = \Delta\sigma \qquad (0 \le \theta-\phi \le \pi) \tag{17}$$

The geometrical method is also very similar to the previous case. Draw a plane at a distance equal to $\Delta\sigma$, and find the junction P between the plane and the $\sigma_{\alpha\beta}$ sphere (Fig. 4). Now locate the center, O_a, of a sphere of radius, $k\sigma_{\alpha\beta}$ $(= a\sigma_{\alpha\beta}/r^*)$, on the line parallel to the y-axis, passing through P and heading toward the side of the origin, O. Then the spherical portion of radius $\sigma_{\alpha\beta}$ outside the sphere of $k\sigma_{\alpha\beta}$ is the equilibrium shape. For a proof, select a point Q along the line passing through both O_a and O. Then $\theta = $ <POQ $= $ <POY + <YOQ $= $ <POY + <PO$_a$O $= $ <POY + ϕ. Since $\sigma_{\alpha\beta}$ cos <POY $= \Delta\sigma$, the Gibbs-Wulff construction satisfies Eq. (17). Again, an effective Wulff volume, V_W, can be given by:

$$V_W = \pi\sigma_{\alpha\beta}^3[(2 - 3\cos\theta + \cos^3\theta) - k^3(2 - 3\cos\phi + \cos^3\phi)]/3$$
$$+ \pi\sigma_{\alpha\beta}^3[\cos\theta\sin^2\theta - 2\nu k^2(1-\cos\phi) + k^3(2-3\cos\phi+\cos^3\phi)] \tag{18}$$

where

$$\cos\theta = (k\nu - 1)/(k^2 - 2\nu k + 1)^{1/2} \tag{19}$$

$$\cos\phi = (k - \nu)/(k^2 - 2\nu k + 1)^{1/2} \tag{20}$$

and $\nu = \Delta\sigma/\sigma_{\alpha\beta}$.

Spherical Interphase Boundary

A β particle is being nucleated at a spherical interphase boundary whose radius of curvature is equal to a (Fig. 5). Unlike a substrate case, the growing β particle now invades both sides of the interphase boundary. Let $\sigma_{\alpha\alpha}$ be the interphase boundary energy, and assume that the boundary structure has a torque of a sufficient strength to prevent any puckering at the triple junction during the formation of a new phase. Let $\sigma_{\alpha\beta1}$ and $\sigma_{\alpha\beta2}$ be the interfacial free energies between the matrix grain 1 and the new phase, and between the matrix grain 2 and the new phase, respectively. r_1 is the radius of the new phase facing the matrix grain 1, and r_2 is the particle radius facing the matrix grain 2. If we define the dihedral angle as the sum of θ_1 and θ_2 with $\theta_1 > \phi$ as indicated in Fig. 5, the free energy change for the formation of a β particle becomes:

$$\Delta G = \Delta G_v\pi[r_1^3(2 - 3\cos\theta_1 + \cos^3\theta_1) + r_2^3(2 - 3\cos\theta_2 + \cos^3\theta_2)]/3$$

$$+ 2\sigma_{\alpha\beta1}\pi r_1^2(1 - \cos\theta_1) + 2\sigma_{\alpha\beta2}\pi r_2^2(1 - \cos\theta_2) - 2\sigma_{\alpha\alpha}\pi a^2(1 - \cos\phi) \quad (21)$$

The geometrical constraint requires that $r_1\sin\theta_1 = r_2\sin\theta_2 = a\sin\phi$. The Lagrange method provides the Gibbs-Thomson equation, and a relationship for the dihedral angle:

$$(\sigma_{\alpha\beta1}\cos\theta_1 + \sigma_{\alpha\beta2}\cos\theta_2)\cos\phi = \sigma_{\alpha\alpha} \quad (0 \leq \theta_1+\theta_2 \leq \pi, \ \theta_1 \geq \phi) \quad (22)$$

As the radius a approaches infinity, the angle ϕ decreases and the equilibrium condition of Eq. (22) becomes identical to the one for a planar interphase boundary [9].

From the geometry of Fig. 5, the y coordinate of the junction P, Δy, can be obtained for a given set of $\sigma_{\alpha\beta1}$, $\sigma_{\alpha\beta2}$, $\sigma_{\alpha\alpha}$, a, and ΔG_v [11]. If the center of the r_1 sphere is located on the same side as the center of the interphase boundary, Δy is given by:

$$\Delta y^2 = \sigma_{\alpha\beta1}^2 k^2\nu^2[2k^2-1-q^2 +2\{k^2\nu^2 +(k^2-1)(k^2-q^2)\}^{1/2}]/\{4k^2\nu^2 -(1-q^2)^2\}$$
$$- \sigma_{\alpha\beta1}^2(k^2-1) \quad (23)$$

where $k = a/r_1^* = -2a\Delta G_v/\sigma_{\alpha\beta1}$, $\nu = \sigma_{\alpha\alpha}/\sigma_{\alpha\beta1}$ and $q = \sigma_{\alpha\beta2}/\sigma_{\alpha\beta1}$. The angular relationship of $\theta_1 \geq \phi$ requires $k \geq 1$. The Gibbs-Wulff construction starts with a sphere of radius equal to $\sigma_{\alpha\beta1}$. Locate the junction P by drawing a plane at a distance given by Δy. Draw a second sphere of radius equal to $\sigma_{\alpha\beta2}$, such that it passes through the point P and its center, O_2, is located on the y-axis and away from the origin O. The trace of the interphase boundary can be given by a third sphere with its radius equal to $k\sigma_{\alpha\beta1}$, which meets the junction P and the center, O_a, of which is also on the y-axis and to the side of the origin. An effective Wulff volume, V_W, is equal to:

$$V_W = \pi\sigma_{\alpha\beta1}^3[(2 - 3\cos\theta_1 + \cos^3\theta_1) + q^3(2 - 3\cos\theta_2 + \cos^3\theta_2)]/3$$

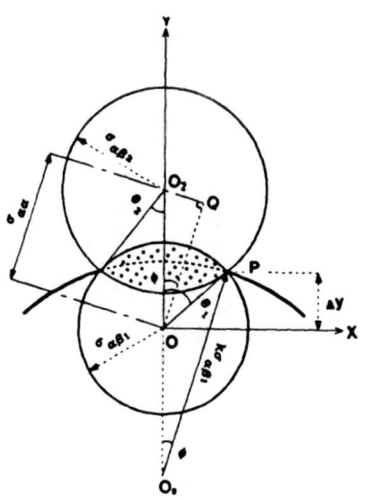

Fig. 5 Interphase Boundary

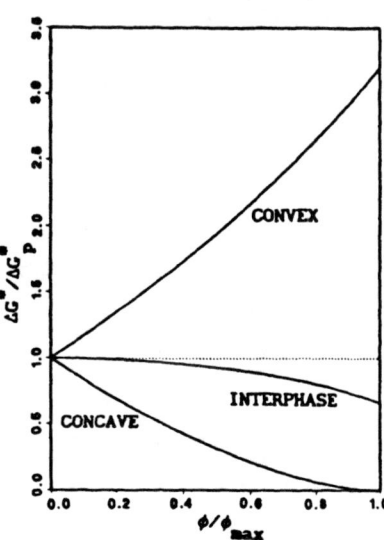

Fig. 6 ΔG^* versus ϕ

$$+ \pi\sigma_{\alpha\beta1}^3 [\cos\theta_1 \sin^2\theta_1 + q^3 \cos\theta_2 \sin^2\theta_2 - 2\nu k^2 (1-\cos\phi)] \qquad (24)$$

where

$$\cos\theta_1 = \Delta y/\sigma_{\alpha\beta1} \qquad (25)$$

$$\cos\theta_2 = \{(\Delta y/q\sigma_{\alpha\beta1})^2 + 1 - q^{-2}\}^{1/2} \qquad (26)$$

The equilibrium condition of Eq. (22) provides $\cos\phi$. A purely graphical method such as Fig. 4 or Fig. 5 has not yet been discovered for this case. When $\sigma_{\alpha\beta2} = \sigma_{\alpha\beta1}$, Eq. (23) becomes:

$$\Delta y^2 = \sigma_{\alpha\beta1}^2 [\{k^2\nu^2 + (k^2-1)^2\}^{1/2} - (k^2-1)]/2 \qquad (27)$$

As k goes to infinity in Eq. (27), Δy approaches $\nu\sigma_{\alpha\beta1}/2 = \sigma_{\alpha\alpha}/2$, resulting in the behavior of a planar grain boundary case [9].

DISCUSSION

It is interesting to compare the activation energy barrier, ΔG^*, for nucleation at a curved interface to the one for a planar interface. Since ΔG^* is proportional to V_w via Eq. (5), a simple way of comparison is to utilize the Gibbs-Wulff construction shown in Figs. 3-5. For the case of a concave substrate (Fig. 3), the critical nucleus volume is seen to decrease with decrease in the radius, a, of the substrate (a = kr*). On the other hand, it increases with decrease in a for the convex substrate case (Fig. 4). For the case of a spherical interphase boundary (Fig. 5), the distance between O and O_2 increases as a decreases. Consequently, both the critical nucleus volume and ΔG^* decrease with decrease in the radius of curvature of the interphase boundary. The behavior of ΔG^* is displayed in Fig. 6, where ΔG^* is plotted as a function of the curvature-related angle, ϕ. Here, $\phi = 0$ indicates a planar interface case, for which ΔG_p^* represents its ΔG^* value. For both the concave and convex substrate case, $\Delta\sigma = 0.5\sigma_{\alpha\beta}$ and $\phi_{max} = 60°$. For a spherical interphase boundary case, $\phi_{max} = 45°$ and $\sigma_{\alpha\alpha} = \sigma_{\sigma\beta1} = \sigma_{\alpha\beta2}$ (marked with INTERPHASE).

Acknowledgments: The authors are indebted to Prof. H. I. Aaronson of Carnegie Mellon for valuable suggestions on this work. The U.S. Dept. of Energy has financially supported this work under Grant No. DE-FG02-87ER45315, for which the authors express their gratitude.

References

1. J. W. Gibbs, On the Equilibrium of Heterogeneous Substances, Collected Works, Vol 1, (Longmans, Green and Co., NY, 1928).
2. K. C. Russell, Phase Transformations, ed., H. I. Aaronson (ASM, Materials Park, OH, 1970), ch. 6.
3. J. W. Christian, Theory of Transformations in Metals and Alloys, Part 1, 2nd ed. (Pergamon Press, Oxford, UK, 1975), p.448.
4. M. Volmer, Kinetik der Phasenbildung, (Steinkopf, Dresden, 1939) quoted in Ref. 3.
5. D. Turnbull, J. Chem. Phys., 18, 198 (1950).
6. H. I. Aaronson and C. Wells, Trans. AIME, 205, 1216 (1956).
7. W. L. Winterbottom, Acta Met., 15, 303 (1967).
8. J. W. Cahn and D. W. Hoffman, Acta Met., 22, 1205 (1974).
9. J. K. Lee and H. I. Aaronson, Acta Met., 23, 799 (1975).
10. J. W. Cahn and J. E. Taylor, Phase Transformations '87, ed., G. W. Lorimer (Inst. of Metals, Cambridge, UK, 1988), p.545.
11. J. K. Lee, J. H. Choy and Y. Choi, Surface Science, in press.
12. J. K. Lee and A. Hellawell, research in progress, Michigan Tech. (1991).
13. J. K. Lee and H. I. Aaronson, Scripta Met., 8, 1451 (1974).

PART III

Semiconductor-Metal Reactions

Interfacial Reactions between In/Pd and GaAs

Z. Ma[a], L.H. Allen[a], B. Blanpain[b], Q.Z. Hong[c], J.W. Mayer[c] and C.J. Palmstrom[d]
[a] Dept. of Materials Science and Engineering, University of Illinois at Urbana-Champaign, Urbana, IL 61801
[b] MTM Dept., Kuleuven, Belgium
[c] Dept. of Materials Science and Engineering, Cornell University, Ithaca, NY 14853
[d] Bellcore, Red Bank, NJ 07701

ABSTRACT

Interfacial microstructure of In/Pd ohmic contacts to n-GaAs was studied by various X-ray diffraction techniques and secondary ion mass spectroscopy (SIMS). Analysis of this interface after various annealing showed that $In_{1-x}Ga_xAs$ compounds are formed at the interface and the composition of these compounds depends upon the annealing temperature. As the temperature increases, the stoichiometry of the In-rich compounds tends toward higher concentrations of Ga. The low contact resistance is achieved by dividing the Schottky barrier between metal and GaAs into two barriers due to metal/$In_{1-x}Ga_xAs$ and $In_{1-x}Ga_xAs$/GaAs. The barrier due to $In_{1-x}Ga_xAs$/GaAs is believed to be the main limiting factor in lowering of contact resistance. The observed ohmic behavior for sample annealed at 500°C for 20 s is attributed to the further reduction of this barrier.

I. INTRODUCTION

The continuing interest in high speed low power field effect transistor (FET) devices based on GaAs requires a further development of shallow and low resistance ohmic contacts. Several successful efforts have been made to achieve such low resistance ohmic contacts to n-GaAs using simple metallization techniques. In various contact metal systems, In-based metallization ohmic contacts to n-GaAs have received considerable attention due to the possibility of the reduction of Schottky barrier via the formation of $In_{1-x}Ga_xAs$ compound between metal and n-GaAs. The typical contact metal systems of this class are In[1], In/Pt[2], In/Pd[3], MoGeInW[4], GeInW[5], and NiInW[6]. In our previously reported work[3], we showed that an ohmic contact to n-GaAs could be formed using In/Pd metallization followed by annealing at 500°C for 20 s. From previous analysis, it is believed that the reaction first occurs between In and Pd to form In_3Pd, then excess In diffuses to GaAs interface through this solid intermediate layer and results in a thin interfacial layer of $In_{1-x}Ga_xAs$. This interfacial layer is suggested to be responsible for the observed electrical behavior.

The objective of this study is to characterize the interfacial microstructure changes during the transition from rectifying behavior to ohmic contact and to correlate the changes in microstructures and contact resistance measured by using the transmission line method (TLM). Various X-ray diffraction techniques and secondary ion mass spectroscopy (SIMS) were used to explore the microstructural evolution and composition depth of the contact , and the property of an $In_{1-x}Ga_xAs$ interfacial layer. By comparing the differences in the interfacial microstructure and crystallographic orientation of the compounds formed at the interface after various heat treatments, the variation in the measured contact resistances is explained in terms of the microstructural and compositional changes.

II. EXPERIMENTAL PROCEDURES

The samples used in the present study were prepared by sequential E-beam evaporation of Pd (100 Å) and In (3000 Å) onto (100) GaAs substrate in a system with base pressure of 1 X 10^{-8} Torr. Layers were deposited at a rate of ≈ 5 Å/s. The GaAs substrate material was bulk Si-doped n-GaAs (N_d ≈ 5 X 10^{17} cm^{-3}). Before deposition, the GaAs wafers were chemically cleaned using an etch in a 1:1 HCl/H_2O solution followed by an etch in a 1:15 NH_4OH/H_2O solution. The samples were annealed at 300°C, 400°C and 500°C, respectively, for 20 s using a Heatpulse™ 410 flash lamp rapid thermal annealer (RTA). During annealing the system was purged (20 cc/s) with either Ar/H_2 (9:1) or N_2.

Microstructure developed at the metal/GaAs interface was characterized by various X-ray diffraction (XRD) techniques. For a powder scan, a computer-controlled Rigaku D/Max IIIA X-ray diffractometer was used with Cu Kα radiation as the X-ray source and a curved graphite (006) monochromator. Both divergent slit (DS) and scattering slit (SS) of 1°, and a receiving slit (RS) of 0.15° were used. Finer slits and a Rigaku RU-200 high resolution double crystal diffractometer with a Ge (111) single crystal as a beam conditioner were used for high resolution X-ray analysis and rocking curve measurements. In order to study the crystallographic orientation of the compounds formed at the interface, a pole figure geometry X-ray diffraction technique was also used. Similar results were obtained by all of these systems. Because of the easier access to the Rigaku D/Max IIIA system in our laboratory and its computerized feature, all X-ray results shown here are gained from that system. To obtain the information about the layer sequence of the compounds formed at the interface, the sample annealed at 400°C was chemically treated by dipping in a solution prepared by dissolution of $HgCl_2$ in organic solvent. The etched sample was studied by X-ray diffraction analysis.

The interface was also probed for the as-deposited sample and the sample after 500°C anneal via SIMS analysis. Because of the nonuniform surface morphology of the sample, which was observed using a scanning electron microscope (SEM), front side depth profiling was not useful. So the technique of the back side depth profiling was used. Special samples were made which consisted of a GaAs epilayer grown on an intermediate AlGaAs buffer layer and a GaAs substrate acting as a carrier substrate for the layers to produce a structure of epi-GaAs/AlGaAs/GaAs. After metallization and annealing the thin GaAs, reacted layer and remaining metal film were removed intact from the supporting substrate using a lift-off technique.

III. RESULTS AND DISCUSSION

X-ray diffraction Analysis

All samples were analyzed by X-ray powder diffraction measurements. The spectra taken from as-deposited sample and the samples annealed at 300°C, 400°C and 500°C for 20 s, respectively, are shown in Figure 1. To emphasize the major observations in these samples, only part of the spectra around a GaAs (200) reflection are shown here. As seen from the spectrum taken from the as-deposited sample, an In_3Pd compound formed even during deposition and there is excess In left. In addition

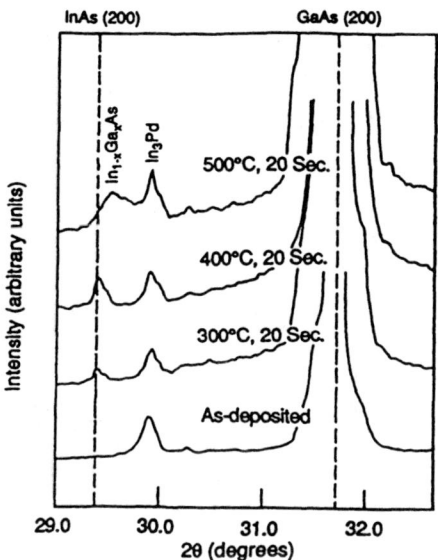

Fig.1. X-ray spectra taken from as-deposited sample and the samples annealed at various temperatures for 20 s.

to the original reflections from excess In and In_3Pd compound, the samples after annealing treatment at 300°C, 400°C and 500°C exhibited two additional peak close to the GaAs (200) and (400), respectively. The positions of these peaks lie between those positions corresponding to InAs and GaAs. They are attributed to be due to the formation of $In_{1-x}Ga_xAs$ compound. When the temperature of annealing increases, the positions of these peaks shift from InAs toward GaAs and the intensity of these peaks becomes larger. Simultaneously the peak gets broader. The sample annealed at 300°C results in an $In_{1-x}Ga_xAs$ compound with stoichiometry very close to that of InAs. The sample after 500°C anneal exhibits a fairly broad peak the composition of which is close to $In_{0.93}Ga_{0.07}As$.

To understand the orientation relations between these compounds and GaAs substrate, the samples were studied by tilting the specimen slightly off the GaAs pole. Such X-ray spectrum taken from the sample after 300°C anneal is shown in Figure 2. The reflections resulting from both GaAs substrate and $In_{1-x}Ga_xAs$ compound disappeared after tilting the specimen 1° or 2° off. This implies that the $In_{1-x}Ga_xAs$

compounds have strong preferred orientation with respect to the GaAs substrate. On the other hand, the peak intensity of In and In₃Pd remain almost the same, which indicates that both are randomly oriented and polycrystalline. This is also confirmed by many other peaks from elemental In and In₃Pd compound. The same phenomena were observed in the sample annealed at 500°C and is shown in Figure 3. The peak

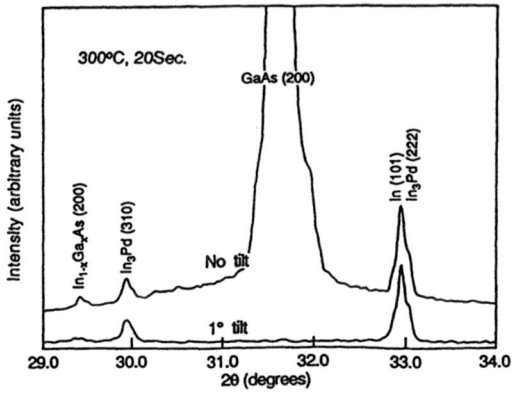

Fig.2. X-ray spectra before and after tilting, taken from the sample annealed at 300°C for 20 s.

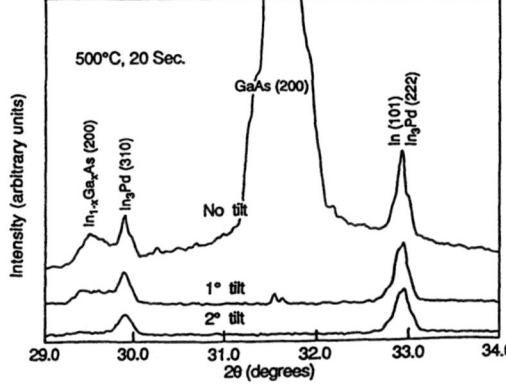

Fig.3. X-ray spectra before and after tilting, taken from the sample annealed at 500°C for 20 s.

broadening is attributed to the partial grading of the composition of this compound layer.

The layer sequence of the compounds at the interface is revealed by the removal of excess In and In₃Pd compounds via chemical etching of the sample annealed at 400°C. Figure 4 shows X-ray spectra taken from the sample before and after etching treatment. It is obvious from the spectra that most In and In₃Pd were removed while $In_{1-x}Ga_xAs$ still remained. This indicates that this compound layer was formed underneath In and In_3Pd.

SIMS analysis

The SIMS depth profiles from both as-deposited sample and the sample annealed at 500°C are shown in Figure 5. The as-deposited sample exhibits a rather

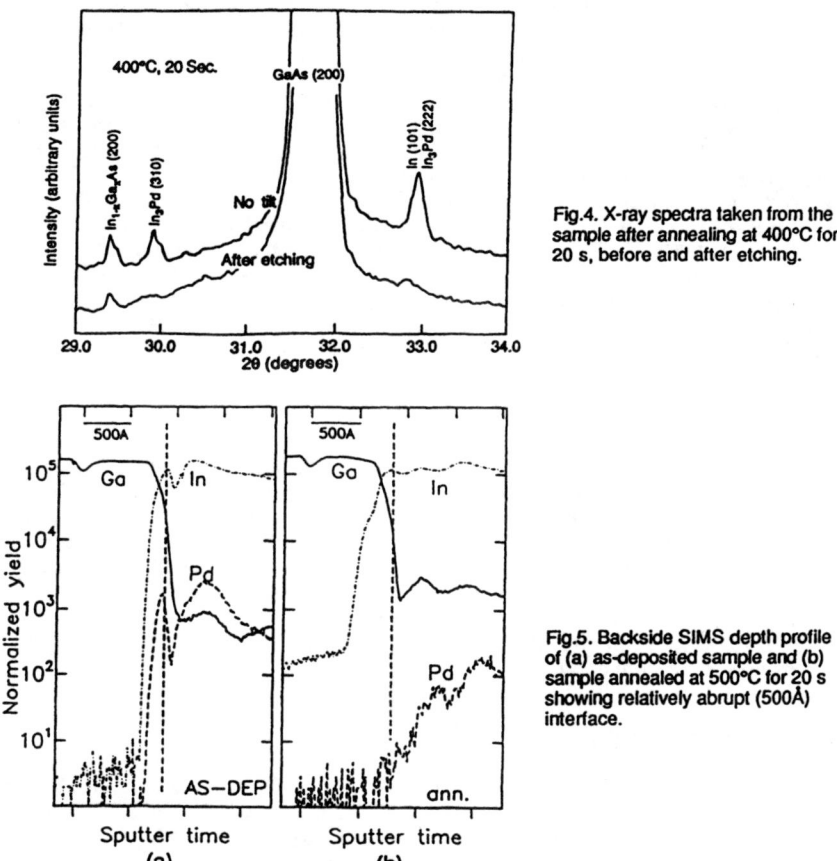

Fig.4. X-ray spectra taken from the sample after annealing at 400°C for 20 s, before and after etching.

Fig.5. Backside SIMS depth profile of (a) as-deposited sample and (b) sample annealed at 500°C for 20 s showing relatively abrupt (500Å) interface.

abrupt transition between In/Pd contact metal layer and GaAs substrate. After annealing at 500°C, the interaction between excess In and GaAs is observed. This interaction seems to be limited by solid state interdiffusion through In_3Pd layer, which resulted in a more uniform and shallow contact. The entire contact layer is on the order of 1000 Å. The leading tail on the In depth profile is attributed to microcracks in the film and not to the In concentration in the bulk GaAs.

In summary, we have observed several key results regarding the formation of ohmic contacts to n-GaAs with In/Pd metallization. First of all from SIMS data we conclude that the metallurgical reaction between the metal and GaAs substrate is relatively limited with the transition region between metal and substrate of thickness on the order of 500 Å.

Furthermore from X-ray diffraction measurements we have identified that during thermal annealing In-rich $In_{1-x}Ga_xAs$ compounds are formed between metal and GaAs substrate. The composition of this compound is dependent upon the temperature of annealing and can be determined from X-ray diffraction analysis. Both tilting and rocking curve measurements confirmed that the $In_{1-x}Ga_xAs$ compounds are polycrystalline but have a strong preferred orientation with respect to the substrate. The electrical measurements[3][7] show that the contact resistance varies from the value greater than 1×10^7 to 1×10^{-2} (Ω) after annealing in a range of 300°C to 500°C for 20 s. The initial decrease in the contact resistance of the metal/n-GaAs contact is due to the formation of $In_{1-x}Ga_xAs$ compound between metal and GaAs. The appearance of this compound divides the original Schottky barrier due to metal/GaAs into two barriers due to metal/$In_{1-x}Ga_xAs$ and $In_{1-x}Ga_xAs$/GaAs, respectively. The latter is believed to be the main factor in controlling the contact resistance[8] and depends on the composition of this compound. The ohmic behavior for the sample annealed at 500°C for 20 s is thus attributed to the further reduction of this barrier.

IV. CONCLUSIONS

Interfacial reactions between In/Pd and GaAs have been characterized. The transition region within the contact is relatively thin and on the order of 500 Å. After annealing at various temperatures $In_{1-x}Ga_xAs$ compounds are formed at the metal/GaAs interface and have a strong preferred orientation with respect to GaAs substrate. The composition of this compound is dependent on the temperature of annealing. As the temperature increases, the composition of $In_{1-x}Ga_xAs$ compound shifts toward higher concentration of Ga. Reduction of the contact resistance of the metal/GaAs contact is correlated with the formation of $In_{1-x}Ga_xAs$ compounds of increasing Ga concentration.

V. ACKNOWLEDGEMENT

We would like to express our thanks for the use of microanalysis facilities in the Center for Microanalysis in the Materials Research Laboratory at the University of Illinois, which is sponsored by the Department of Energy.

VI. REFERENCES

1. A.A. Lakhani, *J. Appl. Phys.*, **56**(1984)1888.

2. D.C. Marvin, N.A. Ives, and M.S. Leung, *J. Appl. Phys.*, **58**(1985)2659.
3. L.H. Allen, L.S. Hung, K.L. Kavanagh, J.R. Phillips, A.J. Yu, and J.W. Mayer, *Appl. Phys. Lett.*, **51**(1987)326.
4. M. Murakami, W.H. Price, Y.C. Shih, N. Braslau, K.D. Childs, and C.C. Parks, *J. Appl. Phys.*, **62**(1987)3295.
5. M. Murakami, Y.C. Shih, W.H. Price, E.L. Wilkie, K.D. Childs, and C.C. Parks, *J. Appl. Phys.*, **64**(1988)1974.
6. M. Murakami, W.H. Price, *Appl. Phys. Lett.*, **51**(1987)664.
7. L.H. Allen, to be published.
8. J.M. Woodall, G.D. Pettit, T.N. Jackson, C. Lanza, K.L. Kavanagh, and J.W. Mayer, *Phys. Rev. Lett.*, **51**(1983)1783.

INITIAL EVOLUTION OF COBALT SILICIDES IN THE COBALT/AMORPHOUS-SILICON THIN FILM SYSTEM

HIDEO MIURA*, EN MA** AND CARL V. THOMPSON**
*MERL, Hitachi, Ltd., 502 Kandatsu, Tsuchiura, Ibaraki 300, Japan and Massachusetts Institute of Technology, Department of Materials Science and Engineering.
**Massachusetts Institute of Technology, Department of Materials Science and Engineering, 77 Mass. Ave. Cambridge, MA 02139.

ABSTRACT

The initial phase formation sequence for reactions in cobalt/amorphous-silicon multi-layer thin films has been investigated using a combination of differential scanning calorimetry, thin film X-ray diffraction, and transmission electron microscopy. Multilayer thin films with various overall atomic concentration ratios and various bilayer thicknesses were used in this study. It was found that crystalline CoSi is always the first phase to nucleate in the interdiffused amorphous layer which preexisted at the as-deposited cobalt/amorphous-silicon interface. The CoSi nucleates at temperatures as low as about 530 K, but does not grow until the next phase, which is Co_2Si when excess Co is available, starts to nucleate and grow. The activation energy of the CoSi nucleation was found to be 1.6±0.1 eV.

INTRODUCTION

Refractory and near-noble metal silicides are widely used in VLSI devices for interconnects and for ohmic contacts between metal interconnects and silicon substrates. Silicides are usually formed by means of reactions between a deposited metal film and a silicon substrate or a silicon film. The first phase to form at the metal/silicon interface is usually not the silicide phase which has the largest negative formation energy. It is generally believed that the phase formation sequence in thin film reactions is determined by kinetic constraints, rather than by the immediate minimization of free energy. There are several phenomenological models which deal with phase formation sequences during interface reactions [1]. However, these models do not permit a priori predictions of phase formation sequences, or the kinetics of phase formation.

Among the metal silicides, cobalt disilicide has attractive properties such as a low resistivity and a good lattice match with silicon. Therefore a large number of studies have been performed on the kinetics of cobalt silicide formation. However, it remains controversial as to which phase forms first at the cobalt/silicon interface. Some researchers report that the first phase is crystalline Co_2Si when the substrate is single crystalline silicon[2]. Others found that both CoSi and Co_2Si form simultaneously upon heating [3]. Crystalline $CoSi_2$ was also observed on the single crystalline silicon [4] or CoSi was observed to appear first on amorphous silicon [5]. Recent TEM studies revealed that there exists a very thin amorphous layer of cobalt silicide at the as-deposited interface [6]. The initial evolution of cobalt silicides is therefore still open to question.

Recently, Clevenger et al. demonstrated that differential scanning calorimetry can be used to study the nucleation as well as the growth of silicides in nickel/amorphous-silicon and vanadium/ amorphous-silicon systems [7]. Calorimetry has two major advantages compared with conventional analysis techniques such as Rutherford backscattering and X-ray diffraction analysis. One advantage is that calorimetry can be more sensitive to both compositional and structural changes than conventional methods. The other is that both kinetic and thermodynamic data for a variety of transitions in a wide temperature range can be obtained in one experiment.

In this paper, the phase transformation sequence in the cobalt/amorphous-silicon thin film system, especially the first phase to form at the cobalt/amorphous-silicon interface, is discussed. The use of multilayer films enables the simultaneous observation of the heat released as the reaction occurs at many reacting interfaces. This allows for the study of the early stages of silicide formation and growth. In addition to calorimetry, cross-sectional transmission electron microscopy and thin film X-ray diffraction were employed to systematically characterize the reaction kinetics in cobalt/amorphous- silicon multilayer thin films.

EXPERIMENT

Free-standing multilayer films were deposited on microscope slides coated with photoresist to a thickness of 1 μm. Cobalt and silicon were alternately electron-beam deposited at room temperature. The background pressure during the film deposition was never higher than 10^{-7} Torr. The multilayer film consisted of 5 bilayers. The bilayer thicknesses were varied from 14 nm to 100 nm. In the following discussion, we use h to designate the total thickness of the multilayer and λ to denote the bilayer thickness. The overall atomic concentration ratios were designed to be 2Co:1Si. After the evaporation, the slides were soaked in acetone to dissolve the photoresist and remove the multilayer films.

DSC samples consisted of pieces of the multilayer films with a total weight of 1 to 2 mg. These were heated in flowing N$_2$ in a Perkin-Elmer DSC2 at heating rates (H) ranging from 2.5 to 20 K/min from 320 K up to 900 K. The sensitivity range used for the calorimetric measurements was 1 mcal/s. After the first heating, samples were quenched from 900 K to 320 K at 320 K/min, and the second heating was performed without disturbing the samples and with the same heating rate used in the first runs. The DSC traces of the second runs were used as the baseline and subtracted from the first traces.

Glancing angle thin film X-ray diffraction for phase identification was performed on the samples annealed in the DSC. The incident X-ray beam angle was fixed at 5 degrees. The multilayer films were also deposited on thermally-oxidized silicon wafers simultaneously with the films deposited on photoresist-coated slides. Cross-sectional samples were used for structural analysis in a JEOL-200CX transmission electron microscope.

RESULTS

DSC traces for multilayer films with atomic concentration ratios of 2Co:1Si are shown in Fig. 1. In this figure, four traces for

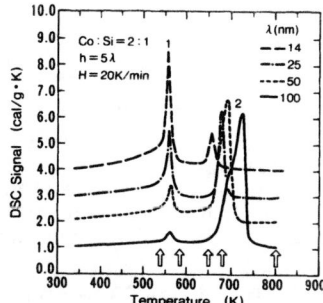

Fig. 1 DSC traces for Co/a-Si multilayer thin films with atomic concentration ratio of 2Co to 1Si

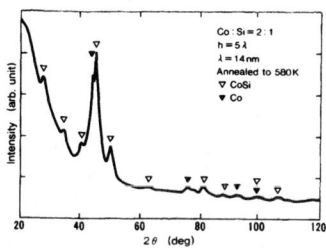

Fig. 2 Thin film X-ray diffraction spectrum for a Co/a-Si multilayer thin film heated to 580 K

different bilayer thickness, λ = 14, 25, 50, and 100 nm, are compared. The heating rate was fixed at 20 K/min. The vertical axis shows the DSC signal, which is the energy change normalized by the sample weight and the heating rate. Two exothermic peaks, labeled as peak 1 and peak 2 in Fig. 1, were observed in every trace. The first peak of each trace appeared at the same temperature, about 560 K. The second peaks, except that for films with λ=14 nm, started at about 660 K and terminated at increasingly higher temperatures as the bilayer thickness was increased. This shift is known to be due to the larger attainable thicknesses of the reactant layers [7]. The peak height of the first peak increased as the bilayer thickness was decreased. This is because there are more Co/Si interface per unit weight in the samples with smaller bilayer thicknesses. Thus, when reaction occur at the Co/amorphous-silicon interfaces, a stronger DSC signal can be detected in smaller bilayer thickness samples.

To correlate thermal changes with structural changes, cross-sectional transmission electron microscopy was performed on samples with 100 nm bilayer thicknesses. It was clearly seen that a crystalline phase nucleated and grew at the cobalt/amorphous-silicon interface, when the sample was heated to 580 K, just after the first peak. The thickness of this interfacial layer was about 5 nm. Fig. 2 shows a thin-film X-ray diffraction spectrum for a multilayer film with a 14 nm-bilayer thickness which was heated to 580 K. This spectrum clearly indicates that the new polycrystalline layer is CoSi. It is interesting to note that this nucleated CoSi layer did not thicken even after the sample was heated up to 650 K (just before the second peak).

The thin-film X-ray diffraction spectrum of the sample heated to 670 K (the middle of the second peak) is shown in Fig. 3. It indicates that there are three crystalline phases: Co, CoSi, and Co_2Si. In view of the composition of the phases, the spatial sequence of the phases should be Co/Co_2Si

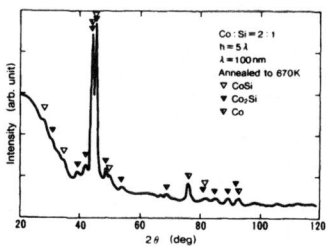

Fig. 3 Thin film X-ray diffraction spectrum for a Co/a-Si multilayer thin film heated to 670 K

CoSi/amorphous—silicon. Such a phase order was confirmed by energy dispersive X—ray analysis using a VG HB5 scanning transmission electron microscope.

The thickness of the CoSi layer was about 5 nm and that of the Co_2Si layer was about 15 nm. This results indicates that the during the second peak, the Co_2Si layer nucleated, coalesced, and thickened, while the thickness of the CoSi layer remained almost constant, even though both excess cobalt and silicon still existed.

The final structure of the multilayer film heated to 800 K consisted of only one polycrystalline phase, Co_2Si. It was confirmed using thin—film X—ray diffraction and it was consistent with the overall atomic concentration of 2Co:1Si. Thus, it is concluded that during the second exothermic peak, both nucleation and growth of Co_2Si and transformation of the CoSi layer to Co_2Si occurred. Silicon is fully consumed first, followed by the reaction of CoSi with excess Co to transform the multilayer into Co_2Si.

DISCUSSION

As shown in Fig. 2, the first phase that formed at the cobalt/amorphous—silicon interface was crystalline CoSi. This result differs from most previous studies except Nathan's [6], which reports that CoSi is always the first phase to form at the cobalt/amorphous—silicon interface. To further investigate the CoSi formation, the interface structure of the multilayer film in the as—deposited state was observed using TEM. Fig. 4 shows a bright—field TEM micrograph of the as—deposited sample. There is an interfacial layer about 10 nm thick between the Co layer and the amorphous silicon layer. The interlayer appears to be amorphous, with indistinct boundaries on both ends. This interface structure remained unchanged even after the sample was heated up to 540 K, i.e., just before the first peak. The observation of this amorphous interlayer is in agreement with other recent studies [8]. EDX analysis using scanning TEM was performed to analyze this layer. Fig. 5 shows the atomic compositional profiles for cobalt in the interface region of an as—deposited film and a film annealed to 540 K. As shown in this figure, in both cases there is a gradual compositional change across the cobalt/amorphous—silicon interface.

The thickness of the interdiffusion layer over which the Co composition varies was about 10 nm. In addition to the gradual composition change in this amorphous layer, we note that the interface between this layer and the amorphous silicon layer shown in Fig. 5 is rather ill—defined. This is quite different from the very sharp amorphous—silicide/amorphous—silicon interfaces frequently observed in other metal/ silicon systems [7]. We thus suspect that the interfacial layer is largely composed of amorphous silicon with an appreciable concentration of cobalt in

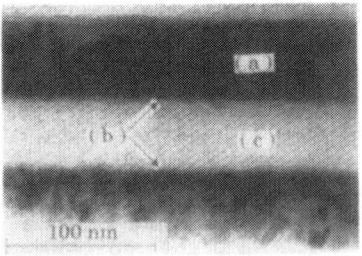

Fig. 4 Bright—field TEM image of an as—deposited Co/a—Si multilayer thin film (λ =100 nm) (a)Co ; (b)amorphous interlayer ; (c)a—Si

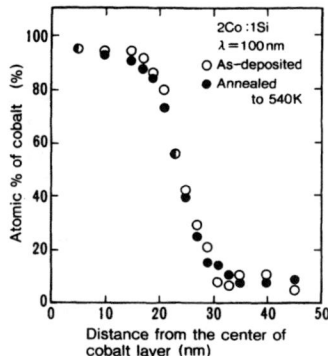

Fig. 5 Concentration profiles for Co at the Co/a–Si interface in the as–deposited state and after being heated to 540 K

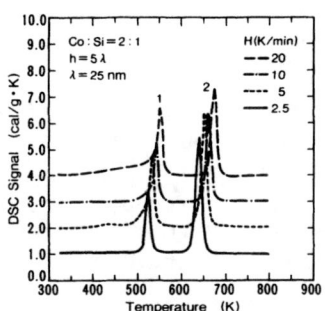

Fig. 6 DSC traces for Co/a–Si multilayer thin films with λ =25 nm, heated at 2.5, 5, 10 and 20 K/min

solution, rather than an amorphous silicide with a nearly fixed composition.

Unlike many other metal/amorphous–silicon diffusion couples where formation of a new amorphous silicide phase occurs, the amorphous interlayer (interdiffusion layer) at the cobalt/amorphous–silicon interface did not grow upon annealing. This behavior has been observed before by Holloway et al. in this system [8]. It was proposed that the inability of the element to diffuse through the amorphous interlayer is responsible for the absence of the growth of the amorphous phase. Our results suggests that the interdiffusion of cobalt and silicon through the interlayer is indeed difficult below and at the temperature at which crystallization of CoSi occurs. In fact, the preexisting amorphous layer did not grow at all. CoSi forms only in this interlayer, and the nucleated CoSi always stops growing at about 5 nm thickness. Thus the kinetic advantage that solid–state amorphization has in some other system does not seem to be present in this cobalt/amorphous–silicon system.

The question why CoSi nucleates first remains to be answered. In systems where new amorphous phases grow, the crystalline phase that subsequently forms is usually not the phase with the largest thermodynamic driving force. Nucleation of the crystalline phase takes place at the interface between the amorphous silicide and either silicon or metal. In the case of cobalt/amorphous–silicon reactions, however, the nucleation may occur within the amorphous interlayer. In such a case crystallization within the interlayer can occur without significant long range diffusion, so that free energy minimization may more readily determine the first phase to form. Indeed, CoSi has the lowest free energy in the cobalt–silicon system.

Activation energies for the formation of CoSi and Co_2Si were measured using an analysis originally developed by Kissinger [9]. When the heating rate was decreased during the DSC experiments, every exothermic peak shifts down to lower temperatures. This peak shift can be related to the activation energy of the phase transformation by the following equation.

$$\ln (H/Tp^2) = C - Q/(k \cdot Tp), \qquad (1)$$

where H is the heating rate, Tp is the peak temperature, C is a constant, k is Boltzman's constant, and Q is the activation energy. Fig. 6 shows examples of the peak shifts with variation of the heating rates. The first peak which corresponds to the nucleation and growth of CoSi went down from about 560 K to about 530 K, and the peak temperature of the second peak which corresponds to the nucleation and growth of Co_2Si also went down from about 680

K to about 640 K when the heating rate of the multilayer film was decreased from 20 K/min to 2.5 K/min. Activation energies were determined by averaging the Q values obtained for multilayer films with different bilayer thicknesses. The activation energy for the nucleation and growth of CoSi was 1.6±0.1 eV, and that of Co_2Si

nucleation and growth was 2.0±0.1 eV.

CONCLUSION

We have applied techniques sensitive to structural and compositional changes, such as differential calorimetry, cross-sectional transmission electron microscopy, and thin-film X-ray diffraction, to obtain a clear picture of the silicide phase formation sequence and the kinetics of reactions in cobalt/amorphous–silicon multilayer thin films. It is found that the first intermetallic phase to form at the cobalt/amorphous–silicon interface is always crystalline CoSi. The nucleation of CoSi occurs at a low temperature, about 550 K, only in an amorphous interdiffused layer which already existed at the interface after deposition of the films. This CoSi layer, about 5 nm in thickness, does not thicken until the next phase starts to nucleate and grow. If excess cobalt exists, Co_2Si

starts to nucleate and grow at the Co/CoSi interface at about 650K with the coexistence of CoSi. The activation energies obtained using a Kissinger analysis of our calorimetric data are 1.6±0.1 eV for the nucleation and growth of CoSi, and 2.0±0.1 eV for the nucleation and growth of Co_2Si.

ACKNOWLEDGEMENTS

This work was supported in part by International Business Machines and by Hitachi, Ltd.. The authors would like to thank J. Carter for assistance in sample preparation.

REFERENCES

1. U. Gosele and K.N. Tu, J. Appl. Phys., 53, 3252 (1983).
2. S.S. Lau, J.W. Mayer, and K.N. Tu, J. Appl. Phys., 49, 4005 (1978)
3. C.D. Lien, M-A. Nicolet, C.S. Pai, and S.S. Lau, Appl. Phys. A. 36, 153 (1985).
4. J.Y. Veuillen, J. Derrien, P.A. Badoz, E. Rosencher, and C.d' Anterroches, Appl. Phys. Lett., 51, 1448 (1987).
5. M. Nathan, J. Appl. Phys., 63, 5534 (1988).
6. P. Ruterana, P. Houdy, and P. Boher, J. Appl. Phys., 68, 1033 (1990).
7. L.A. Clevenger and C.V. Thompson, J. Appl. Phys., 67, 1325 (1989).
8. K. Holloway, R. Sinclair, and M. Nathan, J. Vac. Sci. Technol. A. 7, 1479 (1989).
9. H.E. Kissinger, Analyt. Chem., 29, 1702 (1957).

DIFFUSIONAL PHASE TRANSFORMATION UNDER INDUCED THERMAL STRESS.

E C ZINGU AND B T MOFOKENG, Department of Physics, Medunsa, 0204, South Africa

ABSTRACT

When thin films are deposited on substrates or when compound films are formed through interdiffusion of multi-film structures, intrinsic stress develops in the various films. Thermal mismatch between the expansion coefficients of the substrate and films in multi-film structures gives rise to extrinsic stress at elevated temperatures

By using Si$\langle 100 \rangle$ and rolled Al foil substrates supporting the same multi-film structure $SiO_2/Si/Co$, the effect of extrinsic stress on interdiffusion of thin films is isolated.

Silicide growth is found to be inhibited (delayed) when formed on Al substrates compared to that formed on Si substrates. The delay in silicide growth is ascribed to delamination caused by large tensile stress prior to silicide formation. The growth rate of Co_2Si is found to be similar on both Al and Si substrates

INTRODUCTION

Thin films deposited on substrates are usually in a state of stress which has been established during the deposition process or during the subsequent annealing or cooling. The stress in the film is said to be a combination of intrinsic stress (i.e. stress due to the deposition process, and phase or structural transformations of the deposited film) and extrinsic stress (i.e. thermal stress due to the thermal expansion difference between the film and the substrate). At temperatures well above or below the temperature of deposition thermal stress usually dominates. [1]

In multi-layered films the thermal stress in the individual films is independent of the physical properties of the interposing or covering layers. [2] Consider the thin film structure $Al/SiO_2/Si/Co$ or $Si\langle 100 \rangle/SiO_2/Si/Co$ in which SiO_2, Si and Co films have been deposited sequentially on Al and Si substrates respectively. Each of the three films will be in a state of thermal stress at elevated temperatures due to the difference between the film expansion and that of the substrates. See Table I for the expansion coefficients and the nature of the stress for various films and substrates. The nature of the stress (i.e. tensile or compressive) and its magnitude will be determined by the difference between the expansion coefficient of the film and the substrate

At temperatures around 400°C Co and Si interdiffuse to form Co_2Si. The aim of this investigation was to determine the effect which thermal stress in the Co and Si films has on the formation of Co_2Si. In order to vary the degree of thermal stress in the films, different substrates namely Si and Al were used

Mat. Res. Soc. Symp. Proc. Vol. 230. ©1992 Materials Research Society

Table I : Expansion coefficient and the nature of the stress in various films

Film	Substrate	α_f $(\times 10^{-6} \ K^{-1})$ (film)	α_s $(\times 10^{-6} K^{-1})$ (substrate)	Nature of Stress ($\pm 400°C$)
Co	Si	12	2,3	Compressive
Si	Si	2,3	2,3	0
SiO₂	Si	0,5	2,3	Tensile
Co	Al	12	22	Tensile
Si	Al	2,3	22	Tensile
SiO₂	Al	0,5	22	Tensile
Co₂Si	Si	9,4	2,3	Compressive
Co₂Si	Al	9,4	22	Tensile

Values of α_f and α_s from reference [3]

EXPERIMENTAL

As the main thrust of this investigation was a comparison of the behaviour of thin film structures on different substrates, thin film deposition and heat treatments were identical for samples which were to be compared

The specimens were prepared by depositing thin films of SiO_2, Si and Co sequentially on substrates consisting of 1cm² Si⟨100⟩ and rolled Al foil in an electron-beam evaporator at a base pressure of approximately 5×10^{-7} Torr. The samples were annealed in a vacuum furnace at temperatures around 360°C for various periods of time. The SiO_2 layer served as a barrier to diffusion between Si and the substrate

Rutherford backscattering spectrometry employing 1 MeV alphas was used to characterise the samples with respect to thickness and composition of the various layers

RESULTS

Figure 1 shows typical Rutherford backscattering spectra of Si⟨100⟩/SiO₂/Si/Co and Al/SiO₂/Si/Co samples with identically deposited thin films, prior to annealing. The only noticable difference is found in that part of the spectra (low energy) which originates from the substrates (Al and Si respectively)

The effect of annealing the specimens at 360°C for 40 minutes can be observed in figure 2. The change in the Co and Si layer signals (step formation) indicates that cobalt silicide has formed. The composition of the silicide phase Co_2Si, has been determined by comparing the heights of the steps in the Si and Co signals. It is evident from the widths of the step in the Si and Co signals, that the thickness of the Co_2Si layer formed on the Si<100> substrate exceeds that of the Co_2Si layer formed on Al

By measuring the thickness of Co_2Si formed after isothermal anneals of increasing duration, the rate of Co_2Si formation on both Si and Al substrates was determined. The (thickness)² of the Co_2Si layer formed on Si and Al substrates, plotted against the annealing time at 360°C is shown in Figure 3. The linearity of these curves indicates that silicide formation is limited by the diffusion process. It is observed that the slope of the two curves are comparable which suggests the same diffusion rate for both specimens. The shift in the growth curve of the specimen on the Al substrate suggests a delay in the diffusional growth process

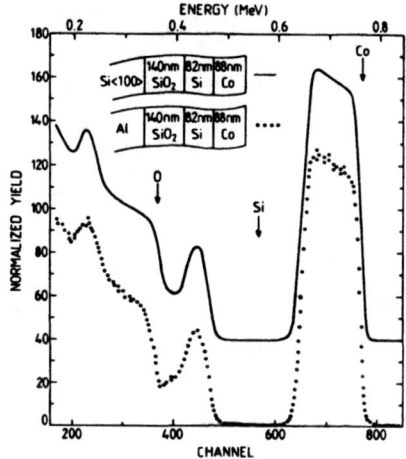

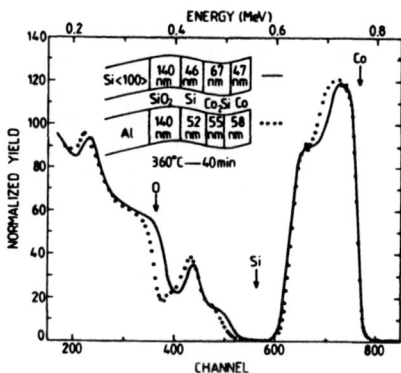

Fig 2 Rutherford Backscattering spectra of the samples in figure 1 after vacuum annealing

Fig 1 Rutherford Backscattering spectra of Si<100>/SiO₂/Si/Co and Al/SiO₂/Si/Co samples before annealing

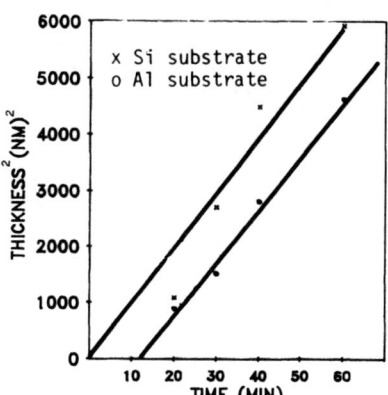

Fig 3 The (thickness)² of the Co_2Si layer plotted against the annealing time. The diffusivities (slope) for specimens on different substrates are similar

DISCUSSION

During the formation of Co_2Si, Co diffuses through the silicide layer and enters the Si layer where it forms new Co_2Si at the Si/Co_2Si interface [4]. This represents a volume increase of the original Si layer which results in a compressive stress in the silicide layer. A theoretical model proposed for the oxidation of silicon suggests that the compressive stress which develops in the growing oxide film reduces the effective diffusion constant [5]. A similar effect in Co_2Si specimens should result in enhanced diffusional growth of the silicide layer for specimens on Al substrates compared to those specimens on Si substrates since the induced tensile stress (i.e. thermal mismatch between the films and the Al substrate) would relieve the compressive stress (due to the volume increase). The similar slope of the curves in figure 4 ((thickness)^2 against annealing time) suggests that the effective diffusivity of the Co_2Si layer is not influenced by the induced thermal stress.

It has been reported that silicide layer growth is inhibited when the metallized silicon wafer is bent forwards or backwards during annealing [6]. Since both types of stress (tensile and compressive) give rise to a similar effect (i.e. inhibiting the growth of the silicide layer), it is concluded that the induced stress does not influence the activation energy of the diffusivity. The inhibited silicide growth on Al substrates as observed in figure 2 can therefore not be ascribed to a change in the activation energy but rather a delay in the commencement of the silicide growth process

The delay in silicide formation as observed in Fig 3 can be explained as follows. On annealing, large tensile stresses develop in the films on the Al substrate, also at temperatures below Co_2Si formation. Such interfacial stresses agravated by contamination at the Si/Co interface, lead to delamination, thus reducing the intimate contact between the Si and Co layers. During the delay time the films undergo sress relaxation via plastic flow, the interfacial contact is restored and silicide formation commences. Due to the statistical nature of the relaxation process, the growth of Co_2Si does not commence uniformly over the surface of the specimens. The Si/Co_2Si interface will therefore be non-uniform. A tail of the low energy edge of the Co signal of the Rutherford Backscattering spectra of annealed $Al/SiO_2/Si/Co$ specimens would be indicative of a non-uniform Si/Co_2Si interface, in agreement with the prediction above. Such tails have been observed in the spectra of samples which had been annealed for very short periods of time. Delay in silicide growth has also been observed when the formation temperature was relatively low [7]. The temperature of 360°C at which this investigation was conducted is well below the lowest temperature which has been reported for Co_2Si growth on amorphous Si [8]. Due to the limited sensitivity of Rutherford Backscattering it is not practical to investigate the non-uniform growth of the very thin layer of Co_2Si during the delay period. It can therefore not be stated conclusively whether there is any silicide growth during the delay period

CONCLUSION

The effect of induced stress through thermal mismatch of the deposited films and substrate on the thin film interdiffusion has been demonstrated. Induced thermal stress is found to influence the onset of silicide formation. Contrary to the predicted enhancement of silicide formation due to induced stress, the growth rate of Co_2Si is found to be similar for specimens on Si<100> and Al substrates.

ACKNOWLEDGEMENT

The authors wish to express their appreciation to T Marais for his technical assistance, the National Accelerator Centre for making their facilities available and the Foundation for Research Development for their financial support

REFERENCES

1. K L Chopra, Thin film Phenomena, (New York, McGraw Hill, 1969)

2. P H Townsend, D M Barnett and T A Brunner, J. Appl. Phys. 62 (1987) 4438

3. M A Nicolet and S S Lau in VLSI Electronics : Microstructure Science, edited by N G Einspruch and G B Larrabee,(Academic Press, New York, 1983)

4. G J van Gurp, W F van der Weg and D Sigurd, J. Appl. Phys. 49, 4011 (1978)

5. A Fargeix and G Ghibaudo, J. Appl. Phys. 54, 7153 (1983)

6. M von Allman, S S Lau and M Nicolet in Thin Film Interfaces and Interactions, edited by J Baglin and J Poate, (Electrochem. Soc., Manchester, 1980) 364

7. A Martinez, D Esteve, A Guivarch, P Henoc and G Pelosis, Solid State Electronics, 23, 55 (1980)

8. C D Lien, M A Nicolet, C S Pai and S S Lau, Appl. Phys. A 36, 153 (1985)

COMPOUND FORMATION IN Pd METALLIZED
STRAINED LAYERS OF SiGe on Si

A. BUXBAUM*, M. EIZENBERG*, A. RAIZMANN** AND F. SCHAFFLER***
*Department of Materials Engineering and Solid State Institute, Technion-Israel Institute of Technology, Haifa, Israel.
**Soreq Nuclear Research Center, Yavne, Israel.
***Daimler-Benz AG, Forschungsinstitut Ulm, Germany.

ABSTRACT

In this paper we report on the formation of compounds following the interaction of Pd with strained epitaxial layers of $Si_{1-x}Ge_x$ (x=0.18) MBE grown on Si(100), in the temperature range of 200 to 650°C. Compositional and structural analyses show that the dominant compound formed was an hexagonal ternary phase, $Pd_2Si_{1-y}Ge_y$ where the value of y is lower than of x, and varies with the reaction temperature. In addition to the ternary phase, the binary phase PdGe was also detected. The hexagonal compound grew in a textured manner with its c-axis oriented along the [100] direction of the $Si_{1-x}Ge_x$ film. High temperature anneals (T≥550°C) resulted in the formation of a double layered structure, with the silicide/germanide compound layer on the surface, and below it a Ge rich epitaxial $Si_{1-x}Ge_x$ layer. Strain relaxation of the unreacted layer occurred for specimens in which the double layer structure appeared (annealing temperature ≥550°C). A mechanism for the formation of the double layered structure is proposed.

INTRODUCTION

Advances in growth techniques of thin epitaxial strained layers of $Si_{1-x}Ge_x$ on Si substrates have allowed for the modification of the optical and electrical properties of the semiconductor, while using standard Si processing technology [1]. The interest in the $Si_{1-x}Ge_x$ layers has only recently prompted investigations of their interaction with thin metal films [2]. An understanding of the interaction with metals is essential, if the strained layers are to be used in device applications. In this work we investigated the interaction of Pd with strained $Si_{1-x}Ge_x$ epilayers on Si(100) as a function of the applied heat treatment.

EXPERIMENTAL

Strained $Si_{1-x}Ge_x$ epilayers with x=0.18 were grown on Si(100) by Molecular Beam Epitaxy (MBE), to a thickness of 2300Å. The strain in the films was measured by Double Crystal X-ray Diffractometry (DCD). The peak positions of the asymmetric {511} reflections were used to calculate the values for the in-plane and perpendicular lattice parameter mismatch, $f^{\parallel}$ and $f^{\perp}$ [3]:

$$f^{\parallel} = \frac{a^{\parallel} - a_0}{a_0} \qquad (1a)$$

$$f^{\perp} = \frac{a^{\perp} - a_0}{a_0} \qquad (1b)$$

where a_0 is the lattice parameter of the Si substrate, and $a^{\perp}$, and $a^{\parallel}$ are the lattice parameters of the $Si_{1-x}Ge_x$ film in the directions perpendicular and parallel to the surface, respectively. The layers were found to be commensurate with the substrate, with $f^{\parallel} = 8.9 \cdot 10^{-5}$, and strained in the direction perpendicular to the surface, with $f^{\perp} = 1. .0 \cdot 10^{-2}$. Oxide removal prior to metal deposition was carried out by immersion in buffered HF, followed by a deionized water rinse. The samples were then immediately loaded into an e-gun evaporation chamber, and pumped to a pressure of $1 \cdot 10^{-6}$ torr, and Pd was then deposited to a thickness of 1300Å. Samples were annealed in a vacuum furnace (with pressures below $1 \cdot 10^{-6}$ torr) at temperatures ranging from 200 to 650°C.

Structural information on the reaction products was obtained by a combination of electron diffraction and bright field / dark field images in a 200 kV transmission electron microscope (TEM), and by X-ray diffraction. Information on elemental distribution in the reacted region was obtained by Energy Dispersive Spectroscopy (EDS) in the TEM, and by Auger Electron Spectroscopy (AES) depth profiles.

RESULTS AND DISCUSSION

Fig. 1 shows AES depth profiles of (a) an unreacted sample, (b) sample reacted at 250°C for 4 hours, and (c) a sample reacted at 550°C for 4 hours. Fig. 1b shows that the Pd was totally consumed in the reaction, resulting in a reacted region containing Pd, Si and Ge. The X-ray diffraction pattern from this sample showed the presence of a ternary phase $Pd_2Si_{1-y}Ge_y$, and a weak reflection from the PdGe phase. The former phase was determined to be ternary by EDS investigation in the TEM. The EDS spectrum (shown in Fig. 2), was collected from a single grain region of the sample, and shows the presence of all three elements. TEM investigations indicated that the ternary compound grew in a fiberous-columnar manner, and the structure of the compound was like that of hexagonal Pd_2Si, with its c-axis oriented in the direction perpendicular to the surface.

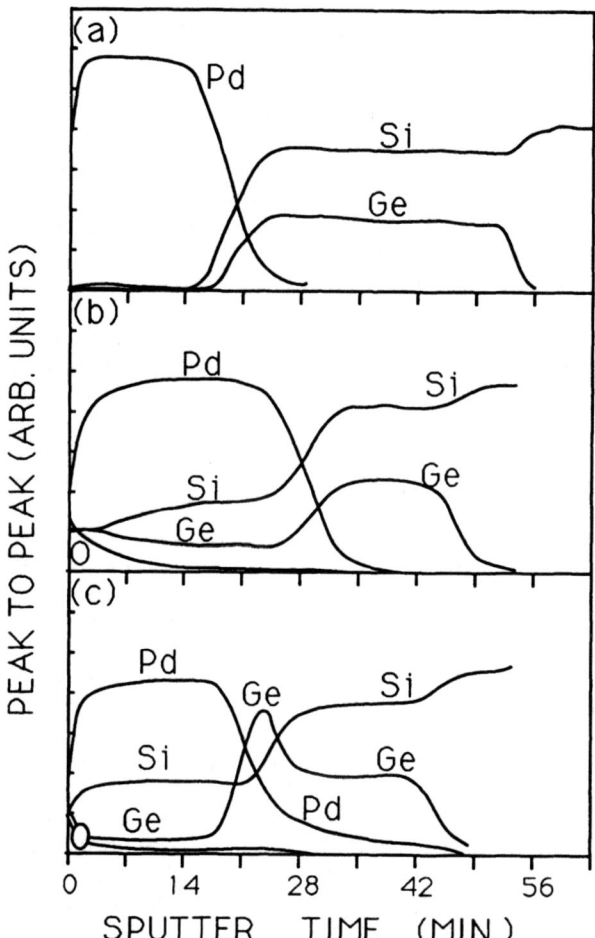

Fig. 1
AES depth profiles of a sample in the as deposited
condition (a), samples reacted for 4 hours at 250°C (b),
and at 550°C (c).

The high temperature interaction between the Pd and the $Si_{1-x}Ge_x$ layers is shown in Fig. 1c, for a sample reacted at 550°C for 4 hours. We note that the Ge signal within the reacted region is lower than that of the sample reacted at 250°C, i.e. the value of y is lowered as a consequence of the high temperature treatment. Furthermore, we note that there is a Ge rich region between the reacted region on the surface and the unreacted $Si_{1-x}Ge_x$ layer below. This double layer structure is shown in a TEM cross-sectional dark field micrograph (Fig. 3), where we see a highly defected layer above the unreacted $Si_{1-x}Ge_x$ layer. Microdiffraction from the Ge rich layer (Fig. 4) reveals the [110] zone axis of the diamond lattice, a pattern identical to that obtained from the unreacted $Si_{1-x}Ge_x$ layer, an indication of the epitaxial relationship. EDS analysis of this region showed that the atomic concentration of Ge in this region was approximately 40%.

Strain in the $Si_{1-x}Ge_x$ layer was measured by DCD. At the low reaction temperature (250°C), the value of $f^{\parallel}$ was low ($1 \cdot 10^{-4}$), and the value of $f^{\perp}$ was high ($1 \cdot 10^{-2}$), indicating a strained layer similar to the as-grown $Si_{1-x}Ge_x$ film. For the sample reacted at 550°C, however, the the value of $f^{\parallel}$ was raised ($1 \cdot 10^{-2}$), and that of $f^{\perp}$ was lowered ($8 \cdot 10^{-3}$). Since $f^{\perp} = f^{\parallel}$, namely the tetragonal distortion has diminished, we conclude that the strain has relaxed.

A Possible mechanisms for the formation of the Ge rich intermediate layer at high temperatures is as follows. At low temperatures only the Pd atoms are mobile, and diffuse into the $Si_{1-x}Ge_x$ layer to form $Pd_2Si_{1-y}Ge_y$ + PdGe (this conclusion was reached based on a TEM analysis). At higher temperatures, most probably the Si and the Ge atoms are mobile as well. The double layer structure can be explained if we assume that the Si diffuses towards the Pd, while the Ge atoms diffuse back to the $Si_{1-x}Ge_x$ layer, and regrow epitaxially, forming the Ge rich layer. This double layer behavior was first observed in the $Pt-Si_{1-x}Ge_x/SiO_2$ system [4], and was used to form $Si_{1-x}Ge_x$ layers when a-Ge was deposited on the epitaxial $Pd_2Si/Si[111]$ system [5]. In the latter work, no Ge rich layer formed in the non-epitaxial $Pd_2Si/Si[100]$ system. In the present work we demonstrate the growth of a Ge rich Si-Ge layer on the $Si_{1-x}Ge_x/Si[100]$ layer. The high defect density in the regrown Ge rich region may be responsible for the strain relaxation observed in the unreacted strained $Si_{1-x}Ge_x$ layer, by providing surface defects which act as nucleation sites for dislocations.

Fig. 2.
EDS spectrum collected
in the TEM, from a single
grain region of the
Pd2SiGe compound. All
elements are present,
thus the compound was
ternary.

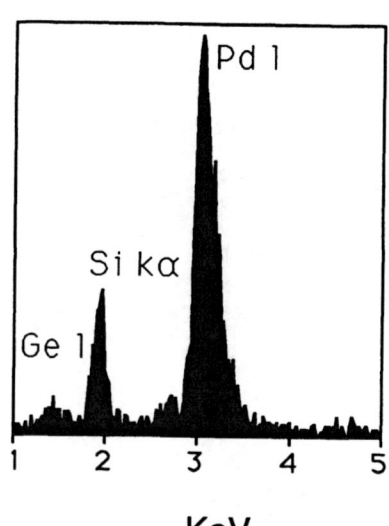

Fig. 3
TEM cross sectional dark field
image of sample reacted at 550°C
for 4 hours, showing the highly
defected Ge rich region. Since
image was obtained by using a
diffracted beam from the $Si_{1-x}Ge_x$
layer, the compound on the
surface is not seen.

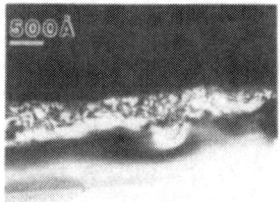

Fig. 4.
Microdiffraction from the Ge rich
$Si_{1-x}Ge_x$ layer of a sample reacted
at 550°C for 4 hours. This has
the diamond structure ([110]
zone axis shown) and is epitaxial
on the unreacted $Si_{1-x}Ge_x$ layer.

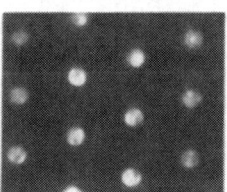

CONCLUSIONS

In this work we found that the low temperature interaction (250°C) of Pd with strained $Si_{1-x}Ge_x$ on Si[100] was dominated by the formation of a ternary compound, $Pd_2Si_{1-y}Ge_y$, and small amounts of PdGe. The $Pd_2Si_{1-y}Ge_y$ grew in a columnar fashion, with its basal plane parallel to the $Si_{1-x}Ge_x$ (100) surface. At high temperatures (550°C) we observed the formation of a Ge rich Si-Ge layer between the silicide/germanide and the unreacted $Si_{1-x}Ge_x$ layer. A suggested mechanism is: Pd and Ge atoms diffuse inwards, while Si diffuses out towards the surface. The Ge rich layer grows epitaxially (with defects) on the unreacted $Si_{1-x}Ge_x$ layer. The presence of the defects may be responsible for strain relaxation in the unreacted $Si_{1-x}Ge_x$ layer.

References:

[1]   D.V. Lang, R. People, J.C. Bean and M. Sergent, Appl. Phys. Lett. 47 (12), 1333 (1985).
[2]   R.D. Thompson, K.N. Tu, J. Angillelo, S. Delage, and S.S. Iyer, J. Electrochem. Soc. 135, 3161 (1988).
[3]   I.B. Bhat, K. Patel, N.R. Taskar, E. Ayers, and S.K. Ghandhi, J. Cryst. Growth, 88, 23 (1988).
[4]   Q.Z. Hong and J.W. Mayer, J. Appl. Phys. 66 (2), 611, (1989).
[5]   Q.Z. Hong, J.G. Zhu, and J.W. Mayer, Appl. Phys. Lett. 55 (8), 747 (1989).

KINETICS OF VACANCY ORDERING IN YSi$_{2-x}$ THIN FILM ON SILICON

T.L. LEE[*], L.J. CHEN[*] and F.R. CHEN[**]

[*]Department of Materials Science and Engineering, National Tsing Hua University, Hsinchu, Taiwan, Republic of China
[**]Materials Science Center, National Tsing Hua University, Hsinchu, Taiwan, Republic of China

ABSTRACT

High resolution and conventional transmission electron microscopy have been applied to study the interfacial reaction of yttrium thin films on Si. Epitaxial YSi$_{2-x}$ film was grown on (111)Si by rapid thermal annealing at 500-1000 ^{0}C. The orientation relationship between yttrium silicide and (111)Si was determined to be [0001]YSi$_{2-x}$//[111]Si and $(10\bar{1}0)$YSi$_{2-x}$//$(11\bar{2})$Si. The vacancies in the YSi$_{2-x}$ film were found to be ordered in the Si sublattice plane and form an out-of-step structure. The range of M values of the out-of-step structure was found to narrow with annealing temperature and time. Defects along specific crystallographic directions were observed and analyzed to be intrinsic stacking faults.

INTRODUCTION

Rare earth (RE) silicide thin films have attracted much attention in the past few years for their potential applications as infrared detectors and ohmic contacts. The compositions previously reported for the stable phases of rare earth silicide films have ranged from RESi$_{1.6}$ to RESi$_2$, with the former representing a vacancy distribution in the Si sublattice. The excellent lattice match between rare-earth silicides and to (111) Si substrates offers a rare opportunity to grow high quality epitaxial silicides on silicon [1,2]. Epitaxial YSi$_{2-x}$ with a χ_{min} as low as 3% was grown on silicon by rapid thermal annealing. Previous studies showed that rare earth silicides on silicon formed by conventional annealing typically have poor surface morphology with pitting in the thin films [1-5]. Some improvement in the surface morphology have been obtained by rapid thermal annealing [1,6,7].

The yttrium silicide is a rare earth silicide with the hexagonal AlB$_2$ structure with the lattice mismatch between (0001) plane of yttrium silicide and (111)Si being nil at room temperature [1,2]. The Schottky barrier height ϕ_b of yttrium silicide is as low as 0.3 to 0.4 eV on n-type silicon [3,8]. The compositions of yttrium silicides, measured by Rutherford backscattering spectrometry (RBS), were found to range from YSi$_{1.7}$ to YSi$_2$ with the former representing the formation of the vacancies in the Si sublattice plane [1,9]. The vacancies in YSi$_{2-x}$ silicides form different ordered superstructures depending on the annealing conditions. Defects along <11$\bar{2}$0> directions were observe to distribute in the YSi$_{2-x}$ films. In this paper, we report the results of a transmission electron microscope study of the kinetics of vacancy ordering in Y/(111)Si system. The superstructures of the ordered vacancies and the nature of the defects in the yttrium silicide thin films were identified by using high resolution and conventional electron microscopy.

EXPERIMENTAL PROCEDURES

Single-crystal, 3-8 cm, 3 or 4 in. in diameter, phosphorus-doped (111) oriented silicon wafers were first cleaned chemically by a standard procedure.

The samples were then dipped into a diluted HF (HF:H_2O =1:50) solution for 3 minutes immediately before loaded into an electron gun evaporation chamber. Thin yttrium films, 30 nm in thickness, were first deposited onto the Si substrates at 350 ^{0}C. The vacuum during deposition was maintained to be better than 2 x 10^{-6} Torr. The deposition rate was about 1 nm/s. An amorphous silicon capping layer, 20 nm in thickness, was subsequently then deposited onto the metal layer to prevent yttrium film from oxidation during heat treatments.

The samples were annealed by the rapid thermal annealing (RTA) in a high-purity argon atmosphere at 500-1200 ^{0}C for various periods of time in a A.G. Heatpulse 310 quartz furnace. The temperature-time profile of the quartz heater was controlled by a digital input. A ramp-up time of a few seconds and a heating rate of about 250 ^{0}C/s were used in a thermal cycle. The temperature of the samples during annealing was measured by a thermocouple attached to the backside of samples.

Both planview and cross-sectional specimens were examined by either a JEOL 200CX for conventional electron microscopy or by a JEOL 4000EX for high resolution electron microscopy. All the diffraction patterns are taken along [0001] axis of YSi$_{2-x}$.

RESULTS AND DISCUSSION

Figs. 1 (a) and (b) show the electron diffraction patterns of the samples annealed at 500 and 600 ^{0}C for 15 s, respectively. For 500 ^{0}C annealed samples, the orientation relationships between epitaxial silicide and (111)Si were found to be [0001]YSi$_{2-x}$//[111]Si and (10$\bar{1}$0)YSi$_{2-x}$//(11$\bar{2}$)Si. After annealing at 600 ^{0}C, additional superlattice spots appear at $\frac{1}{3}$(2$\bar{1}\bar{1}$0) as seen in Fig. 1 (b). A previous low energy electron diffraction (LEED) study

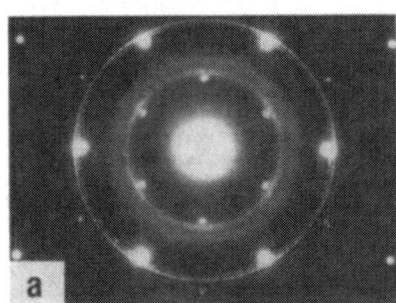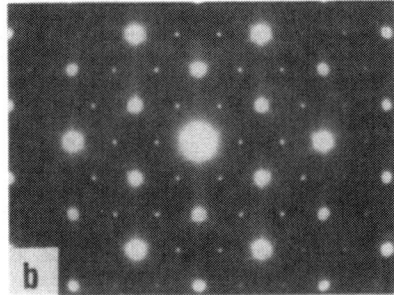

Fig. 1   Diffraction patterns (DPs) of samples annealed at (a) 500 ^{0}C for 15 s and (b) 600 ^{0}C for 15 s.

attributed the appearance of superlattice diffraction spots at $\frac{1}{3}$(2$\bar{1}\bar{1}$0) to the formation of a $\sqrt{3}$x$\sqrt{3}$ R30^0 ordered vacancy mesh in the YSi$_{2-x}$ thin film [10]. Computer simulation using an ordered vacancy structure of YSi$_{2-x}$ as depicted in Fig. 2 (a) generates a diffraction pattern, as shown Fig. 2 (b), with all the essential features of that obtained by the experiment. Fig. 2 (a) illustrates the ordered vacancy structure with vacancies located in the Si sublattice viewed along   the [0001] silicide direction. The diamond indicate the new unit cell for the ordered vacancy structure in YSi$_{2-x}$.

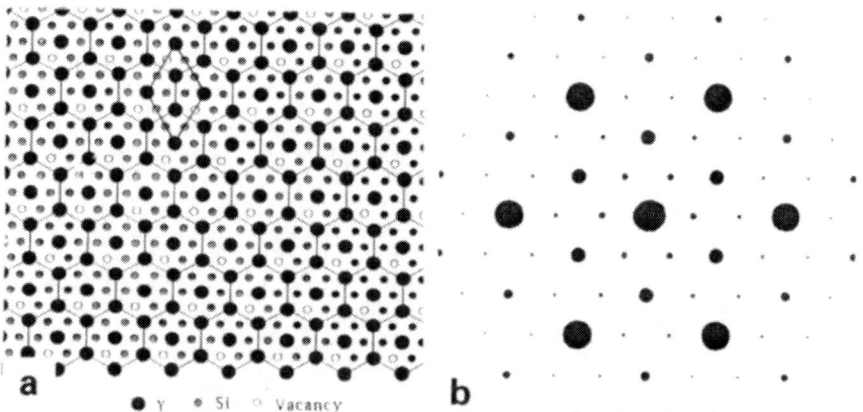

Fig. 2   (a) The structure of YSi$_{2-x}$ with vacancies in Si sublattice plane, viewed along [0001] direction of YSi$_{2-x}$, (b) a calculated diffraction pattern corresponding to the structure of YSi$_{2-x}$ depicted in (a).

The extra superlattice spots at $\frac{1}{3}\{2\bar{1}\bar{1}0\}$ were found to split along the normal of $\{2\bar{1}\bar{1}0\}$ plane for samples annealed at 800 ^{0}C for 5 s. An example is shown in Fig. 3 (a). Streakings were also observed for the split diffraction spots. The extent of streaking was found to depend on the annealing condition. Fig. 3 (b) shows a diffraction pattern of a sample annealed at 1000 ^{0}C for 120 s. The formation of an out-of-step structure, which is analogous to that in the Cu$_3$Au and CuAu ordered alloy systems, in YSi$_{2-x}$ is likely to be responsible for the splitting of superlattice spots [11-13]. The out-of-step structure is characterized by a displacement vector and a parameter M [14]. Here, M is related to the size of the out-of-step structure measured in atomic distance that can be estimated from the separation of split maxima. The value of M is generally not an integer, and the streaking of the split spots in the diffraction pattern, as seen in Fig. 3 suggested that there is a range of M.

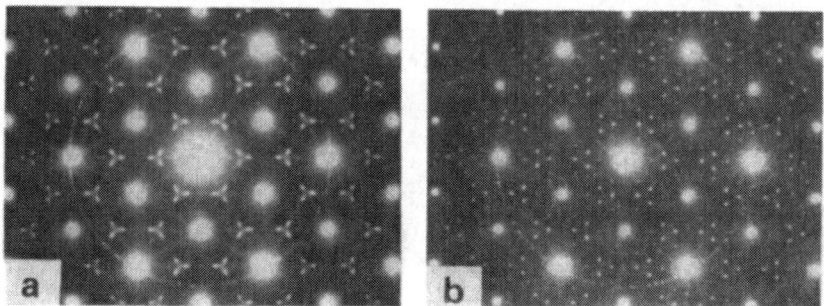

Fig. 3   DP, (a) 800 ^{0}C, 5 s, (b) 1000^{0}C, 120 s.

The displacement vector is a lattice vector of the AlB$_2$ crystal structure, but not a lattice vector of the ordered vacancy structure suggested in figure 2 (a). Figure 4 (a) shows an out-of-step structure for M = 2 which is generated

by shifting every two unit cell as shown in Fig. 2 (a) along [1010] direction
by a [2110] vector. The *parallelgram* outlines the unit cell of the out-of-step
structure of YSi_{2-x} for M = 2. The displacement causes a sinusoidal modulation
for the distribution of vacancy on Si plane in the YSi_{2-x} thin films. The
computed diffraction pattern shown in Figure 4 (b) was obtained by considering
the YSi_{2-x} has a structure shown in Figure 4 (a) and two 60^0 twin symmetry
operations.

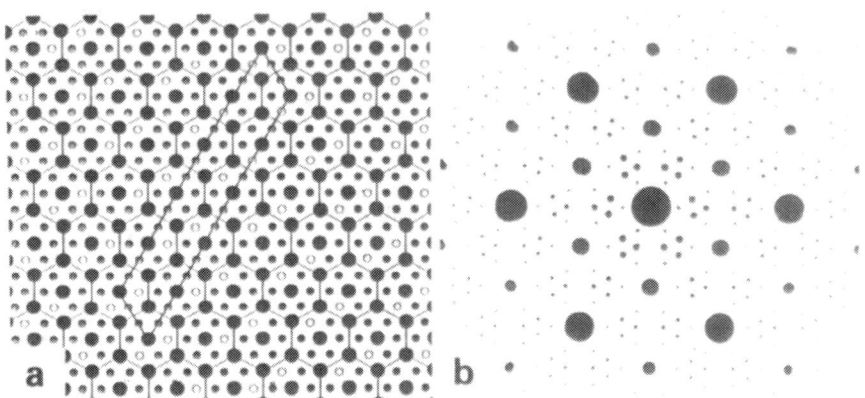

Fig. 4   (a) An out-of-step structure of YSi_{2-x} with M = 2, (b) computed DP
corresponding to the structure of YSi_{2-x} depicted in (a).

The streaking of diffraction pattern shown in Fig. 3 is related to that
there is a range of M. In order to study the kinetics of vacancy ordering,
samples were annealed at 800^0C for 5-120 s. Strakings of split diffraction
spots were found to decrease with annealing time. Figs. 5 (a) to (c) show
diffraction pattern of samples annealed for 20, 40 and 120 s, respectively.
The decrease in the extent of streaking with annealing time implied that M
settled down to an appropriate value for high annealing temperature and/or
long enough annealing time.

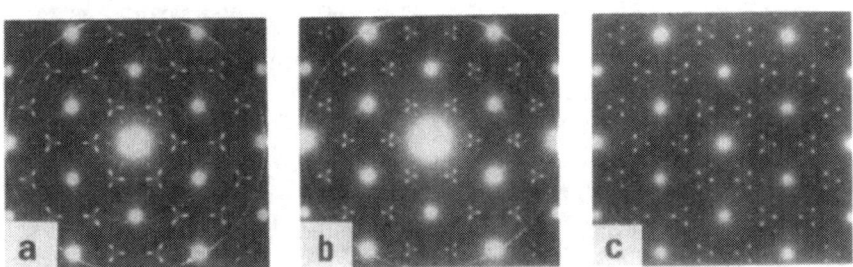

Fig. 5   DP, 800 ^{0}C, (a) 20, (b) 40 and (c) 120 s.

Defects along <1210> were observed in the YSi_{2-x} thin films by high
resolution transmission electron microscopy. An example is shown in Fig. 6.
The defects were believed to be related to the formation of ordered vacancies
in YSi_{2-x} thin films. Figs. 7 (a) and (b) are the dark field images taken with
$\bar{g}$ = (1210) and (1120), respectively. From diffraction contrast analysis, the

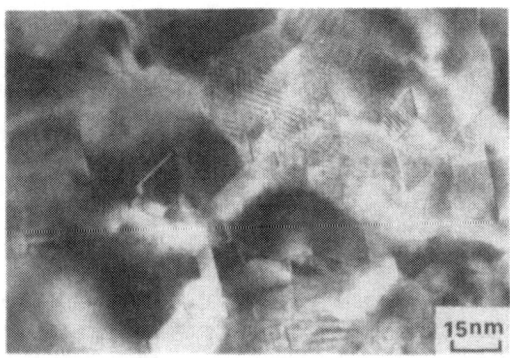

Fig. 6   A HREM image showing the defects along the ⟨1210⟩ direction of
YSi$_{2-x}$.

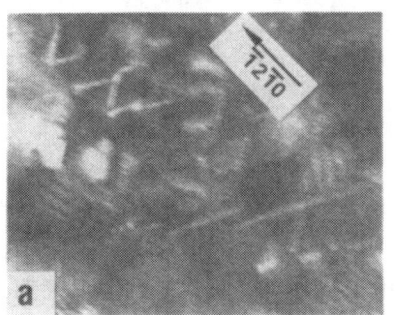
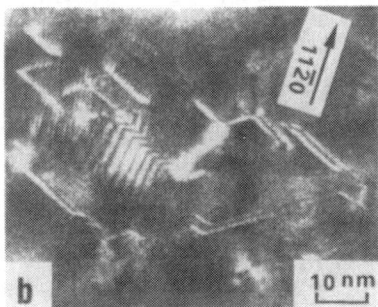

Fig. 7   Dark field images, (a) $\vec{g}$ = (1210), (b) by $\vec{g}$ = (1120).

displacement vector associated with these defects was found to be
perpendicular to the defects. Fig. 8 shows the high resolution image viewed
along the [0001] axis of YSi$_{2-x}$. An atomic displacement across the boundary of
defect was found. The defects were identified to be intrinsic stacking faults.

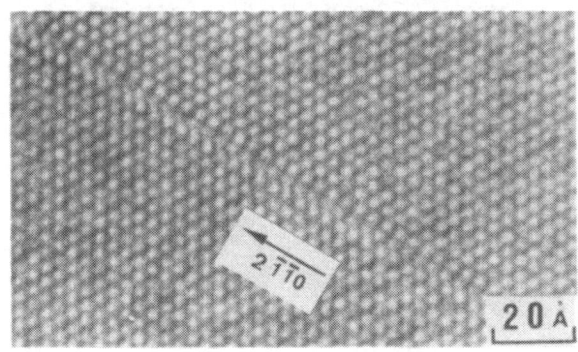

Fig. 8   A HREM image showing atomic displacements across the boundary.

SUMMARY AND CONCLUSIONS

The kinetics of vacancy ordering has been studied by TEM. Structural characterization of YSi_{2-x} thin film has been performed by both high resolution and conventional transmission electron microscopy. Epitaxial YSi_{2-x} with ordered vacancy structure was grown on (111)Si by rapid thermal annealing. The appearance of additional diffraction spots was found to be related to the formation of an ordered vacancy superstructure in the YSi_{2-x} thin films. The splitting of diffraction extra spots was explained in terms of the formation of an out-of-step structure. Streaking of the split diffraction spots in the diffraction pattern is attributed to the presence of out-of-step structures with a range of M values. The M was found to settle down to an appropriate value after high temperature and/or long time annealing. The defects along the <1120> directions of YSi_{2-x} were analyzed to be intrinsic stacking faults.

Acknowledgment
       The research was supported by the Republic of China National Science Council.

References

1.   J.A. Knapp and S.T. Picraux, Appl. Phys. Lett. 48, 466 (1986).

2.   J.E.E. Baglin, F.M. d'Heurle, and C.S. Petersson, J. Appl. Phys. 52, 2841 (1981).
3.   K.N. Tu, R.D. Thomson, and B.Y. Tsaur, Appl. Phys. Lett. 38, 626 (1981).
4.   R.D. Thompson, B.Y. Tsaur, and K.N. Tu, Appl. Phys. Lett. 38, 535, (1981).

5.   E.E. Baglin, F.M. d'Heurle, and C.S. Petersson, Appl. Phys. Lett. 36, 594 (1980).
6.   J.A. Knapp, S.T. Picraux, C.S. Wu, and S.S. Lau, Appl. Phys. Lett. 44, 747 (1984).
7.   J.A. Knapp, S.T. Picraux, C.S. Wu, and S.S. Lau, J. Appl. Phys. 58, 3747 (1985).

8.   H. Norde, J. de Sousa Pires, F.M. d'Heurle, F. Pasavento, C.S. Petersson, and P.A. Tove, Appl. Phys. Lett. 38, 865 (1981).
9.   M. Gurvitch, A.F.J. Levi, R.T. Tung, and S. Nakahara, Appl. Phys. Lett. 51, 311 (1987).
10.   R. Baptist, S. Ferrer, G. Grenet, and H.C. Poon, Phys. Rev. Lett. 64, 311 (1990).
11.   S. Ogawa and D. Watanabe, J. Phys. Soc. Japan, 9, 475 (1954).
12.   S. Ogawa, D. Watanabe, H. Watanabe, and T. Komoda, Acta Cryst. 11, 872 (1958).
13.   K. Fujiwara, J. Phys. Soc. Japan, 12, 7 (1957)
14.   R. Sato and H. Dio, Acta Cryst. 14, 763 (1961).

PART IV

Crystallization, Amorphization and Epitaxy of Semiconductors

A CONTINUOUS HETEROGENEOUS MODEL FOR THE CRYSTALLINE TO
AMORPHOUS TRANSITION IN ION IMPLANTED SEMICONDUCTORS :
RELATIONSHIP TO THE "CRITICAL DAMAGE ENERGY DENSITY" MODEL

C. VIEU, A. CLAVERIE[*,***], J. FAURE[**] AND J. BEAUVILLAIN[***]

LMM/CNRS, 196 avenue H. Ravera, 92220 Bagneux, France
[*]Lawrence Berkeley Laboratory, 1 cyclotron road, Berkeley CA
94720, USA
[**]Laboratoire de Microscopie Electronique, 21 rue C. Ader 51100
Reims, France
[***]CEMES/CNRS, 29 rue J. Marvig, 31055 Toulouse, France

ABSTRACT

A method is presented for calculating amorphization doses
of ion implanted semiconductors, based on a continuous heterogeneous description of damage accumulation. This new approach
is compared to the classical "critical damage energy density"
(CDED) model. For high dose implantations the equivalence of
both descriptions is formally established. It is proposed that
the main limitation of the CDED model lies in the linear
additivity of damage rather than the homogeneous damage build-up.

INTRODUCTION

During the two last decades a large number of papers have
been devoted to the crystalline to amorphous (c-a) transition
in ion implanted semiconductors, dealing both with the
experimental and theoretical point of view. However, the model
proposed by Stein et al. [1] in the early stage of ion
implantation technology, still remains the most universal
approach to the c-a transformation. This approach, often
referred to as the "critical damage energy density" (CDED)
model, assumes that the damage nucleates homogeneously as a
result of Frenkel defect interactions, and that a crystal
containing a critical density of defects will spontaneously
relax into the amorphous state. By coupling experimental studies
on the amorphous phase extension with calculations of damage
depth distributions, the critical damage energy density (E_{dc}),
which corresponds to the concentration of defects inducing the
c-a transformation, has been obtained for several materials.
Severe limitations to this description have been recently
observed for room temperature implantations of light ions in
silicon [2][3]. These deviations are inherent in the basic
assumptions of this model. Indeed, the phenomenological
macroscopic threshold E_{dc} does not take into account the
microscopic nature of damage deposition through collision
cascades, which obviously depends on the incident ion, but
also on ion flux and substrate temperature.

On the other hand, heterogeneous models [4] consider the role of individual ionic impacts in the damage build-up. Although they are expected to give a more realistic description of the c-a transition, they suffer from the lack of available procedures to calculate amorphization doses. The aim of this paper is to show how a continuous heterogeneous description of damage accumulation can be adapted to predict the depth extension of an amorphous layer. This so-called continuous model, which is a generalization of previous discrete heterogeneous descriptions, is applied for the first time to the c-a transition, and is compared to the homogeneous CDED model.

THEORETICAL FRAMEWORK

For implementing an heterogeneous model, we need a full description of the three-dimensional (3D) damage energy deposition. We have proposed a procedure [5] for calculating the spatial distribution $E_d(x,r)$ which represents the damage energy deposited by an incident ion at a depth x and a lateral distance r from its point of impact and its initial direction. This calculation, which is a 3D extension of the earlier work of Brice [6] in many aspects, generates a point response function that describes the spatial damage generated by an "average ion".

Figure 1: *3D damage energy mapping for three systems of implantation : Light (-a-), Intermediate (-b-) and Heavy (-c-) ions incident on silicon. Also shown on the left side are the 1D damage depth profiles obtained after integration of the 3D distributions. Note the difference between the depth position of the 1D damage peak (78nm) and the depth of the core of the 3D mapping (10nm) in the light ion case (-a-).*

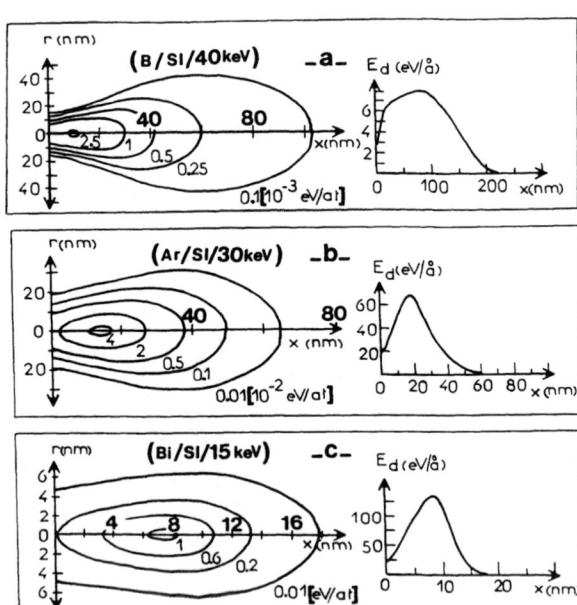

Figure 1 represents a damage energy mapping for three systems (B/Si/40keV) (B$^+$ ion incident on silicon with an energy of 40keV), (Ar/Si/30keV) and (Bi/Si/15keV) representing respectively light, intermediate and heavy ion cases in the LSS classification. Several lines called isodamages connect the points of constant damage energy density. These contours describe the "average cascades", in the sense that they represent a statistic of the individual cascades which can be obtained through Monte-Carlo simulations. Also shown in figure 1 are the one-dimensional (1D) damage depth distributions which are used for the homogeneous CDED model.

Under bombardment the damage level at a point of the target is described by a dimensionless parameter $\theta = E_d(x,r)/\theta_a$, where θ_a stands for the energy density inducing the local c-a transition. As suggested by Thompson and Walker [7] this parameter can be assumed to be the heat of melting of the material (0.8 eV/at for silicon). For $\theta = 0$, the region under consideration is undamaged, for $\theta = 1$, this zone is amorphous and for greater values $\theta > 1$, it is assumed to stay in the amorphous phase. The lateral extent of damage can be accurately described by a Gaussian expression

$$\theta = \theta_{max}(x) \exp - \frac{r^2}{2\sigma(x)^2} \quad (1)$$

where $\sigma(x)$ is the straggling of the lateral damage distribution at a depth x, and $\theta_{max}(x)$ the ratio $E_d(x, r=0)/\theta_a$. These quantities are calculated by our 3D procedure. Because of this Gaussian shape, the total area damaged by one incident ion can be considered as infinite. Let $\rho(x,\theta)$ be the damaging function representing, for a plane parallel to the surface at a depth x, the area of the zone damaged at a level between θ and $\theta + d\theta$. From equation 1) we can deduce the area damaged at a level greater than θ

$$S(x,\theta) = 2\pi\sigma^2(x) \ln\left(\frac{\theta_{max}(x)}{\theta}\right) \quad (2)$$

The damaging function $\rho(x,\theta)$ is then given by the derivative function of $S(x,\theta)$

$$\rho(x,\theta) = \left| \frac{dS(x,\theta)}{d\theta} \right| = \frac{2\pi\sigma^2(x)}{\theta} \quad (3)$$

Finally, the accumulation function $\alpha(x,\theta,\Phi)$, which represents at a depth x the fractional area damaged at a level lying between θ and $\theta + d\theta$ after implantation of a dose Φ, can be described by an evolution equation. When the damage of successive impacts is assumed to be additive, we obtain

$$\frac{\partial \alpha(x,\theta,\Phi)}{\partial \Phi} = \int \rho(x,\theta')\{\alpha(x,\theta-\theta',\Phi) - \alpha(x,\theta,\Phi)\}d\theta' \quad (4)$$

This equation is similar to that proposed by Jimenez-Rodriguez et al. [8]; but with no source term because of an infinite total damaged area. By the Fourier transform technique, we obtain [9]

$$\alpha(x,\theta,\Phi) = \frac{2}{\sqrt{2\pi}\sqrt{\nu\Phi}\left\{1 + erf\left(\frac{\epsilon\Phi}{\sqrt{2\nu\Phi}}\right)\right\}} \exp{-\frac{(\theta - \epsilon\Phi)^2}{2\nu\Phi}} \quad (5)$$

with $\epsilon = 2\pi\sigma^2(x)\theta_{max}(x)$ and $\nu = \pi\sigma^2(x)\theta_{max}^2(x)$

The expression of the fractional amorphous area is then

$$f_a(x) = \int_1^{\infty} \alpha(x,\theta,\Phi)d\theta = \frac{1 + erf\left(\frac{\epsilon\Phi - 1}{\sqrt{2\nu\Phi}}\right)}{1 + erf\left(\frac{\epsilon\Phi}{\sqrt{2\nu\Phi}}\right)} \quad (6)$$

A criterion for the formation of a continuous amorphous layer must be adopted by analogy with discrete models [4]. The amorphization dose Φ_c is obtained when 90% of a layer parallel to the surface belongs to the amorphous phase (f_a=0.9). Under this condition, at each depth within the target Φ_c is calculated through equation (6), and a dose/depth phase diagram can be built.

RESULTS

Figure 2 presents the dose/depth phase diagrams obtained for the three systems presented before. The continuous and CDED models are compared. The CDED calculation was carried out with a critical damage energy density E_{dc}= 0.8 eV/at, identical to the microscopic threshold θ_a used in the continuous procedure. For the light ion case (figure a]) both calculations give the same diagram. For the intermediate system (figure b]) at high doses the CDED and continuous models converge, while around the minimal amorphization dose a slight divergence is observed. The continuous model gives lower amorphization doses. This difference is stressed for the heavy system (figure c]), where the minimal dose for generating a continuous amorphous layer is around 2.5 times lower for the continuous calculation. However, for high doses the continuous and CDED predictions merge. Briefly, we observe that when the average damage density of the cascades is low the two calculations are equivalent. This occurs at any depth for light systems and only at high doses for heavy systems.

Moreover, as previously discussed [5], for light systems the peak of the damage depth distribution does not coincide with the position of the cascades of maximal damage density. For example in figure 1a) the core of the mapping (at a depth x=10nm) is shifted towards the surface compared to the peak of the damage depth distribution (for x=78nm). Nevertheless,

the continuous approach, although based on a 3D calculation
of damage, provides a diagram agreeing with the CDED model
which proceeds from a 1D calculation.

These observations can be qualitatively explained considering that, for cascades of low damage densities, a high
number of overlaps is required to create a continuous amorphous
layer. This results in an integration over numerous impacts
so logically the heterogeneous approach tends to an homogeneous
damage build-up.

Figure   2:
*Dose/Depth   phase
diagrams for light
a], intermediate b]
and heavy c] ions
in silicon. The
amorphous (-a-) and
crystalline   (-c-)
phases are separ-
ated by a curve
which represents
the depth variation
of the amorphiz-
ation dose. The
predictions of the
CDED and continuous
models are pres-
ented. The macros-
copic   critical
damage   energy
density Edc and the
microscopic amor-
phization thresh-
old θa were fixed
at 0.8 eV/at.*

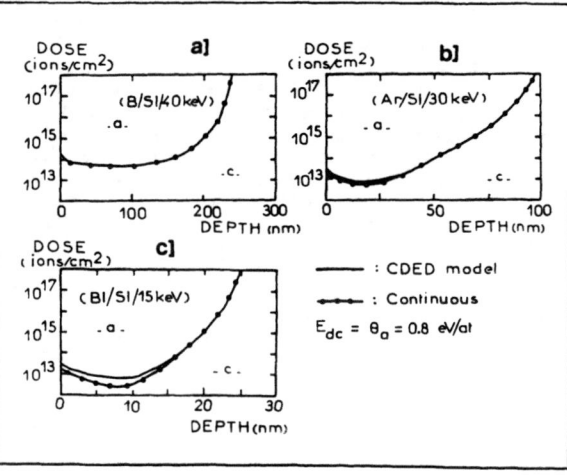

This behaviour can be predicted from equation (6). In the
limit of average cascades of low damage energy density, we
can show that the quantity $(2\nu\Phi)^{1/2}$ is very small compared to
$\epsilon\Phi$. Consequently, the fractional amorphous area tends to a
step function : $f_a=0$ for $\epsilon\Phi<1$; $f_a=0.5$ for $\epsilon\Phi=1$ and $f_a=1$ for
$\epsilon\Phi>1$. According to our conventions the amorphization dose is
then obtained for $\epsilon\Phi_c=1$. This relation leads to

$$\Phi_c 2\pi\sigma^2(x)\theta_{max}(x) = 1 \quad (7)$$

Moreover, the damage depth profile is obtained by integration
of the 3D distribution according to

$$E_d(x) = 2\pi \int_0^\infty rE_d(x,r)dr = 2\pi E_d(x,r=0)\sigma^2(x) = 2\pi\theta_{max}(x)\theta_a\sigma^2(x) \quad (8)$$

Substituting this relation into equation (7) gives

$$\Phi_c(x)E_d(x) = E_{dc} = \theta_a \quad (9)$$

DISCUSSION

At high doses, the mathematical equivalence of the homogeneous CDED and the continuous heterogeneous models has been established, when the critical damage energy density E_{dc} equals the local amorphization threshold θ_a. For low doses the divergence of the two predictions is small. This heterogeneous effect, is noticeable only in the extreme cases of heavy ions at low energies. Regardless of the microscopic model of damage energy deposition, both models, this minor effect aside, produce similar results because the quantity of interest is the amount of defects created on a large scale after a high number of impacts. One can better understand why the very simple CDED model was so successfully applied to many experiments. However, this also sheds light on the limitations of the CDED model. Rather than the homogeneous description, the relevant basic assumption of the CDED model is the linear additivity of damage between successive ionic impacts. This assumption can be easily relaxed in our continuous heterogeneous approach, by a slight modification of the evolution equation (4) [9]. This should explain why the phenomenological amorphization threshold E_{dc} deduced from experiments (4 eV/at for low temperature implantations in silicon), is an order of magnitude higher than the heat of melting of the implanted material (0.8 eV/at for silicon). Moreover, an universal self-consistent model of the c-a transition of implanted semiconductors based on a continuous heterogeneous description of damage accumulation can be planned, taking into account the influence of parameters as complex as temperature and flux. Comparisons with experiments where the classical CDED model fails should be very promising.

REFERENCES

1. H.J Stein, F.L Vook, D.K. Brice, J.A. Borders and S.T. Picraux, in Proceedings of the 1st International Conference on Ion Implantation (Gordon and Breach, London, 1971), P. 17.
2. K.S. Jones, D.K. Sadana, S. Prussin, J. Washburn, E.R. Weber and W.J. Hamilton, J. Appl. Phys. 63 (5), 1414 (1988).
3. A. Claverie, A. Roumili, N. Gessin and J. Beauvillain, MRS Proc. Symp A Boston Fall (1990), in print.
4. G. Carter and R.P. Webb, Radiat. Eff. 42, 159 (179).
5. C. Vieu, A. Claverie, J. Faure and J. Beauvillain, Nucl. Instr. Meth. B 36, 137 (1989).
6. D.K. Brice, J. Appl. Phys. 46 (8), 385 (1975).
7. D.A. Thompson and R.S. Walker, Radiat. Eff. 36, 36 (1978).
8. J.J. Jimenez-Rodriguez, A. Gras-Marti and G. Carter, Phys. Stat. Sol. (a) 81, 267 (1984).
9. C. Vieu, A. Claverie, J. Faure and J. Beauvillain, to be published.

CRYSTAL NUCLEATION IN AMORPHOUS Si FILMS ON   GLASS SUBSTRATE BY
Si$^+$ ION IMPLANTATION

Tomonori Yamaoka, Keiji Oyoshi, Takashi Tagami,
Yasunori Arima, and Shuhei Tanaka
Tsukuba Research Laboratory, Nippon Sheet Glass Co., Ltd.,
5-4 Tokodai, Tsukuba City, Ibaraki, Japan

Abstract

    Crystallization of amorphous Si films on a glass substrate
using Si$^+$ ion implantation is investigated. 100keV and 180keV
Si$^+$ ion implantations into 600nm-thick amorphous Si layers
crystallize half and almost all of the film thicknesses,
respectively. This result demonstrates that crystallization by
ion implantation, which contains both crystal nucleation and
grain growth, is due to ion-solid interaction, and not to "pure"
thermal effect by ion beam heating. Furthermore, two distinct
regions are observed in transmission electron microscopy
investigation of grain size at different depths of crystallized
Si/SiO$_2$ multi-layer specimens. The deep region below the
projected range is composed of grains smaller than in the
shallow region. This result is strongly related with crystal
nucleation and growth kinetics by ion implantation.

Introduction

    Solid phase epitaxy of amorphous Si (a-Si) layers on single
crystal Si using the implantation of high energy ions has been
investigated[1-5]. In these reports, the activation energy of
epitaxial growth was estimated as 0.2-0.3eV[2, 3, 5]. This is
1/10 that found in, so called "pure" thermal annealing. The
substrate temperature needed for growth during ion implantation
was very low (<300°C). It has been considered that some kinds of
defects generated by the ion implantation contribute to these
phenomena. Recently, Yamaoka et al. have reported that
crystallization of a-Si films on glass substrates is achieved
by Si$^+$ ion implantation through crystal nucleation[6]. Im and
Atwater have also reported Xe$^+$ ion beam enhanced nucleation and
estimated the activation energy of the nucleation as 3.9eV[7].
Furthermore, Spinella et al. have reported that only grain
growth under Kr$^+$ ion irradiation is observed in a-Si films
containing pre-existing crystal grains, but no nucleation[8]. It
is not clear what mechanism decides crystal nucleation under
ion irradiation.
    In this paper, to make clear the effect of ion-solid
interaction, accelerated energy dependence of crystallized depth
and grain size of crystallized Si layers in Si/SiO$_2$ multi-layer
specimens are investigated by transmission electron microscopy.

Experiments

Specimens to be crystallized in an experiment in which the acceleration energy was to be varied were prepared through a process described as follows. Poly-Si films of 600-nm-thickness were deposited on silica glass by the atmospheric CVD method and then amorphized by two-step Si^+ implantation (260keV, $5X10^{15}$ions/cm^2 and 100keV, $1X10^{16}$ions/cm^2). Si/SiO$_2$ multi-layer specimens were prepared by the following procedure with a-Si layers and SiO$_2$ layers deposited, one after the other, by a D.C and R.F sputtering method, respectively. The substrate temperature was held at 300 °C during the deposition. Both a-Si and SiO$_2$ layers contained phosphorus as impurity, because the sputtering target used in the deposition contained phosphorus. After the samples were cut, 1cmX1cm, crystallization was performed by Si^+ ion implantation.

In the experiment on the acceleration energy dependence, 600nm-thick a-Si films were irradiated with either 180keV Si^+ at a current density of 10μA/cm^2 or 100keV Si^+ at 20μA/cm^2; the effect of ion beam heating was almost the same under the two conditions. Si/SiO$_2$ multi-layer specimens were irradiated with 180keV Si^+ at 10μA/cm^2. Both experiments were performed without external heating of the substrate.

Plane and cross-section transmission electron microscopy (TEM) observation were carried out for crystallized specimens. Plane TEM observation of multi-layer specimens was carried out, after the samples were obliquely polished.

Results and Discussion

Figs. 1(a) and (b) show cross-sectional TEM images of specimens crystallized by 100keV or 180keV Si^+ ion implantation, respectively. As shown in Fig.1(c) which describes the distribution of the implanted ions accelerated by the respective energies mentioned, the crystallized depths in both cases correspond to the depths where the implanted ions reached with respective accelerated energies. This crystallization must have proceeded through crystal nucleation, because the Si film did not contain any crystal phase in the initial stage.

The temperature rise during the implantation under the condition in Fig. 1(b) (1.8W/cm^2 beam heating) was estimated as 500°C based on the following assumption. The energy loss was due only to thermal radiation from the silica glass substrate with radiation from the Si surface not taken into account. If the crystallization was due to an effect of the "pure" thermal annealing by ion beam heating, the whole thickness of films should be crystallized from the thermal diffusion consideration[6]. This demonstrates that ion-solid interaction plays an important role in crystallization by ion implantation.

The crystallization was completed in a very short time, equal to the implantation time, although the substrate temperature in the crystallization by ion implantation was lower than that in the "pure" thermal annealing (<600°C). The time spent for the crystallization in Fig. 1(b) was 30min in this experiment. Furthermore, it is clear that a larger beam current would have made this time even shorter.

To investigate the grain size at certain depths in the crystallized films, separation of Si films using SiO_2 was proposed. In the multi-layer structure as shown in Fig. 2 (a), the thicknesses of the Si and SiO_2 layers are 80 and 50 nm, respectively. The number of Si and SiO_2 layers are 5 and 4, respectively. The total thickness of the multi-layers is, therefore, 600nm.

Fig. 2(a) shows the cross-sectional TEM image of a crystallized multi-layer specimen under the same condition of ion implantation as in Fig. 1(b). As can be seen, the 1st, 2nd, 3rd and 4th-layer defined in Fig. 2 (a) are crystallized. Only the 5th-layer is not, because there is no ion-solid interaction. Fig. 2(b) shows a respective distribution of implanted 180keV Si^+ ions into either the Si or the SiO_2 films. As can be seen in this figure, the penetration depths of implanted ions are almost the same in both. It is confirmed that the crystallized depth of the Si/SiO_2 multi-layer specimens corresponds to the depth that the implanted ions can reach.

Figs. 3(a), (b), (c) and (d) show plane TEM images of the 1st, 2nd, 3rd and 4th-layer, respectively. These represent the depths through the film. As seen in the figures, the four layers are classified into two distinct regions of grain size. One region, composed of large grains (LG), is observed in the 1st and the 2nd-layer as shown in Figs. 3(a) and (b). They locate above the projected range, as can be seen in Fig. 2. The other region, composed of small grains (SG), is observed in the 3rd and the 4th-layer as shown in Figs. 3(c) and (d). They locate below the projected range, as can be seen in Fig. 2. No difference between the two layers in each region was recognized in this experiment, because observed grain boundaries were not clear. Therefore, to clarify the differences among all four layers, an investigation into how the number and size of grains depend on ion dose would have to be conducted.

The maximum grain size in the LG and the SG regions can be roughly estimated as ~700 and ~70 nm, respectively. Crystallization kinetics in each region would therefore be quite different.

In solid phase crystallization, the final grain diameter d_g can be described as the following equation[9],

$$d_g = [(4 / \pi) A_g]^{1/2} \tag{1}$$

$$A_g = \frac{(\pi / 3)^{1/3}}{\Gamma (4 / 3)} \left(\frac{v_g}{e r_n} \right)^{2/3} \tag{2}$$

where r_n, v_g and e are the nucleation rate, the growth rate and the thickness of a film, respectively. $\Gamma()$ is the general Γ function. The characteristic crystallization time t_c is defined

174

in Ref. 9, as follows,

$$t_c = [(\pi / 3) v_g{}^2 e r_n]^{-1/3}$$

(3)

t_c is qualitatively equal to the time needed for complete crystallization.

Ignoring $\Gamma(\frac{4}{3})$ in equation (2), one order of magnitude difference in d_g is caused by three orders of that in v_g/r_n, as seen from equations (1) and (2). If the grain size 1/10 in the SG region compared with that in the LG region is only due to either 10^3 of r_n or $1/10^3$ of v_g in the SG region compared with that in the LG region, the SG region should have 1/10 or 10^2 of t_c compared with the LG region, respectively, as seen from equation (3). However, the time for complete crystallization may not be so different between each region, because co-existing amorphous and crystal phases were observed in every layer of multi-layer specimens, partially crystallized by the ion implantation in our experiments. Therefore, we believe that the SG region may have high nucleation and low growth rate compared with the LG region.

Grain size generally decreases in standard "pure" thermal treatment, as the treatment temperature increases. This is due to the high activation energy of the nucleation compared with that of the growth. Meanwhile both the nucleation and growth rate increase with rise of temperature. However, as was discussed above, the SG region has high nucleation and low growth rate compared with the LG region. Therefore, the two distinct grain sizes in the multi-layer crystallized by ion implantation may be attributed to quite different kinetics from those of the standard "pure" thermal treatment.

To explain these results, we propose one possibility for the kinetics of nucleation and grain growth by ion implantation, as described below.

In the investigation of defects generated by ion implantation, it was reported that a heavily disordered region existed in the deep region below the projected range[10]. This phenomena was emphasized at relatively high temperature and suggested that the deep region below the projected range may have had different kinetics of defects generation and annihilation from the surface region. Furthermore, it can be assumed that atom movement is enhanced in the disordered region. For instance, the deep disordered region probably has a high concentration of mobile atoms compared with the surface region. On the other hand, since nucleation needs some critical number of atoms to form a stable nucleus, the nucleation rate is very sensitive to concentration of atoms contributing to nucleation. Therefore, a high nucleation rate in the SG region may be attributed to heavy disorder in the deep region, which may result in a high concentration of mobile atoms.

However, it should be considered why the high concentration of mobile atoms in the SG region results in low growth rate. It is generally thought that regrowth rates in solid phase epitaxy enhanced by ion irradiation are decided in competition between amorphization and crystallization[3]. As a result, amorphization proceeds below the critical temperature. Therefore, the

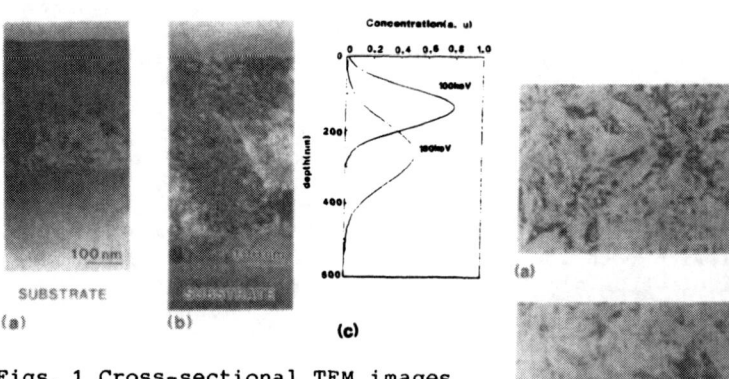

Figs. 1 Cross-sectional TEM images of Si films crystallized by (a) 100keV and (b) 180keV Si$^+$ ion implantation. Calculated distribution of Si$^+$ ions accelerated by respective energy is shown in (c).

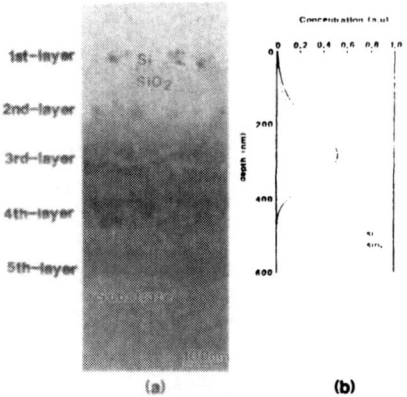

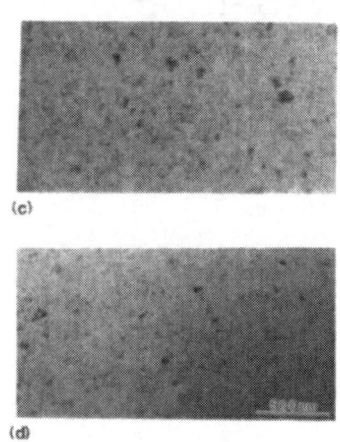

Fig. 2 (a) Cross sectional TEM image of a multi-layer specimen crystallized by 180keV Si$^+$ ion implantation. (b) Calculated distribution of 180keV Si$^+$ ions implanted into Si(solid line) SiO$_2$(dashed line) films.

Figs. 3 Plan TEM images of a multi-layer specimen crystallized by 180keV Si$^+$ ion implantation; (a) 1st-layer, (b) 2nd-layer, (c) 3rd-layer and (d) 4th-layer.

low growth rate in the SG region compared with that in the LG region may be attributed to the large amorphization effect in the SG region, because energy deposition of implanted ions is dominated by elastic collision in the SG region. This amorphization effect may be caused by the disorder discussed above.

The above discussion can be summarized as follows. (1) Since nucleation needs some critical number of atoms to form a stable nucleus, the nucleation rate is very sensitive to concentration of mobile atoms and the SG region has a high nucleation rate compared with the LG region. (2) Since defect generation results in transformation from crystal to amorphous, growth rate is low in a heavily disordered region. The SG region has a lower growth rate than the LG region.

Possible effects of electronic energy deposition should be looked into, because electronic energy deposition must somehow be related with both generation and annihilation of defects. The effects of electronic energy deposition is not clear.

Conclusions

Crystallized depths by Si^+ ion implantation at 100keV and 180keV correspond to the depths that implanted ions can reach with their respective accelerated energies. This demonstrates that crystallization through nucleation by Si^+ ion implantation is strongly related with ion-solid interaction. Furthermore, investigation of grain size along the film depth using Si/SiO_2 multi-layer specimens show two distinct regions. One region is composed of large grains($\sim$700nm) and the other region is composed of small grains ($\sim$70nm). It is suggested that the difference between the two regions may be attributed to the different concentrations of mobile atoms caused by the different energy depositions of the implanted ions.

Reference

1. J. Nakata and K. Kajiyama, Appl. Phys. Lett. 40, 686 (1982).
2. J. S. Williams, R. G. Elliman, W. L. Brown, and
   T. E. Seidel, Phys. Rev. Lett. 55, 1482 (1985).
3. A. Leibrich, D. M. Maher, R. V. Konell, and W. L. Brown,
   Nucl. Instrum. Methods Phys. Res. B19/20, 457 (1987).
4. J. M. Poate, J. Linnros, F. Priolo, D. C. Jacobson, and J. L.
   Batstone, Phys. Rev. Lett. 60, 1332 (1988).
5. N.Kobayashi,H.Kobayashi,and Y. Kumashiro, Proc. 12th
   Symp. Ion Sources Ion-Assisted Technol., Tokyo, 523 (1989)
6. T. Yamaoka, K. Oyoshi, T. Tagami, Y. Arima, K. Yamashita, and
   S. Tanaka, Appl. Phys. Lett. 57, 1970 (1990).
7. J. S. Im and H.A.Atwater,Appl. Phys. Lett. 57, 1766(1990).
8. C.Spinella,S. Lombardo, and S. U. Campisano, Appl. Phys.
   Lett. 55, 109 (1989).
9. R. B. Iverson and R. Reif, J. Appl. Phys. 62, 1675 (1975).
10. L.Csepregi, E. F. Kennedy, S. S. Lau, and J. W. Mayer,
    Appl. Phys. Lett. 29, 645 (1976).

SUBSTRATE EFFECTS ON THE KINETICS OF SOLID PHASE CRYSTALLIZATION IN a-Si

L.HAJI[*], P.JOUBERT[*], M.GUENDOUZ[**], N.DUHAMEL[**] and B.LOISEL[**]
*Laboratoire Composants et Systèmes de Visualisation, Université de Rennes I,
I.U.T. de Lannion, 22302 Lannion Cedex, France
**Centre National d'Etudes des Télécommunications, LAB/OCM,
22301 Lannion Cedex, France

ABSTRACT

The effect of substrate nature on the solid phase crystallization at 600 °C of a-Si deposited by low pressure chemical vapor deposition is investigated by x-ray diffraction and transmission electron microscopy. The nucleation rate varies slightly resulting to a weak variation in the final grain sizes as a function of the substrate type. In all cases the grain growth mode is found to be three dimensional. In contrary, a drastic effect of the substrate is observed for films deposited by plasma enhanced CVD. Fast crystallization is obtained on indium tin oxide (ITO) resulting to small grain poly-Si, whereas the crystallization is retarded on glass leading to an increase in the grain size.

INTRODUCTION

Polycrystalline silicon thin film transistors (TFT's) on glass substrates have attracted a great importance in active matrix liquid crystal displays (AMLCD). Poly-Si offers much higher carrier mobilities, up to 100 $cm^2.V^{-1}.s^{-1}$, than hydrogenated amorphous silicon[1]. Nowadays large grain poly-Si is achieved by thermal crystallization of amorphous silicon (a-Si) at 600 °C. The a-Si starting material is obtained directly by low pressure chemical vapor deposition (LPCVD), plasma enhanced chemical vapor deposition (PECVD) or via implantation of as-deposited poly-Si films.

Crystallization of these a-Si films occurs through nucleation and growth of crystallites during the thermal annealing. In the case where the a-Si film is deposited on patterned substrates, the crystallization process may be different according to the substrate nature. To our knowledge, very little is known so far on the substrate effects on the kinetics of crystallization in a-Si. In this report we will present the kinetics of crystallization of LPCVD and PECVD a-Si films deposited on glass and on glass covered with indium tin oxide or silicon nitride.

EXPERIMENTAL PROCEDURE

Undoped a-Si films 500-nm-thick were deposited by thermal decomposition of pure silane (LPCVD) at 535°C on glass substrate (Hoya NA40) uncovered or covered by a 230-nm-thick ITO film or a 250-nm-thick silicon nitride film deposited by PECVD. 100-nm-thick a-Si:H films were also deposited by PECVD at 180°C in a RF diode-type apparatus on patterned ITO onto glass substrate. These films were heated for dehydrogenation from RT to 500°C with a 1°C min^{-1} heating rate. Crystallization annealing were performed at 600°C in nitrogen ambient.

Kinetic of crystallization and texture of the films were deduced from x-ray diffraction measurements performed in a symmetrical θ-2θ mode. To quantify the texture, the orientational factors O_{hkl} were deduced from the measurements of the intensity of the three peaks {111}, {220} and {311}, as previously reported [2]. The final grain size was measured on transmission electron micrographs by using the procedure described in Ref.[3].

RESULTS AND DISCUSSION

The nucleation in *a*-Si films has been investigated by transmission electron microscopy (TEM). The density of crystallites was measured for films which were subjected to short-time annealing. Figure 1 shows the dependence of the nucleation density on the annealing time for the film deposited by LPCVD on glass substrate and annealed at 600 °C. A transient period 25 minutes is observed for the nucleation regime. This corresponds to the time at which the first crystallites larger than 24 nm were detected by TEM. This suggests that the time before the onset of nucleation is very short. After this transient period, the density of crystallites increases linearly with a steady-state nucleation rate (I_s) of 6×10^6 cm^{-2}.min^{-1}. This rate will be compared below with those obtained for films deposited at the same conditions on other type of substrates. On the other hand, we have found by TEM observation in the cross sectional mode, that the nucleation always starts at the film/substrate interface.

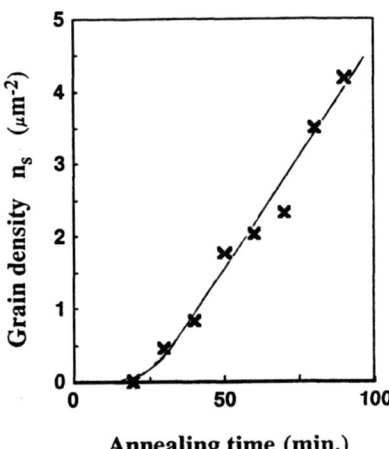

Fig. 1 Grain density (per area unit) vs annealing time for a LPCVD film deposited on glass and annealed at 600°C .

Crystallization kinetics for this film, as deduced from x-ray diffraction analysis, are reported in Fig.2. The variations of the diffracted intensities for the only detected three peaks {111}, {220} and {311} are plotted as a function of the annealing time. As it can be seen, the {111} reflection is the most intense, but the film is found to be not strongly textured as discussed in the following. However, considering that the crystallized volume increases in the same proportion for the three reflections, the crystallization kinetics could be deduced from the only variation of the {111} reflection.

According to Avrami's model [4], for hemispherical shaped grains, the growth mode is determined from the variation of ln { $1/[1-x(t)]$ } with annealing time (Fig.3), where $x(t)$ is the crystallized volume fraction deduced from the ratio of the {111} integrated intensities at annealing time t and at the end of the total crystallization of the film. The slope of such a curve is found to be equal to 4 which corresponds to a three dimensional grain growth mode. In this case the volume

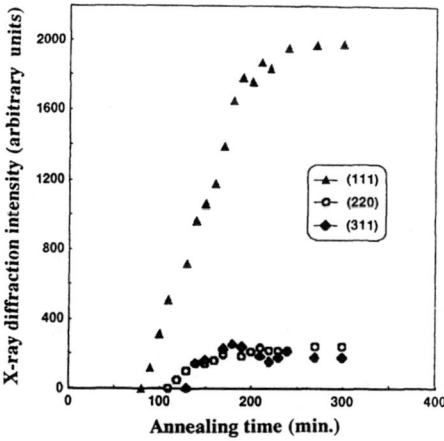

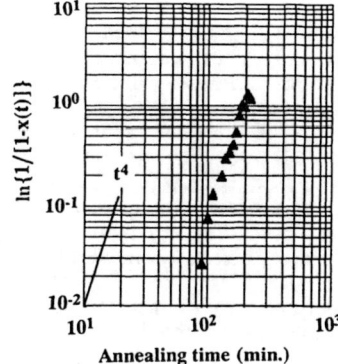

Fig. 2 X-ray diffraction intensities vs annealing time for a LPCVD film deposited on glass and annealed at 600 °C.

Fig. 3 Variation of the crystalline fraction x vs annealing time for a LPCVD film deposited on glass and annealed at 600 °C.

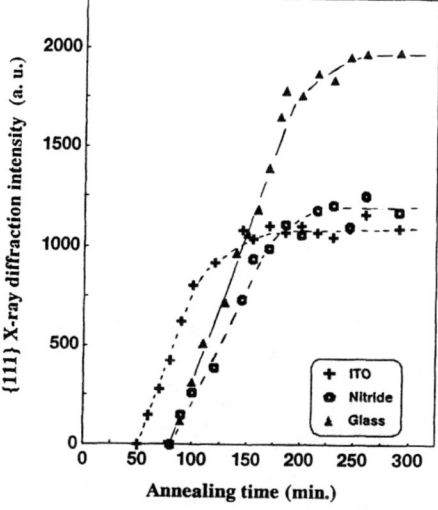

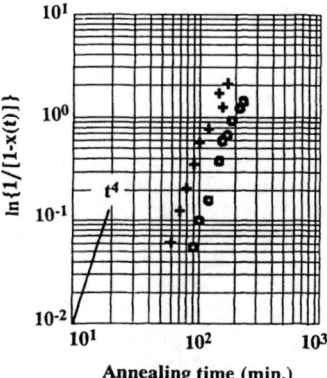

Fig. 4 {111} X-ray diffraction intensities vs annealing time for a LPCVD film deposited on glass (▲), silicon nitride (◓), ITO (+) and annealed at 600 °C.

Fig. 5 Variation of the crystalline fraction x vs annealing time for a LPCVD films deposited on ITO (+) and silcon nitride (◓) and annealed at 600°C.

fraction is expressed by the relation

$$x(t) = 1-\exp[-(\pi/6)(I_s/e)V_g^3t^4] \tag{1}$$

where e is the film thickness, V_g the growth velocity and I_s the nucleation rate per area unit. Thus, we extract a growth velocity (V_g) of 21 A°/min. It should be noticed that this is a mean value which account for the growth along all the crystallographic directions.

In the same manner, we have investigated the kinetics of crystallization of LPCVD a-Si films deposited on glass substrates covered with silicon nitride or indium-tin-oxide (ITO). The {111} x-ray integrated intensities for films crystallized on the different type of substrates are reported on Fig.4. Some differences appear on this figure. The crystallization starts on the ITO earlier than on glass and silicon nitride. This fact was verified by TEM observations. In addition, the integrated intensity of the fully crystallized film on glass is the highest. This could be interpreted as an increase of the grain size or texture of this film. The preferred orientations for the three reflections of the films discussed here are given in Table I. As we have mentioned above, the crystallized films do not show a strong texture; however, we can see that the {111} texture is more pronounced for the film crystallized on glass than on the other substrates. On the other hand, the final grain sizes measured by TEM are quite the same (Table I). Thus, we conclude that the highest {111} intensity observed for film totally crystallized on glass is related to a strongest {111} texture.

Table. I: Orientational factors O_{hkl} (%) and average grain size d_g for fully crystallized LPCVD films on various substrates. (annealed at 600 °C)

Substrate	ITO	Silicon nitride	Glass
O_{111}	22	28	41
O_{220}	26	22	31
O_{311}	52	50	28
d_g (nm)	230	280	300

To determine the dependence of the growth mode on the substrate nature, we plot in Fig.5 the quantity $\ln\{1/[1-x(t)]\}$ versus the annealing time for ITO and silicon nitride cases. As for the glass case (Fig. 3), this quantity varies as t^4. Therefore, the grain growth is independent on the substrate nature and is three dimensional. If we assume that the growth velocity is only dependent on the deposition conditions of the starting a-Si material, we can deduce the ratios of the nucleation rates. The nucleation rate on ITO is 2.5 times greater than on glass; while, on silicon nitride, it is only 1.2 times greater.

The relatively small difference between the final grain sizes (Table I) results from a slight dependence of the nucleation rate on the substrate nature.Indeed, the final grain size could be extrapolated from the experimental nucleation rate in a way similar to the one reported in Ref.[5]. The area (A_g) of the final grain is calculated from the final density of grains and expressed by

$$A_g = \lim_{t\to\infty} [n_s(t)]^{-1} \tag{2}$$

where n_s is the number of grains per area unit. This density is given by

$$n_s(t) = \int_0^t I_s [1 - x(t')] \, dt' \qquad (3)$$

where $x(t')$ follows relation (1). After integrating (3), the final grain size obtained from (2) is

$$d_g = (A_g)^{1/2} = [0.91 \; \pi \; V_g^3/(6e)]^{-1/2} \; I_s^{-3/8} \qquad (4)$$

The experimental grain size measures are in very good agreement with the calculated ones as it is shown in Fig. 6, where the solid line is drawn from Eq. (4) for a V_g value of 21 A°min⁻¹. In addition, this representation shows that to obtain an increase by a factor 10 in the grain size, the nucleation rate should be lowered by unless a factor 500. This is in accordance with the results of Iverson and Reif [5] obtained for Si films amorphizised by self-implantation and annealed at 600°C and reported with our results in Fig. 6. These results have shown that the variation of the grain size of crystallized LPCVD a-Si is slightly dependent on the substrate nature; this is mainly due to the weak differences between the nucleation rates.

In contrast with the above results, the crystallization behavior of a-Si:H films deposited by PECVD is markedly different   . Indeed, the crystallization starts rapidly on ITO during the dehydrogenation annealing at 500 °C leading to a small-grain structure. The grain size is estimated to be less than 30 nm. While, at this step, the film on glass remains amorphous. During the annealing at 600 °C, the PECVD films on glass crystallize more slowly than the LPCVD films. Figure 7 shows a TEM micrograph of a-Si:H film deposited on patterned ITO/glass substrate annealed at 600 °C for 15 hours. On the ITO region, where the crystallization is complete, no further increase in the grain size is observed. On the edge of the ITO pattern large grains have nucleated and grown laterally into the Si/glass region. Far from the ITO pattern, the Si/glass is only partially crystallized with large dendritic grains. The total crystallization of the film is obtained after an annealing as long as 70 hours. The average grain size measured on such a totally crystallized film on

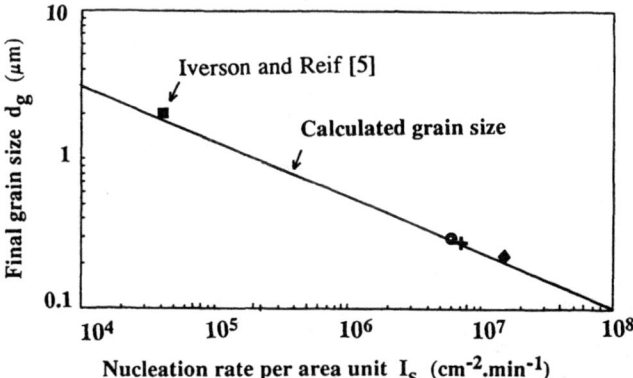

Fig. 6 Final grain size vs the nucleation rate for films annealed at 600 °C. Our data are from films deposited on glass (o), ITO (◆) and silicon nitride (+). The data from Ref. [5] is from self implanted silicon film. The solid line was calculated from Eq. (4) for $V_g = 21$ A°min⁻¹.

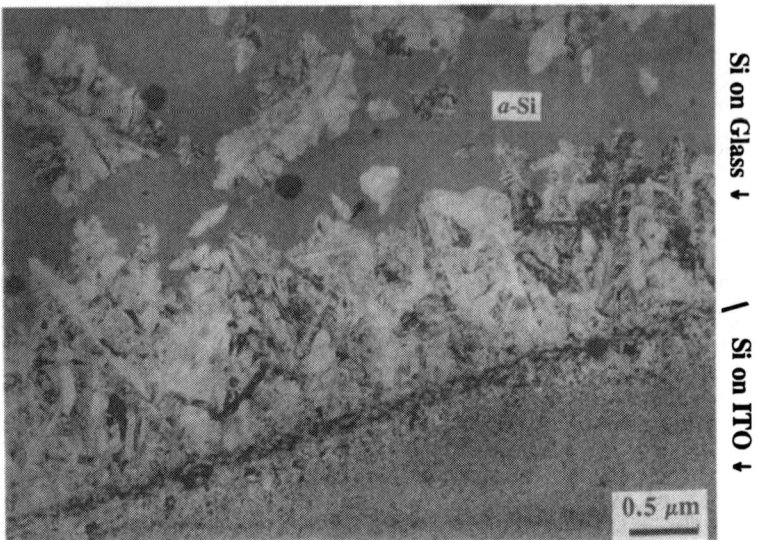

Fig. 7 Bright-field plane view TEM micrograph of PECVD a-Si on ITO patterned glass substrate annealed at 600 °C for 15 hours.

glass is about 1 μm. The nucleation rates in PECVD films on ITO and glass have not been measured, nevertheless, the large difference in grain sizes between the two cases and the lateral over growth result from the fact that the nucleation rate is much higher on ITO than on glass. Indeed, Fig.6 shows that the ratio of the nucleation rates could be estimated to 10^4 for a grain sizes ratio of about 30. The reason why the effect of the substrate nature is dominant in PECVD a-Si:H crystallization is not known; investigations on this subject are currently undertaken.

CONCLUSION

The present work has shown that the substrate nature has a more evident effect on the kinetics of crystallization of PECVD a-Si:H than LPCVD a-Si films. The great difference in nucleation rates observed in the former case results in a marked difference in grain sizes and in a large-grain lateral over growth from the patterned ITO.

REFERENCES

1. I.W. Wu, A. G. Lewis, T. Y. Huang and A. Chiang, 1990 SID International Symposium Digest of Technical Papers, (The Society for Information Display, Playa del Rey, C.A, 1990), pp 307-310.
2. P. Joubert, B. Loisel, Y. Chouan and L. Haji, J. Electrochem. Soc. 134, 2541 (1987).
3. M. K. Hatalis and D. Greve, J. Appl. Phys. 63, 2260 (1988).
4. J. W. Christian, Physical Metallurgy, edited by R. W. Cahn (North-Holland, Amsterdam, 1970), Chap. 10.
5. R. B. Iverson and R. Reif, J. Appl. Phys. 62, 1675 (1987).

CRYSTALLIZATION PROCESSES IN AMORPHOUS HYDROGENATED SILICON BASED ALLOYS

F.DEMICHELIS, C.F.PIRRI, E.TRESSO
Dip. Fisica Politecnico - C.so Duca degli Abruzzi 24 10129 Torino (Italy).

L.BATTEZZATI, E.GIAMELLO
Dip. Chimica Inorganica, Chim. Fisica e Chim. Materiali Universita' di Torino - Via Pietro Giuria 7 10125 (Italy)

P.MENNA
ENEA, Centro Ricerche Fotovoltaiche - 80055 Portici (Italy)

ABSTRACT

Amorphous and microcrystalline films of silicon and silicon carbide have been deposited by means of PECVD at low substrate temperature (200°C), with reactive gases highly diluted in H_2. Devices quality a-Si:H films have been annealed in vacuum at temperatures in the range 200- 1000 °C. The transition from amorphous to crystalline structure was studied by X-ray, Raman and I.R. spectroscopies, optical analysis in UV-VIS-NIR region, Transmission Electron Microscopy, Differential Scanning Calorimetry and Electron Spin Resonance measurements. By comparing the results of the two methods to obtain microcrystalline films we have deduced information on the process of growth of Si and SiC microcrystals.

INTRODUCTION

In the last years amorphous silicon (a-Si:H) has been the subject of considerable research and development leading to a number of applications in photovoltaic solar cells, thin-film field-effect transistors for liquid crystals displays, logic circuits, electrophotography and so on [1].

Recently [2-5] amorphous and microcrystalline silicon carbon alloys (a-SiC:H and μc −SiC:H) have got interest for their applications, for instance the use of μc −SiC:H films as heterojunction window material. Nevertheless up today the processes of microcrystal growth and the phase transition from amorphous to microcrystalline state have not been completely studied.

Using high H_2 dilution, μc −Si:H and μc −SiC:H can be grown at temperature as low as 200°C from SiH_4 and SiH_4+CH_4 by Plasma-Enhanced Chemical Vapor Deposition (PECVD). The microcrystallization process depends on the H_2 dilution, on the power density and on the substrate temperature. On the other hand microcrystallization can be obtained by thermal treatments of amorphous films.

We have realized amorphous and microcrystalline films of silicon and silicon carbide by PECVD. Furthermore amorphous films were annealed at high temperatures in order to obtain crystallization. The phase transition were studied by I.R. and Raman spectroscopies, Transmission Electron Microscopy (TEM), X-ray diffraction, Differential Scanning Calorimetry (DSC), Optical spectroscopy, Electron Spin Resonance (ESR) and D.C. conductivity.

By comparing the results of the two methods for obtaining microcrystals we have deduced information on the transition from amorphous to crystalline structure.

EXPERIMENTAL

a-Si:H, μc −Si:H, a-SiC:H and μc −SiC:H were deposited by Plasma Enhanced Chemical Vapor Deposition (PECVD) of SiH_4+H_2 and $SiH_4+CH_4+H_2$ mixtures. The flow rates were for H_2 in the range 180-200 sccm, for SiH_4 3-13 sccm, and for CH_4 1-4 sccm. The pressure during film growth was varied in the range 30-40 Pa, the substrate temperature from 200 to 260°C and the power density was 0.15 W cm^{-2}. The film growth rate was about 0.3 Å s^{-1} for μc −films and about 1.5 Å s^{-1} for amorphous films.

The amorphous or microcrystalline structure of the films was determined by X-ray diffraction analysis, Raman scattering and I.R. absorption.

Mat. Res. Soc. Symp. Proc. Vol. 230. ©1992 Materials Research Society

The Raman spectra were measured using 514.5 nm (Ar$^+$-ion laser) excitation with triple spectrometer and optical multichannel analizer, I.R. spectroscopy was performed by a Bruker FS85 single beam spectrometer in the range 400 - 4000 cm^{-1}.
Grain size and distribution was observed by a Transmission Electron Microscope (TEM) Siemens EmisKop 1A at 80 keV.
Optical transmittance and reflectance were measured by a Perkin Elmer UV-visible-NIR Lambda 9 Spectrophotometer in the wavelength 0.2-3 μm. ESR measurements were performed at room temperature by a Varian EPR 109 spectrometer. DC electrical conductivity was measured in vacuum (10^{-2} Pa) in planar configuration by using a Hewlett-Packard high resistance meter 4329A.
The phase transitions were studied by a Dupont Thermal Analyzer in the range of temperature 50 to 725°C.
Isochronal annealing of amorphous films was carried out in vacuum (10^{-5} Pa) for 1 hour at temperature in the range 200-1000°C so to reach the crystallization of the samples.

RESULTS AND DISCUSSION

Fig. 1 shows the X-ray diffraction patterns of two typical amorphous and μc −SiC:H samples. We observe the (111) and (200) diffraction peaks in microcrystalline sample and a broad line in the amorphous one.
The Raman spectra of amorphous and microcrystalline Si and SiC are shown in Fig.2 and 3. The Si films remain amorphous (Fig. 2a) for SiH$_4$ concentration in the gas mixture above 5%. In fact we observe a TO-like (transverse optical phonon like) peak at an energy of about 480 cm^{-1}. As SiH$_4$ concentration decreases a microcrystalline phase appears in the amorphous matrix and a new TO peak at higher energy (about 520 cm^{-1}) is detected (Fig. 2b). The addition of a CH$_4$ flow (≤1 sccm) introduce C without changing microcrystallinity (Fig. 3d). It is interesting to note that we obtain microcrystalline films with about 15% of C, while in previous reported experiments [4] the crystalline phase disappeared for C content higher than 6%. As CH$_4$ flow exceeds 1 sccm, we observe the decrease of microcrystallinity (Fig. 3c) and then the desappearence (Fig. 3b). In Fig. 4 TEM pictures are reported. We can observe in Fig. 4a a completely amorphous structure corresponding to Raman spectra of Fig. 3a. In Fig. 4b (Raman Fig. 3c) microcrystalline islands are embedded in an amorphous matrix. Microcrystals, having a size of about 200 Å, connect together in Fig. 4c (Raman Fig. 3d).
Fig. 5 show the SiH vibrational band at 2000 cm^{-1} and 2100 cm^{-1} as appears in IR spectra for typical amorphous and microcrystalline SiC samples.
In amorphous samples, H is bonded predominantly as monohydride having the 2000 cm^{-1} SiH stretching and 630 cm^{-1} rocking modes. In crystalline phase the SiH stretching mode shifts to 2100 cm^{-1} and the rocking mode disappear. Since the 2100 cm^{-1} mode has been identified as (SiH$_2$)$_n$, we can deduce that in the microcrystalline samples higher hydrides dominate.
The optical properties (absorption coefficient as a function of photon energy in the range 1.5 - 3.5 eV) of μc −deposited films are similar to those of amorphous films, even with the presence of a big amount of microcrystals of size ~200 Å.
Electron Spin Resonance (ESR) signal with a g-factor of 2.005, a line width (ΔH_{pp}) of 7 G and spin density (N$_s$) of about 2 10^{18} cm^{-3} were found in a-Si$_{70}$C$_{15}$:H$_{15}$ samples. For μc −SiC:H films of about similar Si and C amount the g-factor and ΔH_{pp} do not change significantly. Only N$_s$ decreases to 5 10^{17} cm^{-3}.
The values of the D.C. conductivity were measured at room temperature for films prepared at different reactive gas concentrations. The conductivity increases from 10^{-10} $\Omega^{-1}cm^{-1}$ to 10^{-6} $\Omega^{-1}cm^{-1}$ as the structure of the films change from amorphous to crystalline.

Thermal treatments at high temperature (above 700 °C) constitute another approach to crystallization processes. a-Si:H films were annealed in the range of temperature 200-1000 °C. Fig. 6 shows the absorption coefficient α as a function of photon energy for some annealing temperatures. A sharp change in slope is evident between 600 and 800 °C, characteristic of a transition from amorphous to crystalline material (as shown in the inset of Fig. 7). These results, compared with the ones obtained for microcrystalline depodited films, indicate that the absorption coefficient strongly depends on crystallization process.

185

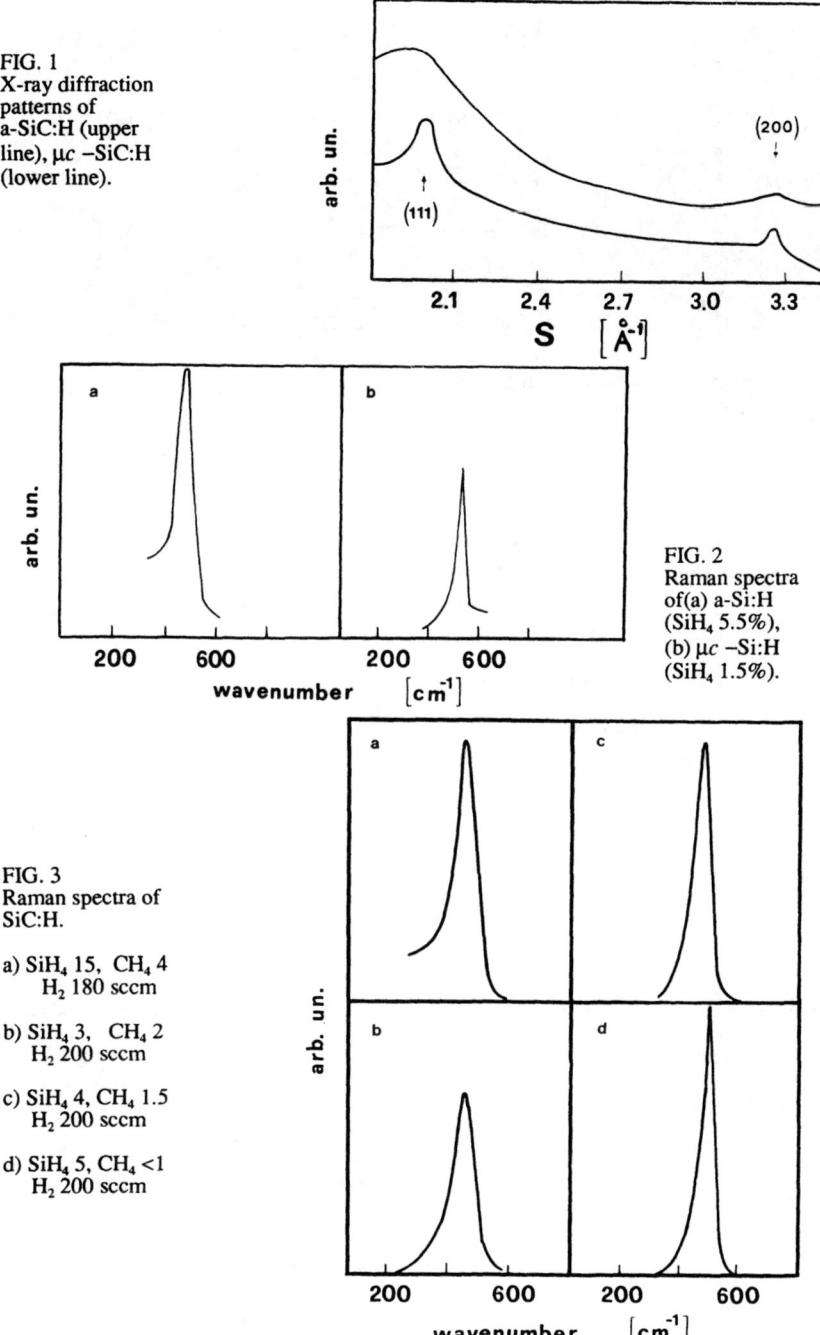

FIG. 1
X-ray diffraction
patterns of
a-SiC:H (upper
line), μc –SiC:H
(lower line).

(200)

(111)

S [Å⁻¹]

arb. un.

2.1   2.4   2.7   3.0   3.3

a

b

arb. un.

200   600   200   600

wavenumber   [cm⁻¹]

FIG. 2
Raman spectra
of(a) a-Si:H
(SiH₄ 5.5%),
(b) μc –Si:H
(SiH₄ 1.5%).

FIG. 3
Raman spectra of
SiC:H.

a) SiH₄ 15, CH₄ 4
   H₂ 180 sccm

b) SiH₄ 3, CH₄ 2
   H₂ 200 sccm

c) SiH₄ 4, CH₄ 1.5
   H₂ 200 sccm

d) SiH₄ 5, CH₄ <1
   H₂ 200 sccm

a   c

arb. un.

b   d

200   600   200   600

wavenumber   [cm⁻¹]

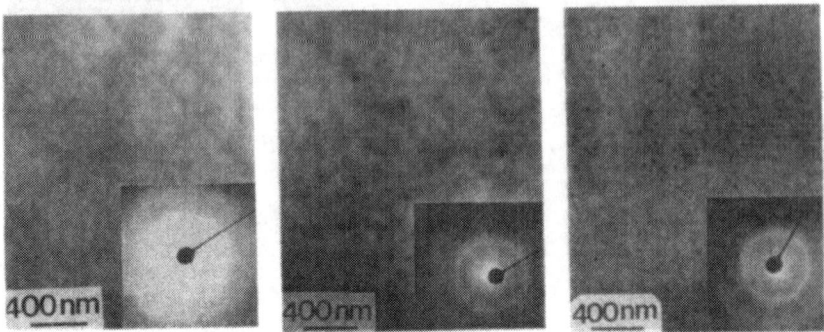

FIG. 4 Transmission electron micrographs
(a, b, c) of the SiC:H samples as
in Fig. 3a, 3c, 3d respectively.

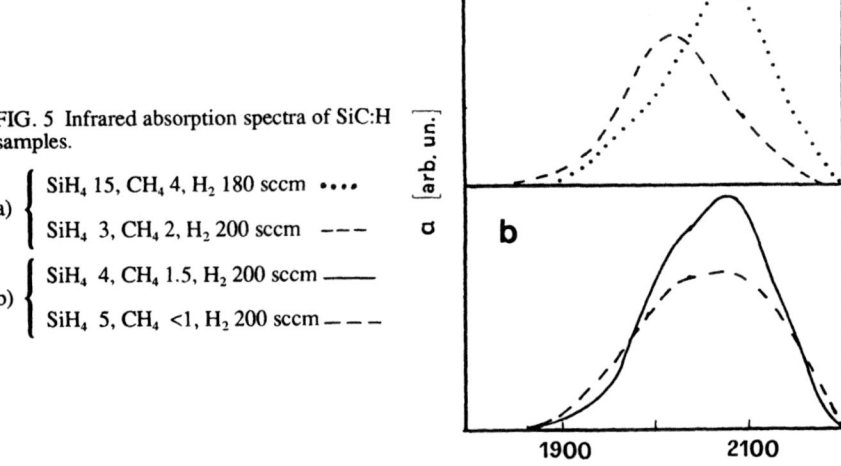

FIG. 5 Infrared absorption spectra of SiC:H
samples.

a) $\begin{cases} \text{SiH}_4 \ 15, \text{CH}_4 \ 4, \text{H}_2 \ 180 \ \text{sccm} \ \cdots \\ \text{SiH}_4 \ 3, \text{CH}_4 \ 2, \text{H}_2 \ 200 \ \text{sccm} \ \text{---} \end{cases}$

b) $\begin{cases} \text{SiH}_4 \ 4, \text{CH}_4 \ 1.5, \text{H}_2 \ 200 \ \text{sccm} \ \text{———} \\ \text{SiH}_4 \ 5, \text{CH}_4 \ <1, \text{H}_2 \ 200 \ \text{sccm} \ \text{— — —} \end{cases}$

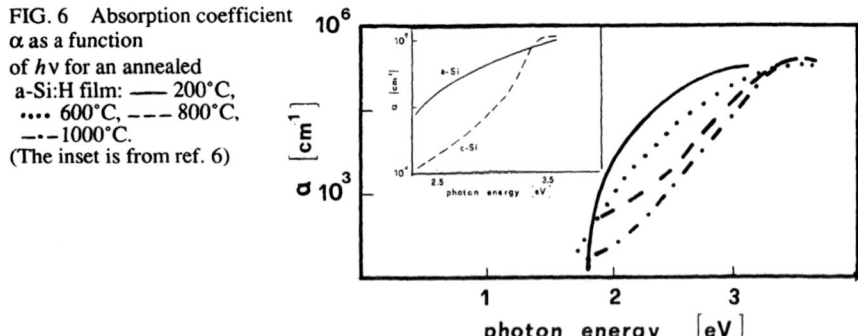

FIG. 6 Absorption coefficient
α as a function
of $h\nu$ for an annealed
a-Si:H film: ——— 200°C,
•••• 600°C, ——— 800°C,
—·—1000°C.
(The inset is from ref. 6)

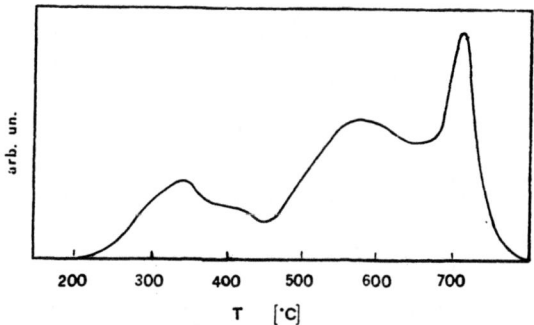

FIG. 7 Subctracted DSC spectrum of a-Si:H powder (weight about 0.25 mg.).
After a first scan, a second one was made to establish a reference state for the previous transformations.

FIG. 8
X-ray diffraction pattern of
a-Si:H powder crystallized
during a DSC measurement.

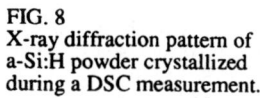
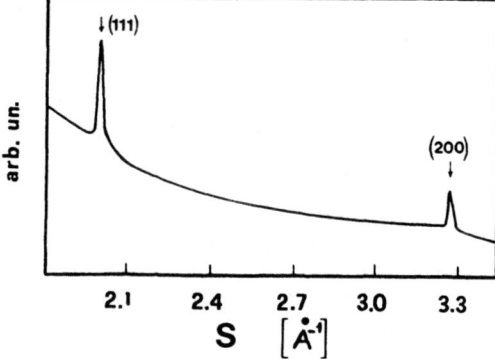

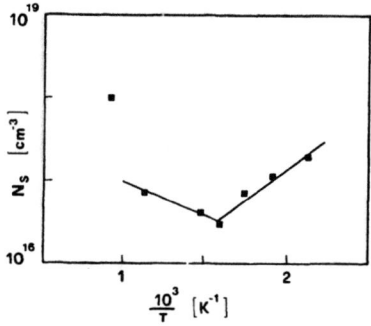

FIG. 9
ESR results of
an a-Si:H
sample
annealed in the
range of tem-
peratures
between 200 -
800 °C.

Finally a differential calorimetry measurement of a typical a-Si:H sample heated at 5 °C min⁻¹ is reported in Fig. 7. The data show a broad exothermic peak in the range 250 - 400 °C corresponding probably to a structural rearrangement in the network. Furthermore we observe a gradually evolution of hydrogen for temperatures higher than 450°C shown by another exthermic process. The strong exothermic peak at about 700 °C is due to crystallization of the sample. X-ray diffraction analysis of the powders after DSC measurements are reported in Fig. 8, where we can observe the typical pattern of crystalline silicon.

From a comparison of X-ray spectra of μc –deposited films and thermally crystallized films it is possible to conclude that thermal processes allow the growth of larger grain size. But the violence of the phoenomenon involves the formation of a greater amount of defects as tested by ESR measurements reported in Fig. 9.

CONCLUSIONS

The experimental results confirm that different mechanisms govern the two crystallization processes: the growth of microcrystals under particular deposition conditions and the annealing of amorphous samples. We will point out that after annealing the sample have lost hydrogen then the comparison is made betwen hydrogenated and non-hydrogenated μ-crystalline films.

To explain the growth of microcrystalline Si by PECVD at temperatures as low as 200 °C, two interpretations have been suggested. In one [7] the radicals, by covering more of growing surface with hydrogen, enhance the surface diffusion of the growth species; in the other [8] H radicals govern the film formation process by promoting etching. The addition of C drastically influences the amorphous-microcrystalline transition; we observed a decrease both of grains size and of the amount of microcrystalline phase.We suppose that the growing surface of the samples is submitted to the attack of hydrogen radicals which remove weak-bonded hydrogen from the surface by etching reaction and promote the formation of non-hydrogenated Si and SiC clusters.

The reported results on the annealing of device quality a-Si:H and expecially the presence of only a crystallization DSC peak suggest that overall crystallization in tethraedral configuration occurs for temperatures higher than 700 °C. This is in agreement with results recently reported [9] which show the existence of a two-phase network and consequently a two stage crystallization only in a poor quality material.

From the above results it is possible to conclude that Si atoms in the as deposited a-Si:H films have a tendency for chemical ordering and that when the films are annealed they segregate into crystalline clusters.

Further experiments and more detailed models would clarify both the mechanisms.

REFERENCES

1) P.G. Le Comber - Proceedings of the 8-th E.C Photovoltaic Solar Energy Conference , Kluwer Acad. Publ. p. 1229 (1988)
2) K. Tanaka and A. Matsuda - Jap. Ann. Rev. Electron Comp. and Telecomm. 6, 161 (1983)
3) S. Usui and M. Kikuchi - J. Non-Crys. Solids 34, 1 (1979)
4) B. Goldstein, C.R. Dickson and P.M. Fauchet - Appl. Phys. Lett. 52, 2672 (1988)
5) G. Ganguly, S.C. De, S. Ray and A.K. Barna - J. Non-Cryst. Solids 114, 822 (1989)
6) G. Willeke and R. Martins - Phil. Mag. B 63, 79 (1991).
7) A. Matsuda - J. Non-Cryst. Solids 59-60, 767 (1983)
8) C.C. Tsai in "Amorphous Silicon and Related Materials" - Ed. H. Fritzsche, World Scient. Publ. Co. p. 123 (1988)
9) W. Paul, J.J. Scott, W.A.T. Turner - Phil. Mag.B 63, 247 (1991)

CRYSTALLIZATION OF AMORPHOUS Si IN Al/Si MULTILAYERS

Toyohiko J. Konno and Robert Sinclair
Department of Materials Science and Engineering, Stanford University, Stanford, CA 94305

ABSTRACT

The crystallization of amorphous Si in a Al/Si multilayer (with a modulation length of about 120Å) was investigated using transmission electron microscopy, differential scanning calorimetry and X-ray diffraction. Amorphous Si was found to crystallize at about 175°C with the heat of reaction of 11±2(kJ/mol). Al grains grow prior to the nucleation of crystalline Si. The crystalline Si was found to nucleate within the grown Al layers. The incipient crystalline Si initially grows within the Al layer and then spreads through the amorphous Si and other Al layers. Because of extensive intermixing, the original layered structure is destroyed. The Al(111) texture is also enhanced.

INTRODUCTION

The stability of amorphous Si (a-Si) in contact with Al is an important issue in view of recent developments in semiconductor electronics and electrophotography.[1,2] For example, the temperature rise in a-Si photoreceptor drum can crystallize the photoreceptor layer and destroy the drum.

Apart from its technological significance, the crystallization kinetics of a-Si and/or amorphous Ge (a-Ge) have also attracted a number of researchers from the scientific point of view. The nucleation of crystalline Si (c-Si) in a-Si is one of the issues still under investigation.[3] Although the crystallization temperature is known to be about 600-700°C,[4,5] this temperature is strongly structure dependent. For instance, an a-Si whisker is reported to crystallize at about 900°C.[6] On the other hand, a-Si and a-Ge crystallize at surprisingly low temperatures when they are in contact with metals with which they form eutectic systems.[7-12] This phenomenon is sometimes called Metal Contact Induced Crystallization (MCIC).

The mechanism of MCIC is still not well known. Early RBS studies on an a-Si/metal/c-Si structure indicated that the a-Si phase dissolves into the metal phase prior to the nucleation of c-Si on the substrate c-Si.[10,11] This technique can be used to induce solid phase epitaxy. Hung et al., in their study on an a-Si/Al system, observed that small crystalline phases appeared in the amorphous region at 200°C.[13] Homma et al. suggested that Pb grains might act as templates for crystalline Ge from their X-ray study on crystallization in a Pb/Ge multilayer system.[14]

In the present study, crystallization of a-Si in an Al/Si multilayer system was investigated using transmission electron microscopy(TEM), calorimetry, and X-ray diffraction. The kinetics of nucleation of c-Si was examined by in-situ as well as ex-situ experiments.

EXPERIMENTAL

Al/Si multilayers were deposited by RF/DC magnetron sputtering on the following substrates: (100) Si wafer, slide glasses, and slide glasses coated with photo-resist. These substrates were placed on a rotating table in the chamber. The power of the guns and the speed of the rotation of the table were adjusted to obtain the desired modulation length and composition. The base pressure was less than 3×10^{-7}torr, and the sputtering was done in 3mtorr Ar atmosphere. A calorimetric study was performed using a Perkin-Elmer DSC-7 on a free standing multilayer film of modulation length of 100Å(total thickness: 1000nm). The approximately 0.5mg specimen was heated at 10°C/min in an Ar atmosphere. Structural analysis was performed using a Philips 430ST microscope operated at 300kV and a Philips X-ray diffractometor on multilayer films with a modulation length of 120Å(total thickness: 400nm). For ex-situ experiments, the samples were annealed on a hot plate at 175°C and subsequently quenched in de-ionized water. Cross-section samples of as-deposited and

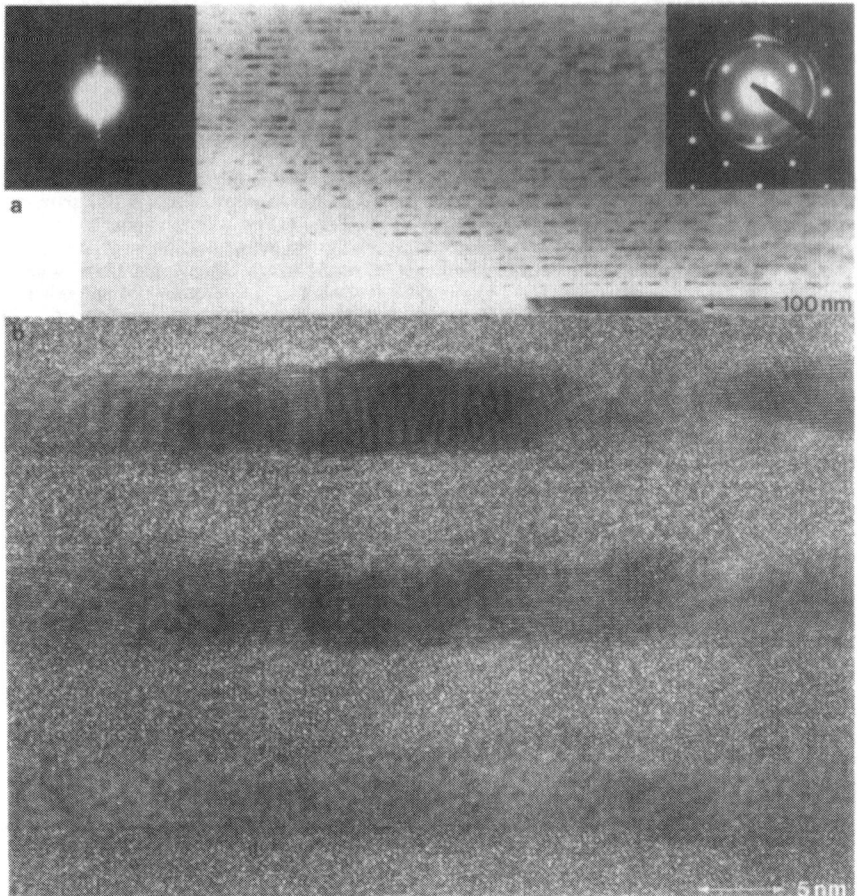

Fig.1 Cross-section TEM micrographs of as-deposited sample. (a) BF image, small-angle(left), and conventional diffraction pattern, (b) high-resolution image.

annealed films were made by the standard procedure.[15] Ion-milling was done at 3kVx4mA using a liquid nitrogen stage in order to avoid any reaction during sample preparation. In-situ heating experiments were also carried out using a single tilt heating holder. The reaction was directly recorded through a Gatan image pick-up system. Texture evolution was measured by X-ray diffraction.

RESULTS

A cross-section TEM micrograph of the as-deposited multilayer is shown in Fig.1-a. Inserted in the picture are both small-angle and normal electron diffraction patterns. Layers containing strongly diffracting grains are Al. It is seen from the diffraction pattern that these Al grains possess a (111) texture. On the other hand, the inner "halo" caused by Si layers indicates that the Si is indeed amorphous. A high-resolution micrograph of the same film is shown in Fig.1-b. Amorphous Si layers and textured Al grains can be clearly seen.

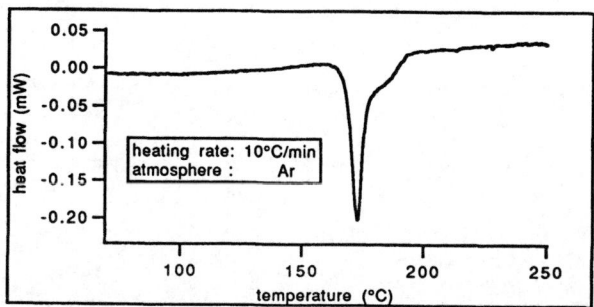

Fig.2 DSC scan of Al/Si multilayer: Tpeak = 173°C, heat of reaction = 11±2 kJ/mol.

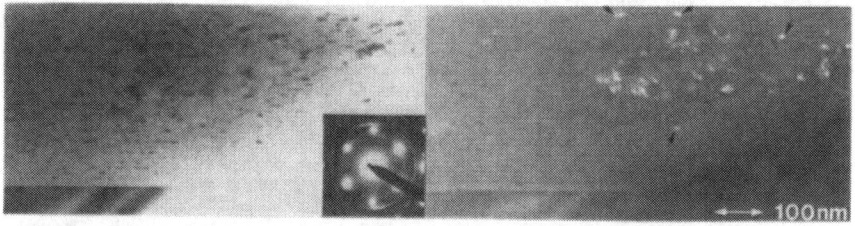

(a)   (b)
Fig.3 Cross-section TEM micrographs of the Al/Si multilayer, annealed at 175°C for 40 seconds: (a) BF image and diffraction pattern, (b) DF(Si) image.

Fig.2 shows the result of the DSC scan. The curve was obtained by subtracting the heat flow of the second run from that of the first run. An exthothermic reaction started at about 170°C. The composition of the film was deduced from a cross-section TEM picture and from density data of c-Si. The resultant composition of the film was used to obtain the heat of the reaction per mole of Si (11±2 kJ/mol-Si). This value can be compared with the heat of crystallization of amorphous Si obtained by Donovan et al. (11.9 kJ/mol). [5]

Fig.3 shows the bright field(BF), dark field(DF) images, and diffraction pattern of the multilayer annealed at 175°C for 40 seconds. The Si(111) ring in the diffraction pattern indicates that crystallization of a-Si has already started. The DF image was taken with the aperture at the Si(111) ring position. An intermixed region containing relatively large Si grains can be seen in the BF and DF pictures. Around this intermixed region in the DF image, small strongly diffracting grains can be found to exist within the Al layer (indicated by arrows).

High-resolution images of the annealed sample taken from different positions in the film are shown in Fig.4. Away from the intermixed region, growth of Al(111) grains in their [111] direction is frequently observed(Fig.4-a). Si(111) fringes appear among the grown Al grains(Fig.4-b). A relatively large Si grain is found in the Al layers(Fig.4-c). This grain corresponds to one of the strongly diffracting grains visible in low magnification, such as those shown in Fig.3. The stacking faults and micro-twins shown in this picture are frequently found during the crystallization of pure amorphous Si.[16,17] In this picture the Si stacking faults in the upper Al layer are not continuous with those in the lower layer, indicating that this crystal grew independently into each Al layer.

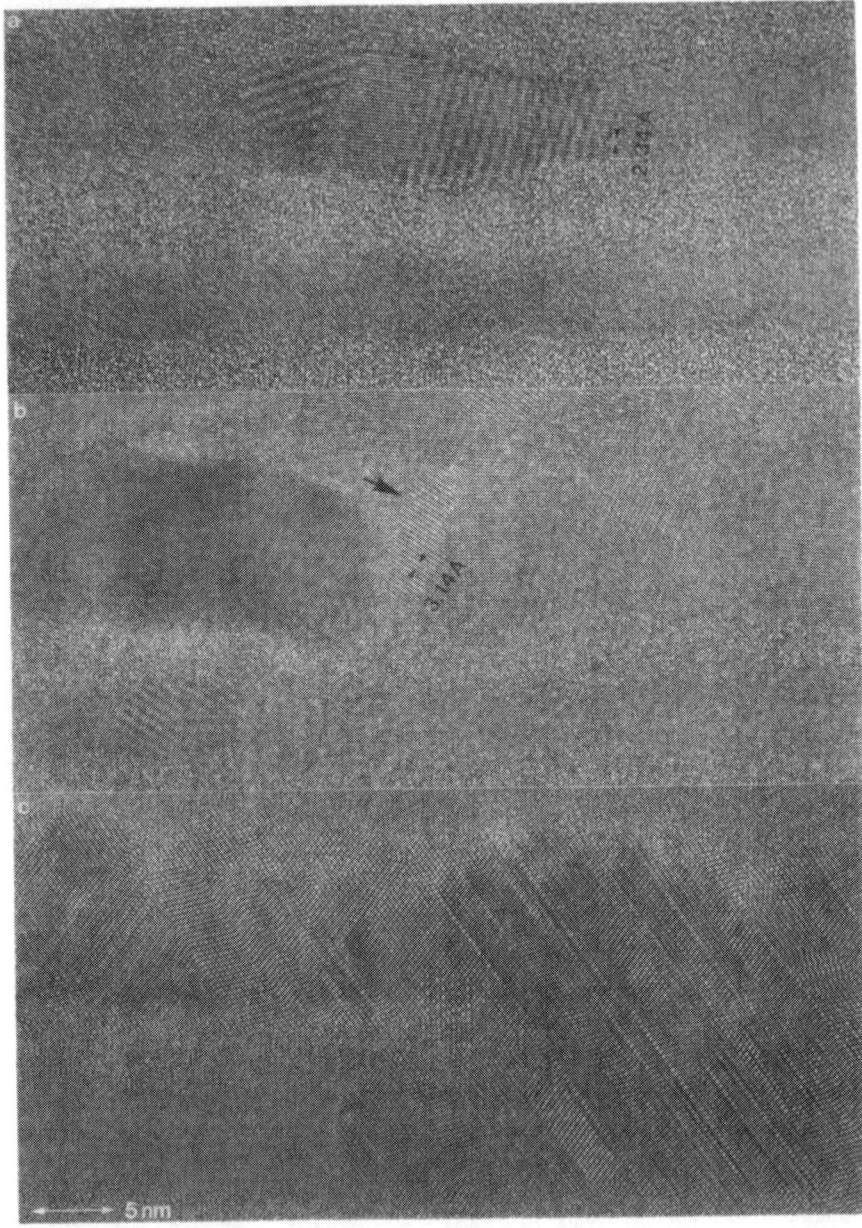

Fig.4 Cross-section TEM micrographs of the Al/Si multilayer, annealed at 175°C for 40 seconds: (a) growth of Al(111) grain, (b) Si(111) fringes appear in the Al layer, and (c) Si grain grows into Al layers. Note that the Si grain in the lower Al layer possess different stacking faults from those found in the Si grain in the upper Al layer.

Fig.5 shows BF and high-resolution images of the multilayer annealed at 175°C for 90 seconds. In this sample, the reaction is almost complete and the layered structure is destroyed.

In-situ TEM heating experiments were carried out at the nominal temperature of 170°C. Fig.6 shows the bright field image of the annealed sample. The corresponding image of the same sample before the annealing is the one shown in Fig.1. The morphology of the film after the anneal is same as that of the sample obtained by the ex-situ experiments. Further examination of the in-situ recordings at high resolution is expected to reveal the associated mechanisms more clearly.

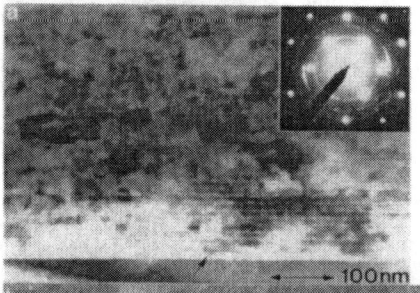

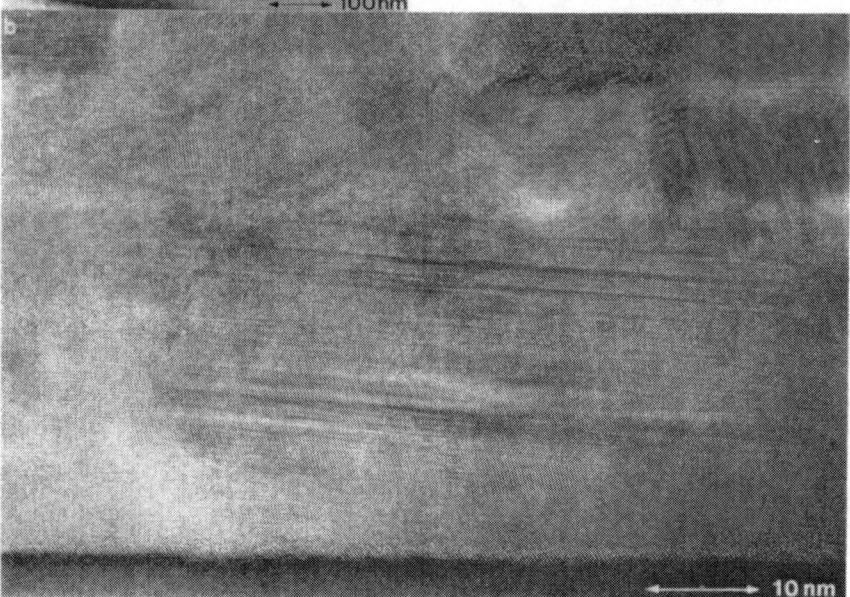

Fig. 5 Cross-section TEM micrographs of Al/Si multilayer, annealed at 175°C for 90 seconds.
(a) BF image and diffraction pattern.
(b) high-resolution image of the area shown by the arrow in (a).

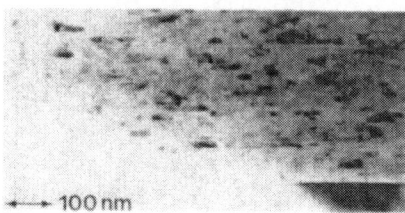

Fig. 6 Cross-section TEM micrograph of Al/Si multilayer, annealed in-situ at a nominal temperature of 170°C.
Corresponding BF picture of the same sample before the annealing is shown in Fig.1.

194

The evolution of the texture of Al grains is also confirmed by X-ray diffraction. The multilayer film deposited on a glass slide was heated at about 200°C for 5 seconds in the diffractometor. The diffraction patterns before and after annealing are shown in Fig.7.

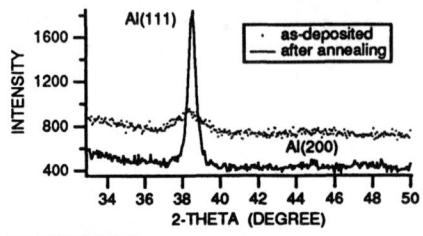

Fig. 7 X-ray diffraction pattern before and after annealing.

CONCLUSIONS

We have shown that amorphous Si crystallizes at about 175°C in an Al/Si multilayer system, consistent with previous results for MCIC. The heat of reaction was 11±2 (kJ/mol). Prior to the nucleation of c-Si, significant grain growth on the part of Al grains was observed, which led to the enhancement of Al(111) texture. c-Si nucleates within the Al layers. During the incipient stage, the c-Si grows into the Al layer and then spreads into the a-Si layer and other Al layers. The original layered structure is destroyed as a result of the extensive intermixing.

ACKNOWLEDGEMENTS

We would like to thank Professor Robert Waymouth and Ms.Anne-Lise Mogstad, Department of Chemistry, Stanford University, for permission to use the Perkin-Elmer DSC-7 and for help during the operation. This work is supported by the National Science Foundation (Grant number DMR 8902232).

REFERENCES

1. Y.Kuwano and M.Ohnishi, JARECT, ed.Y.Hamakawa, Ohmsha and North-Holland Publishing Co., 6 "Amorphous Semiconductor Technologies & Devices" (1983) 204
2. T.Kawamura, N.Yamamoto and Y.Nakayama, ibid. 6 (1983) 325
3. S.Roorda, D.Kammann, W.C.Sinke, G.F.A.Van De Walle and A.A.Van Gorkum, Mat.Lett. 9 (1990) 259
4. K.Zellama, P.Germain, S.Squelard, J.C.Bourgoin and P.A.Thomas, J.Appl.Phys. 50 (1979) 6995
5. E.P.Donovan, F.Spaepen, D.Turnbull, J.M.Poate and D.C.Jacobson, ibid. 57 (1985) 1795
6. Y.Tatsumi, M.Hirata and K.Yamada, J.Phys.Soc.Jpn., 50 (1981) 2288
7. F.Oki, Y.Ogawa and Y. Fujiki, Jpn.J.Appl.Phys., 8 (1969) 1056
8. J.R.Bosnell and U.C.Voisey, Thin Solid Films, 6 (1970) 161
9. S.R.Herd, P.Chaudhari and M.H.Brodsky, J.Non-Cryst.Solid, 7 (1972) 309
10. D.Sigurd, G.Ottaviani, V.Marrello, J.W.Mayer and J.O.McCaldin, ibid., 12 (1973) 135
11. G.Ottaviani, D.Sigurd, V.Marrello, J.W.Mayer and J.O.McCaldin, J.Appl.Phys. , 45 (1974) 1730
12. L.Hultman, A.Robertsson, H.T.G.Hentzell, I.Engstrom and P.A.Psaras, ibid., 62 (1987) 3647
13. L.S.Hung, S.H.Chen and J.W.Mayer, Mat.Res.Soc.Proc., 25 (1984) 253
14. H.Homma, I.K.Schuller, W.Sevenhans and Y.Bruynseraede, Appl.Phys.Lett. 50 (1987) 594
15. J.C.Bravman and R.Sinclair, J.Electron Microsc.Tech. 1 (1984) 53
16. J.Morgiel, I.W.Wu, A.Chiang and R.Sinclair, Mat.Res.Soc.Proc., 182 (1990) 191
17. A.S.Kirtikar, J.Morgiel, R.Sinclair, I.W.Wu and A.Chiang, Mat.Res.Soc.Proc., in press (1991)

NUCLEATION AND CRYSTALLIZATION OF AMORPHOUS SILICON-ALUMINUM THIN FILMS

F. Lin*, M. K. Hatalis*, S. Girginoudi**, D. Girginoudi**, N. Georgoulas** and A. Thanailakis**.
*Lehigh University, Dept. of Computer Science and Electrical Engineering, Bethlehem, PA 18015.
**Democritus University of Trace, Dept. of Electrical Engineering, 67100 Xanthi, Greece.

ABSTRACT

The phase transformation of amorphous silicon-aluminum thin films with 16 at.% and 30 at.% aluminum was characterized by TEM during an in-situ anneal at 500°C. No crystallization was observed in the samples having 16 at.% Al even after a 10 hour anneal at 500°C. In contrast, nucleation of crystallites was observed in the samples having 30 at.% Al after a 40 min anneal. The size of these crystallites grew during annealing. Single crystal regions were also observed and were identified as aluminum having <111> orientation.

INTRODUCTION

Amorphous silicon films have received considerable interest for several applications including solar cells and flat panel displays [1]. For the latter application the main advantage of amorphous silicon is the low processing temperature which makes this material compatible with low cost glass substrates. The electron mobilities in amorphous silicon is, however, very low and this limits the speed of the devices fabricated using this material. In contrast, polycrystalline silicon films offer one to two orders of magnitude higher electron mobility, but the deposition of polycrystalline silicon takes place at temperatures higher than 600°C. By depositing the silicon in the amorphous phase and annealing the amorphous films at low temperatures polycrystalline silicon can be obtained at temperatures lower than 600°C. Amorphous silicon crystallizes at temperatures above 500°C but its crystallization time can be very long at temperatures lower than 600°C [2].

The crystallization of amorphous silicon films in contact with certain metals has been studied and it was shown that the crystallization process is enhanced at temperatures even far below 500°C [3,4,5]. Few studies on the crystallization of certain amorphous silicon-metal alloys have also been reported and enhancement of the crystallization process was also observed [6,7]. A study on amorphous silicon-aluminum alloys has shown that the crystallization of the amorphous phase depends upon the aluminum concentration, and for concentrations of aluminum up to 20% polymorphous transformation takes place, i.e. crystallization without any change in concentration. Since the solid solibility of aluminum in crystalline silicon is smaller than 1 at.% [8,9], the polymorphous transformation is an attractive way to synthesize artificially crystalline silicon-aluminum alloys.

In this work, Transmission Electron Microscopy was used to study the in-situ crystallization of amorphous silicon-aluminum alloys.

Mat. Res. Soc. Symp. Proc. Vol. 230. ʿ 1992 Materials Research Society

EXPERIMENTAL

Two different concentrations of amorphous silicon-aluminum films were prepared and studied. The aluminum concentration was either 30 at.% or 16 at.%. The 30 at.% alloy was prepared by co-evaporation of aluminum and silicon on a glass substrate. The 16 at.% alloy was prepared by e-beam evaporation of a sintered aluminum-silicon material, which was prepared by mixing predetermined amounts of silicon and aluminum and heating them in a rotating crucible at 1600°C in a helium atmosphere. The vacuum during evaporation was 10^{-7} Torr. The thickness of the as-deposited films was in the 80-200 nm range.

The TEM samples were prepared either by lifting the thin films from the glass substrates or by thinning the silicon substrates. The lifting of the thin films was accomplished by immersing the glass substrates in a 10% HF: 90% H_2O mixture. Small pieces of the silicon-aluminum film were lifted by using copper grids. The thinning of silicon substrates was accomplished by first mechanical grinding and polishing the wafers from the back, then breaking the substrates in small 2 mm x 2 mm squares and selectively etching the silicon from the back in a HNO_3:CH_3COOH:HF (5:2:1) mixture.

TEM examination was performed in a Philips 400T electron microscope operating at 120 kV. A hot stage holder was used in order to study the in-situ crystallization of the samples. The annealing temperature was 500°C. The composition of the samples conatining 30 at.% Al was determined by both XPS and EDAX, while of the 16 at.% Al was determined by an Electron Probe Analyzer.

RESULTS AND DISCUSSION

Crystallization of the amorphous samples having 30 at.% aluminum was observed even after a relative short anneal at 500°C but we did not detect any crystallization in the samples having 16 at.% aluminum even after several hours of anneal. In the following, we discuss in detail the structural characteristics observed during the in-situ annealing of the amorphous silicon-aluminum films.

Silicon-Aluminum with 30 at.% Al.

In **Fig. 1** we show a TEM micrograph of the as-deposited film and its diffraction pattern. The observed diffused rings of the diffraction pattern indicate that the as-deposited material is amorphous. Nucleation of crystallites was observed in this material after a 40 min anneal at 500°C. **Figure 2** shows a micrograph of a thin film annealed for 45 min at 500°C. Growth of these crystallites was observed during annealing. The average size reached ~10 nm after 2 hours and ~18 nm after 7 hours. TEM micrographs after the 2 and 7 hours anneals are shown in **Fig. 3** and **Fig. 4** respectively.

We have also observed abnormal growth in this material. Unlike most abnormal or secondary grain growth in which large grains grow by consuming the small grains around them, the result of growth in this film is the formation of large single crystal-like region as shown in **Fig. 5**. By using electron diffraction we have identified that the single crystal-like region consisted primarily of aluminum having <111> orientation, as is shown in **Fig. 6**.

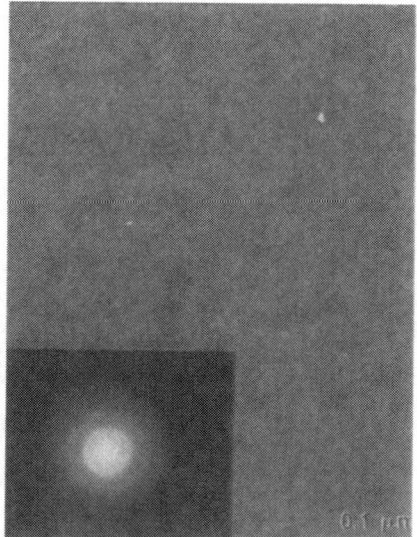

Fig. 1 TEM micrograph of an as-deposited $Si_{0.7}Al_{0.3}$ alloy.

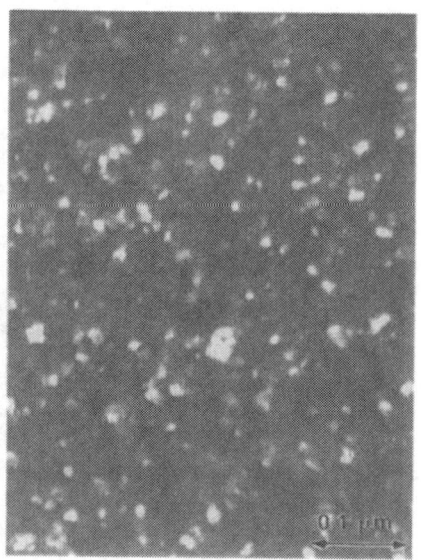

Fig. 2 TEM image of the 30 at.% Al sample after it was annealed at 500°C for 45 minutes.

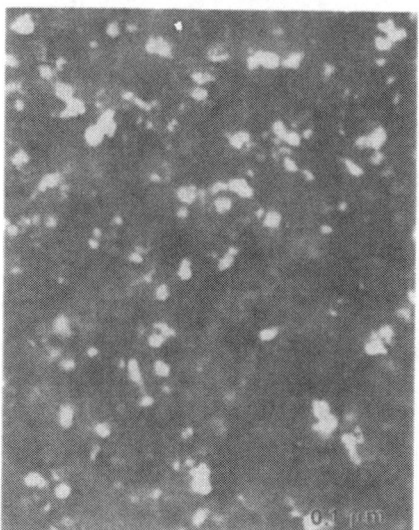

Fig. 3 TEM image of the 30 at.% Al sample after it was annealed at 500°C for 2 hours.

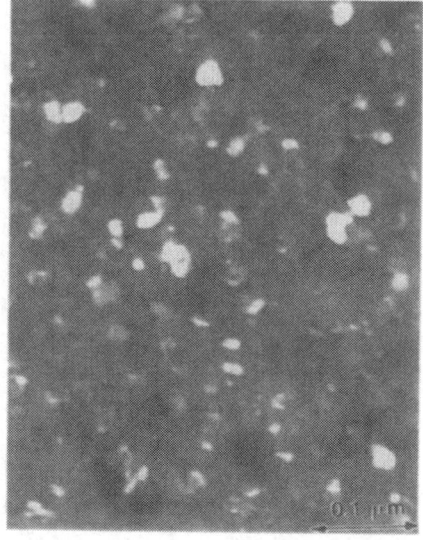

Fig. 4 TEM image of the 30 at.% Al sample after it was annealed at 500°C for 7 hours.

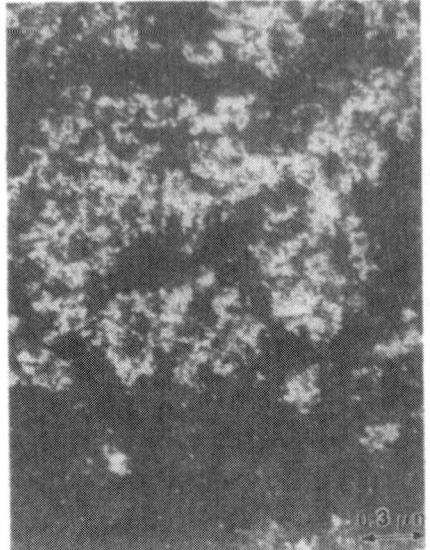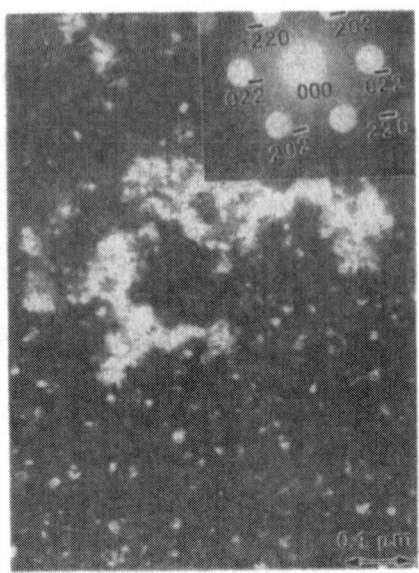

Fig. 5 Single crystal regions formed in $Si_{0.7}Al_{0.3}$ alloy after it was annealed at 500°C for 3.5 hours.

Fig. 6 Diffraction pattern of the single crystal regions formed in $Si_{0.7}Al_{0.3}$ alloy after annealing.

Fig. 7 TEM micrograph of an as-deposited $Si_{0.84}Al_{0.16}$ alloy.

Fig. 8 $Si_{0.84}Al_{0.16}$ alloy after it was annealed at 500°C for 10 hours.

The formation of single crystal regions has been reported in amorphous silicon-nickel alloys in which $NiSi_2$ precipitates were first formed and then catalyzed the formation of the single crystal regions [7]. We have observed previously the formation of large single crystal like region in CdSe thin films that were annealed in-situ [10]. The formation of the single crystal regions in our silicon-aluminum alloys do not resemble any of the previously reported works [7,10]. In the case of silicon-nickel it was the migration of the $NiSi_2$ that caused the formation of the single crystal regions, whereas in the case of CdSe the formation occurred gradually as certain grains, having <111> orientation, grew and consumed all the other grains.

In a previous work [6], amorphous silicon-aluminum alloys containing 30 at.% aluminum crystallized by an eutectic reaction, i.e. separate crystalline silicon and aluminum phases were formed, with"spherulitic" morphology. In our samples we did not observe any "spherulitic" morphology, neither did we obtain strong evidence from the diffraction patterns that crystalline silicon phase is formed.

Silicon-Aluminum with 16 at.% Al.

Similar in-situ experiments were carried out for the samples containing 16 at.% aluminum. However, we did not observe any crystalline phase even after a 10 hour anneal at 500°C. In Fig. 7 and Fig. 8 we show the microstructure of the film as-deposited and after a 10 hour anneal at 500°C respectively. As it can be seen from Fig. 7 and 8 there is no change in the film structure. Using electron diffraction we determined that the structure of the film is amorphous before and after the annealing.

Our results indicated that no polymorphous transformation takes place when the concentration of aluminum is 16 at.%. In fact as we mentioned above we did not detect any phase transformation at all in this material. This is in contrast with a previous report that indicated that polymorphous transformation of amorphous $Si_{1-x}Al_x$ alloy having $x \leq 0.20$ takes place, and for $x=0.15$ this occured even after a 1 min anneal at 500°C [6].

SUMMARY

The crystallization of amorphous silicon-aluminum alloys was studied by TEM during an in-situ anneal at 500°C. It is shown that the phase transformation of the film depends upon the aluminum concentration. For films containing 30 at.% Al, nucleation was observed after 40 min at 500°C while no nucleation was observed in samples with 16 at.% Al even after 10 hours anneal at 500°C. The size of the crystallites in the samples having 30 at.% Al grew during annealing, while in some areas aluminum single crystal regions were formed having <111> orientation.

ACKNOWLEDGEMENTS

Two of the authors FL and MKH would like to acknowledge support from DARPA under contract MDA972-91-K-0001.

REFERENCES

1. C.R. Wronski, Solid State Tech., 131, 113 (1988).
2. M.K. Hatalis and D.W. Greve, J. Appl. Phys., 63, 2260 (1988).
3. S.R. Herd, P. Chaudhari and M.H. Brodsky, J. of Non-Cryst. Solids, 7, 309 (1972).
4. D. Sigurd, G. Ottaviani, V. Marrello, J.W. Mayer and J.O. McCaldin, J. of Non-Cryst. Solids, 12, 135 (1973).
5. J. Stoemenos et al, Appl. Phys. Lett., 58, 1196 (1991).
6. U. Koster and P. Weiss, J. of Non-Cryst. Solids, 17, 359 (1975).
7. R.C. Cammarata, C.V. Thomson, C. Hayzelden and K.N. Tu, J. Mater. Res., 5, 2133 (1990).
8. F.A. Trumbore, Bell System Tech. J., 39, 205 (1960).
9. R.P. Elliot, Constitution of Binary Alloys, first supplement (McGraw-Hill, New York, 1965).
10. M.K. Hatalis, F. Lin and M.R. Westcott, MRS Symposium Proc., 164, 87 (1990).

SECONDARY GRAIN GROWTH IN HEAVILY DOPED POLYSILICON DURING RAPID THERMAL ANNEALING

S. BATRA, K. PARK, M. LOBO AND S. BANERJEE

The University of Texas at Austin, Austin, TX - 78712.

ABSTRACT

To successfully implement Silicon-on-Insulator (SOI) technology using polysilicon-on-oxide, it is necessary to maximize the grain size such that the active devices are entirely within very large single crystal grains. A drastic increase in grain size in polysilicon has been reported due to secondary grain growth in ultra-thin, heavily n-type doped films upon regular furnace annealing. Very little work has been undertaken, however, to study secondary grain growth during Rapid Thermal Annealing (RTA).This paper is a study of the grain growth mechanism in heavily P-doped, amorphous silicon films during RTA. Secondary grains as large as 16 μm have been obtained in 160 nm thick films after a 180 s RTA at 1200 °C, representing a grain-size-to-film-thickness-ratio of 100:1. This is the largest secondary grain size and grain-size-to-film-thickness reported in the literature. A detailed analysis of negatively charged silicon vacancies has also been employed to explain the lower activation energy (1.55 eV) of secondary grain growth compared to that of normal grain growth (2.4 eV).

INTRODUCTION

Polysilicon-on-oxide is a promising implementation of SOI technology towards achieving three-dimensional integration in integrated circuits [1]. It is a simple, inexpensive approach with the additional benefit of being compatible with bulk silicon technology. Polysilicon, however, is composed of small grains with random orientations separated by grain boundaries, which degrade the electrical performance of polysilicon MOSFETs due to the presence of a large number of electrically active traps. There are two approaches to reduce the number of traps at the grain boundaries. One of them is to increase the grain size in polysilicon to the extent that the active devices would be entirely within very large, single crystal grains and thus be devoid of grain boundaries altogether. However, normal grain growth in polysilicon upon high temperature annealing is limited by the film thickness due to a "thermal grooving" effect where the grain boundaries intersect the surface when adjacent grains impinge [2,3]. It has been reported that it is possible to achieve extremely large grains by exploiting a "secondary" grain growth phenomena in ultra-thin, heavily n-type polysilicon films upon furnace annealing [4]. The other approach is an interim solution in which the grain boundary potential barrier height is reduced by using grain boundary passivation. In this paper, we will discuss secondary grain growth in heavily P-doped, as-deposited amorphous silicon-on-oxide ultra-thin films upon RTA. Extensive experimental data about the mechanism is presented for a comprehensive time-temperature RTA matrix. Although, negatively charged silicon vacancies have been known to induce secondary grain growth by enhancing grain boundary mobility [5], we show how the relative change of the total vacancy concentration as a function of doping and temperature lowers the total activation energy of grain boundary mobility during secondary grain growth compared to that in lightly doped polysilicon.

EXPERIMENTAL

Amorphous silicon films, 30 to 300 nm thick, were deposited at 580°C by Low Pressure Chemical Vapor Deposition (LPCVD) on silicon substrates with a 200 nm oxide layer on top. The films were implanted/co-implanted with various doses of P or B resulting in doping concentrations ranging from 3×10^{16} cm^{-3} to 1.25×10^{21} cm^{-3}. To prevent dopants from escaping during annealing, the samples were capped with 20 nm of thermally grown oxide. The films were then subjected to RTA in an AG Associates annealer at temperatures between 950°C to 1200°C for various times between 3 s and 180 s. Plan-view as well cross-sectional Transmission Electron Microscopy (TEM) was used to examine the film morphology.

THEORY

From a phenomenological viewpoint, the grain growth rate G is proportional to the grain boundary mobility M and a driving force term F, where the mobility has an Arrhenius form, characterized by an activation energy Q and pre-exponential factor M_0.

$$G = M.F = M_0 \exp(-Q/kT).F \tag{1}$$

Since the grain boundary velocity also has the same activation energy Q, measurement of activation energy of grain growth should thus provide important clues about the growth mechanism. The standard model for grain growth cannot explain the lower activation energy of secondary grain growth in heavily n-type, doped amorphous silicon films, as compared to the activation energy of normal grain growth in lightly P-doped films, because it does not take into account the role of the negatively charged silicon vacancies at grain boundaries. From an atomistic picture of grain boundary motion, grain growth involves Si atoms in a higher potential grain jumping to vacancies at the grain boundary, and subsequently to sites in a lower chemical potential grain. Therefore, to include the silicon vacancies involved in the process, equation (1) should be modified as

$$G = M \left(\frac{C_v}{N_{Si}}\right) F = M' F \tag{2}$$

where C_v is the total vacancy concentration, N_{Si} is the Si atomic density and M' is the modified grain boundary mobility in the new model for grain growth.

Two mechanisms, dislocation glide which is a non-diffusive process independent of the silicon vacancy concentration, as well as dislocation climb which is a diffusive mechanism proportional to the vacancy concentration, are responsible for the grain boundary motion. Dislocation climb, however, can be considered as the dominant mechanism for secondary grain growth in heavily doped films, where the charged silicon vacancy concentrations are high. Hence, in lightly doped films where neutral vacancies are the dominant vacancy species, the grain boundary mobility M_l can be expressed as

$$M_l = KC_l \exp\left(\frac{E_{ml}}{kT}\right) = K\left[C_{lo}\exp\left(-\frac{E_{vl}}{kT}\right)\right]\exp\left(\frac{E_{ml}}{kT}\right) \tag{3}$$

where K is a proportionality constant, C_l is the total vacancy concentration in lightly doped polysilicon, C_{lo} is a prefactor, E_{vl} is the activation energy of formation of a vacancy at a grain boundary (=1.8 eV), and E_{ml} is the activation energy of migration of a grain boundary (=0.6 eV) in lightly doped polysilicon [5]. Therefore, $Q = E_{vl} + E_{ml}$ (= 2.4 eV) should be the activation energy for grain boundary mobility for normal grain growth in lightly doped polysilicon. For heavily P-doped polysilicon the grain boundary mobility M_h during secondary grain growth is also primarily due to dislocation climb, which is proportional to the total silicon vacancy concentration.

$$M_h = KC_h \exp\left(\frac{E_{mh}}{kT}\right) \tag{4}$$

C_h is the total (neutral plus charged) vacancy concentration in heavily doped polysilicon, and E_{mh} is the activation energy of migration of a grain boundary in heavily doped polysilicon. Assuming the activation energy of grain boundary migration to be the same in heavily and lightly doped polysilicon, the following relation is developed

$$M_h = M_l \left(\frac{C_h}{C_l}\right) \cong M_l \left(\frac{C_h}{C_{vo}}\right)$$

(5)

where C_l has been approximated as the neutral vacancy concentration in lightly doped polysilicon film, C_{vo}, and the total vacancy concentration for the heavy n-type doping, C_h, is the sum of the neutral and the singly and doubly negatively charged species, i.e. C_{vo}, C_{v-} and $C_{v=}$, respectively.

RESULTS AND DISCUSSION

The dependence of grain growth on P concentration is discussed first. Fig. 1 shows the grain growth of 160 nm polysilicon thin films with a P doping concentration of 3×10^{16} cm^{-3}. As can be seen from the figure there is no enhancement in grain growth for such low dopings. For heavier P-doping (1.25×10^{21} cm^{-3}), however, there is an enhanced secondary grain growth mechanism and secondary grains as large as 16 μm are observed in 160 nm polysilicon films after 180 s RTA at 1200 °C (Fig. 2). Different methods have been reported in the literature to distinguish secondary grains from normal grains and to analyze secondary grain growth. One method has been to consider grains greater than 2-4 times the film thickness as secondary grains and apply the 2-D Johnson-Mehl-Avrami analysis [4]. We have adopted the Wada-Nishimatsu technique [6], to determine the size of the 10 largest secondary grains in each sample, where each of these grains is typically much larger (greater than 2 times depending upon the annealing temperature) than the film thickness. A very large area of approximately 1mm² was scanned under plan-view TEM to determine the 10 largest secondary grains. This approach has been adopted because the secondary grains have large variations in grain sizes upon completion of the crystallization process, which precedes secondary grain growth in as-deposited amorphous silicon films. The crystallization process could, therefore, complicate the physical interpretation of the grain growth results if the Johnson-Mehl-Avrami analysis were employed. Also, this analysis is strictly valid only for 3-D grain growth because it is based on the growth impingement of various grains in an aggregate and the calculation of an "extended volume" had grain growth been unimpeded by impingement. Extension of this technique to 2-D for thin films is of doubtful validity.

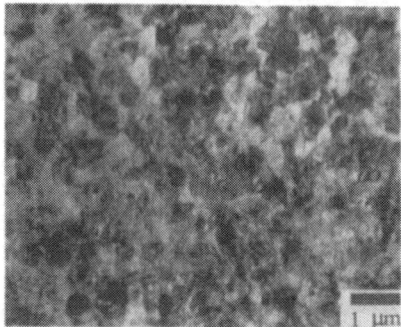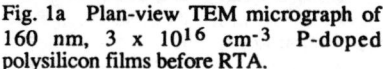

Fig. 1a Plan-view TEM micrograph of 160 nm, 3×10^{16} cm^{-3} P-doped polysilicon films before RTA.

Fig. 1b Normal grain growth in 160 nm, 3×10^{16} cm^{-3} P-doped, as-deposited amorphous silicon films after 60 s RTA at 1150 °C.

Using the above technique, the average secondary grain size in 160 nm thick films doped to 1.25×10^{21} cm^{-3} with P increases very rapidly in the first 3 - 10 s, and subsequently, increases linearly with anneal time (see Fig. 3). The films were also subjected to a 60 s isochronal RTA in the temperature range of 1000 - 1200 °C, in order to determine the activation energy of grain boundary mobility during secondary grain growth. Fig. 4 shows an Arrhenius plot obtained from the isochronal RTA data which determines the activation energy to be 1.55 eV for secondary grain growth. The fact that the grain size increases linearly with time confirms that we have secondary grain growth, and that we do not have impingement of secondary grains which would tend to give erroneously low activation energies. The initial non-linear growth rate is not clearly understood at present but may be related to the rapid grain growth in amorphous silicon during recrystallization.

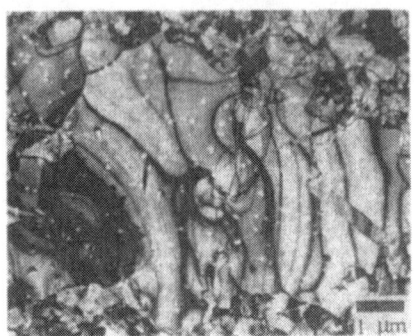

Fig. 2   Secondary grain growth in 160 nm, 1.25 x 10^{21} cm^{-3} P-doped, as-deposited amorphous silicon films after 180 s RTA at 1200 °C.

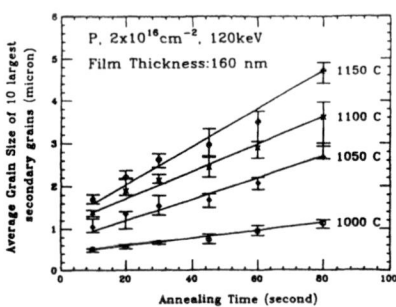

Fig. 3   Average grain size of 10 largest secondary grains in 160 nm, 1.25 x 10^{21} cm^{-3} P-doped, as-deposited amorphous silicon films for RTAs between 1000-1150 °C.

Understanding the reasons behind the lower activation energy of grain boundary mobility during secondary grain growth (1.55 eV) in heavily n-type, as-deposited amorphous silicon films compared to that for normal grain growth in lightly doped films (2.4 eV), is extremely important as it provides insight into the secondary grain growth mechanism. It was modelled earlier in equation (5) that the grain boundary mobility in highly n-type doped films is dependent on the concentration of the negatively charged silicon vacancies, which, in turn, is dependent on doping through the Fermi level [7]. The concentration of these silicon vacancies increases with higher doping concentrations as the Fermi level moves past the singly (E_c - 0.44 eV) and doubly negatively (E_c - 0.11 eV) charged vacancy levels, thereby increasing grain boundary motion via dislocation climb [5]. We have calculated the ratio of the total vacancy concentration to the neutral vacancy concentration as a function of temperature (950 to 1200 °C) and P doping concentration (10^{19} to 1.25 x 10^{21} cm^{-3}). The temperature dependence of the band edges and the various charged vacancy levels, along with Fermi statistics for donor ionization and bandgap narrowing at high doping have been included in the calculation. It is interesting to note that the doping in these films is slightly below the solid solubility limit of P in this temperature range (~1.6 x 10^{21} cm^{-3}), and hence all the P should be substitutionally dissolved. Also, we do not expect any significant loss of P during the anneals because, from cross-sectional TEM micrographs, the 20 nm cap oxide was found to remain intact. The total vacancy concentration at 1150 °C increases by more than two orders of magnitude when the P doping is increased from 3 x 10^{16} cm^{-3} to 1.25 x 10^{21} cm^{-3} because of increased negatively charged vacancies, thereby enhancing grain boundary mobilities for during secondary grain growth (Fig. 5). However, the ratio of total to neutral vacancies decreases from 940 to 450 when the temperature increases from 1000 to 1200 °C. As shown in Fig. 5, the activation energy associated with the ratio of the total to neutral vacancy concentration in heavily n-type films, C_h/C_l, is -0.71 eV which reduces the activation energy of grain boundary mobility during secondary grain growth in heavily n-type polysilicon from 2.4 eV for normal grain growth in lightly doped polysilicon to (2.4 - 0.71 =) 1.69 eV. This value agrees fairly well with the experimental value of 1.55 eV (Fig. 4).

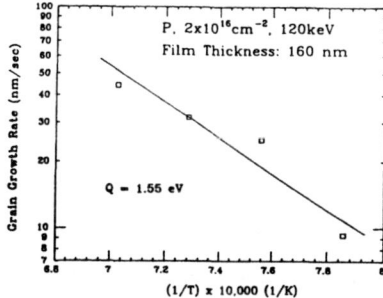

Fig. 4   Arrhenius plot of secondary grain growth rate in 160 nm, 1.25 x 10^{21} cm^{-3} P-doped, as-deposited amorphous silicon films versus RTA temperature.

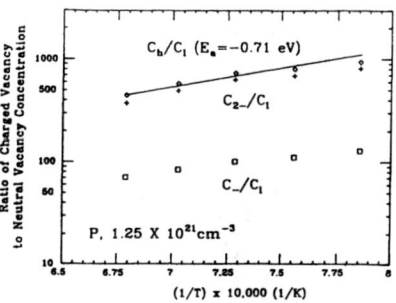

Fig. 5   Calculated ratio of total-to-neutral silicon vacancy concentration as a function of anneal temperature in 1.25 x 10^{21} cm^{-3} P-doped silicon.

If the negatively charged vacancies do, in fact, lower the activation energy of grain boundary mobility during secondary grain growth, then we should not observe significant enhancement of grain boundary mobility in B-doped films because, the positively charged vacancy level is extremely close to the valence bandedge ($E_v + 0.6$ eV) and therefore the total vacancy concentration should not exhibit a drastic change with p-type doping. Indeed, negligible grain growth enhancement is observed in heavily B-doped polysilicon for 60 s RTA at 1150 °C (Fig. 6). Detailed calculations, taking into account the solid solubility limit of B at 1150 °C (~ 2.5 x 10^{21} cm^{-3}), show that the total vacancy concentration increases by less than an order of magnitude for B doping of 1.25 x 10^{21} cm^{-3} compared to that in intrinsic silicon at 1150 °C. Fig. 7 also shows that the grain growth enhancement upon RTA for heavy P doping can be quenched by counter-doping with B, whereby the Fermi level moves closer to midgap.

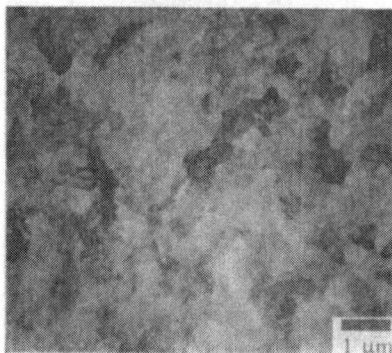

Fig. 6   Negligible grain growth in 160 nm, 1.25 x 10^{21} cm^{-3} B-doped as-deposited amorphous silicon films after 60 s RTA at 1150 °C.

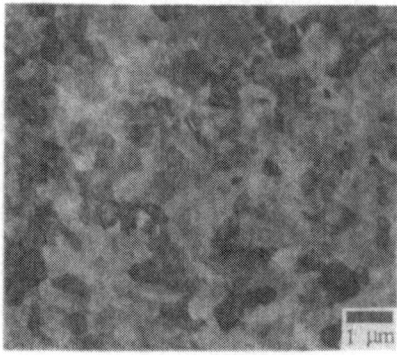

Fig. 7   Quenching of secondary grain growth in 160 nm, 1.25 x 10^{21} cm^{-3} P-doped, as-deposited amorphous silicon films after 1150 °C, 60 s RTA by 1.25 x 10^{21} cm^{-3} B counter-doping.

As reported by us elsewhere [8], a grain size to film thickness ratio of over 100:1 has been obtained after 60 s RTA at 1150 °C in 30 nm thick, 1.25×10^{21} cm^{-3} P-doped super-thin films, while in 160 nm films a ratio of 20:1 is obtained. This observation, coupled with the observed linear increase of grain size with time, is consistent with earlier reports in the literature [9] that secondary grain growth is driven by surface energy anisotropy, and that the driving force is inversely proportional to the film thickness.

CONCLUSIONS

The variation of the negatively charged silicon vacancy concentration with doping and temperature has been used to explain the enhanced grain boundary mobility in heavily P-doped films. In particular, the lower activation energy of grain boundary mobility during secondary grain growth (1.55 eV) compared to that for normal grain growth (2.4 eV) has been shown to result from the variation of the charged vacancy concentration with temperature. Secondary grains as large as 16 µm have been obtained in 160 nm thick films which is the largest secondary grain size reported in the literature. The ratio of 100:1 of resulting grain size to film thickness is to date the largest reported in the literature.

ACKNOWLEDGEMENTS

This work was supported in part by SRC/SEMATECH and in part by the Texas Advanced Research Program.

REFERENCES

1. J. F. Gibbons and K. F. Lee, IEEE Electron Dev. Lett., EDL-1, 117 (1980).
2. W. W. Mullins, ACTA Metallurgica, 6, 414 (1958).
3. T. Nolan, R. Beyers, and R. Sinclair, presented at the 1990 MRS Fall Meeting, Boston, MA, 1990 (unpublished).
4. H. J. Kim and C. V. Thompson, J. Appl. Phys, 67, 757 (1990).
5. H. J. Kim and C. V. Thompson, J. Electrochem. Soc., 135, 2312 (1988).
6. Y. Wada and S. Nishimatsu, J. Electrochem. Soc., 125, 1499 (1978).
7. W. Shockley and J. T. Last, Phys. Rev., 107, 392 (1957).
8. K. Park, S. Batra and S. Banerjee, accepted for publication by Appl. Phys. Lett.
9. C. V. Thompson, J. Appl. Phys., 58, 763 (1985).

EFFECT OF PROCESSING CONDITIONS ON THE SURFACE MORPHOLOGY OF THIN POLYSILICON FILMS USED FOR DRAM CELL CAPACITORS

VIJU K. MATHEWS
Micron Technology, Inc., Boise, ID 83706.

ABSTRACT

The surface texture of polysilicon films, used for capacitor storage nodes, have been investigated for various deposition conditions. An increase in capacitance of approx. 70% has been observed in this work. The reproducibility of the surface texture and the resulting capacitance can be improved by optimizing the deposition pressure. The gain in capacitance was also observed to be affected by the thickness of the film. In the mass flow controlled regime the increase in capacitance was 12%. The increase in surface roughness and the gain in capacitance for the hot wall process provides a simple alternative for satisfying the charge capacity requirements in high density dynamic random access memories.

INTRODUCTION

The reliable operation of a dynamic random access memory (DRAM) device requires a minimum level of stored charge based on design and functional parameters. The reduction in cell sizes associated with high density DRAM's has a direct effect on this charge capacity since it reduces the area available for the storage node of the cell capacitor. The implementation of cell designs utilizing advanced trench [1] or stacked[2] structures, to overcome this problem, usually tends to increase the complexity of the fabrication process. The improvement in capacitance that can be achieved by reducing the thickness of the dielectric film is constrained by tunneling mechanisms that can lead to high leakage current. The application of high dielectric constant materials, like Ta_2O_5 [3], has not yet being realized on a large scale because of several unresolved issues related to leakage currents and reliability. The other potential option involves the "roughening" or "texturization" of the polysilicon film used as the storage plate. These techniques increase the surface to planar area ratio of the films. One approach has been the deposition of standard (smooth) polysilicon films followed by "texturization" using reactive ion etching [4], low temperature oxidation [5] or excimer laser processing [6]. Another approach employs the fact that the surface texture of polysilicon films can be altered by the deposition process [7]. The commonly used deposition conditions generate a smooth surface texture with a a columnar microstructure to satisfy lithographic and film doping considerations. Modification of these process conditions can result in the development of a very rough surface texture with a corresponding enhancement in the cell capacitance [8-10].

The changes in the surface texture of polysilicon films under various deposition conditions have been investigated in this study. The surface roughness of the films were evaluated using reflectance measurements and scanning electron microscopy. Electrical measurements were made to determine the gain in cell capacitance for the rough films.

EXPERIMENTAL

P-type, (100) wafers with 200 nm of silicon dioxide were used as the substrates for the experiments. A vertical hot wall and a cold wall reactor were used for depositing the polysilicon films (approx. 60 nm) using silane and silane/hydrogen mixture. The hot wall reactor was operated at temperatures ranging from 550 to 650C and a minimum pressure of 80 mTorr. The cold wall deposition was at 700C and 100 Torr. The cell dielectric consisted of an 8 nm reoxidized silicon nitride film. A 100 nm, thermally doped, polysilicon film was used as the top plate. The structures used for the capacitance measurements had a planar area of 6.5 E-5 cm2. The relative values mentioned in the paper are with respect to a smooth (625 C) polysilicon film deposited under standard conditions.

RESULTS AND DISCUSSION

The effect of deposition temperature on the surface texture of polysilicon films deposited in a hot wall reactor is shown in Fig. 1. The corresponding change in cell capacitance is shown in Fig. 2. The deposition process in this case is controlled by the reaction rate and is very sensitive to variations in temperature. For the temperature range used in this study, the film goes through an amorphous to polycrystalline transition [7]. Also shown in Fig. 2, is the change in the reflected intensity of the film with deposition temperature. These values correlate very well with the surface roughness observed in the SEM micrographs and the changes in the capacitance. However, if the films are less than 100 nm thick, interference from the underlying films affects the refelected intensity. This phenomena also prevents the accurate measurement of the film thickness using optical measurements. The reproducibility of the the surface texture and capacitance at a given temperature can be improved by varying the pressure. The change in capacitance with pressure is shown in Fig. 3 and the surface texture related to these changes is shown in Fig. 4. It was also observed that the capa-

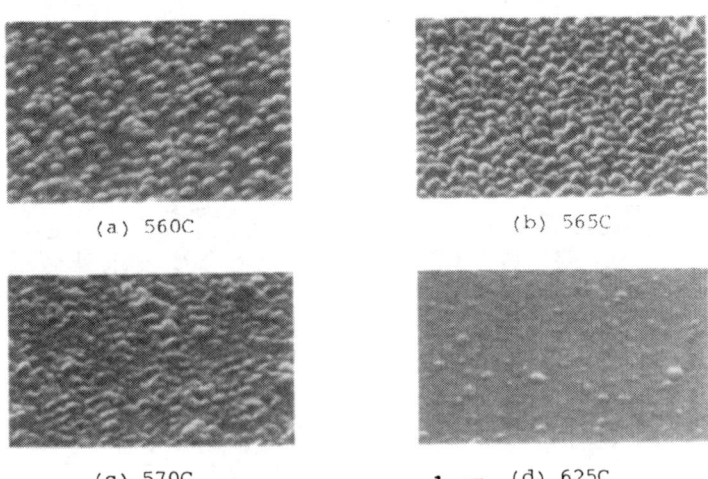

(a) 560C   (b) 565C

(c) 570C   ├─────────────┤ 1 μm (d) 625C

Fig. 1. Effect of deposition temperature on the surface texture of polysilicon films.

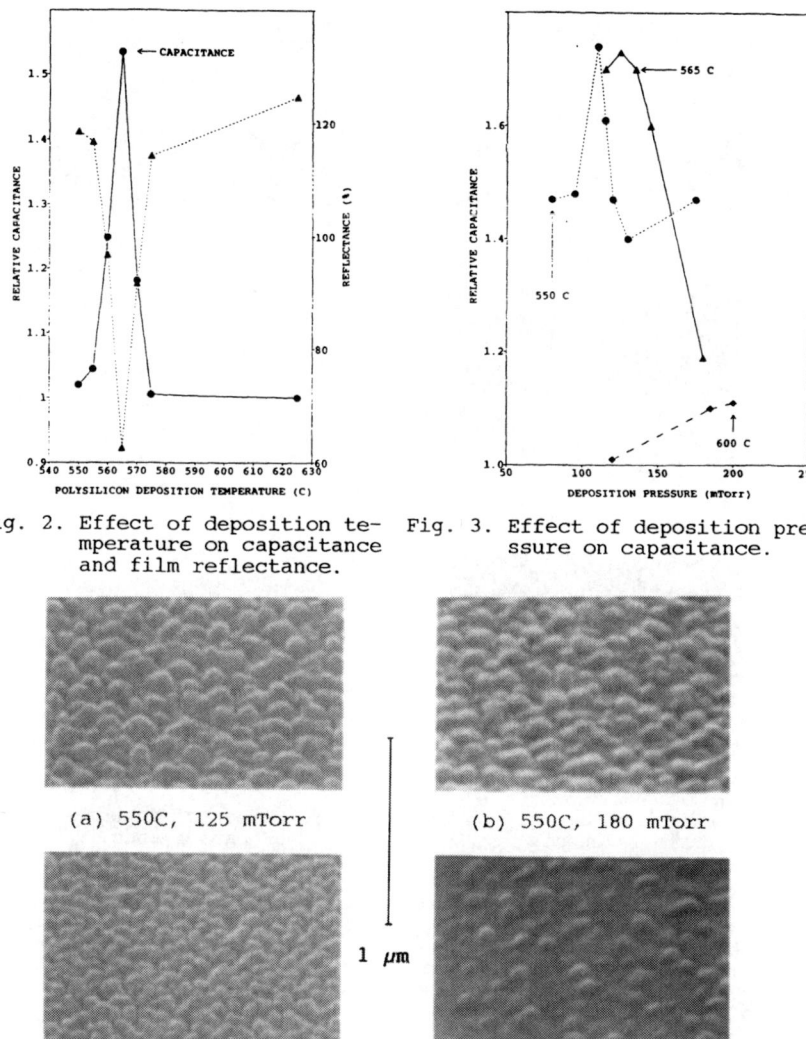

Fig. 2. Effect of deposition temperature on capacitance
and film reflectance.

Fig. 3. Effect of deposition pressure on capacitance.

(a) 550C, 125 mTorr

(b) 550C, 180 mTorr

1 μm

(c) 565C, 125 mTorr

(d) 565C, 180 mTorr

Fig. 4. Effect of deposition pressure on the surface texture.

citance was affected by the thickness of the film. The surface texture for films deposited at 550C with different deposition times
is shown in Fig. 5. As mentioned earlier, the low reflectivity of
the film made it difficult to measure the thickness of the film
using optical techniques. For the 550C film, a deposition time of
15 min corresponds to a thickness of approximately 60 nm (based on
step height measurements). The effect of this texture variation
with thickness on the capacitance is shown in Fig. 6.

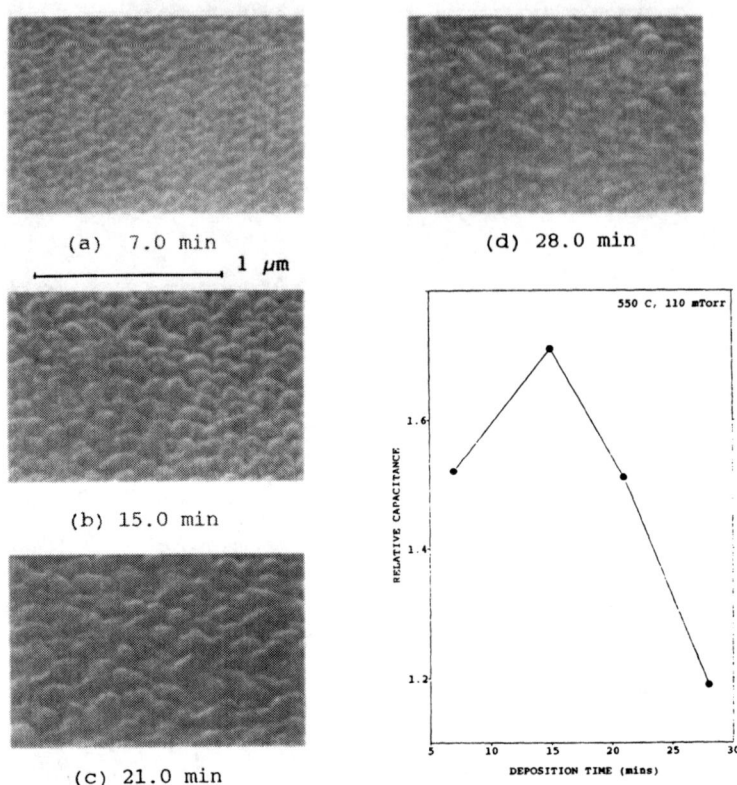

(a) 7.0 min

1 μm

(d) 28.0 min

(b) 15.0 min

550 C, 110 mTorr

RELATIVE CAPACITANCE

DEPOSITION TIME (mins)

(c) 21.0 min

Fig. 5. Effect of deposition time
(film thickness) on surf-
ace texture.

Fig. 6. Effect of deposition time
on capacitance.

Further increase in the deposition temperature and pressure in
a hot wall reactor results in the type of surface texture shown in
Fig. 7A and B. In addition to the increase in the surface roughness, large structures similar to those reported by Joubert [11]
and Kamins [12], are observed on the surface. However, the distribution of these structures is very non-uniform across the wafer.

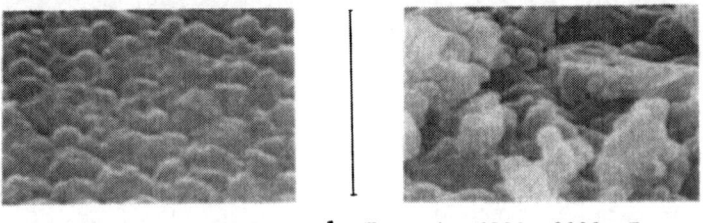

(a) 600C, 1000 mTorr   1 μm   (b) 600C, 2000 mTorr

Fig. 7A. Effect of high pressures on the surface texture at 600C.

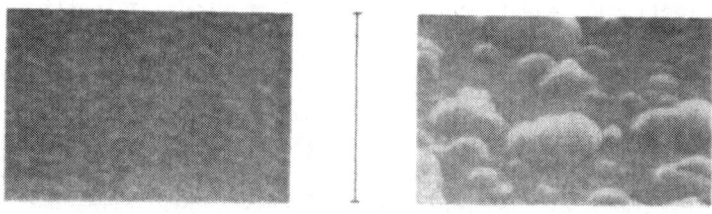

(a) 650C, 1000 mTorr   1 μm   (b) 650C, 2000 mTorr

Fig. 7B. Effect of high pressures on the surface texture at 650C.

The type of surface texture observed for the reaction rate controlled process can also be obtained in the mass flow controlled
regime. In this case the process is less sensitive to temperature
variations. For a given temperature, pressure, and carrier gas flow
rate the surface texture depends on the deposition rate, which in
turn is determined by the flow rate of silane. Micrographs of surface textures as a function of the deposition rate is shown in Fig.
8. The change in capacitance for these films is shown in Fig. 9.

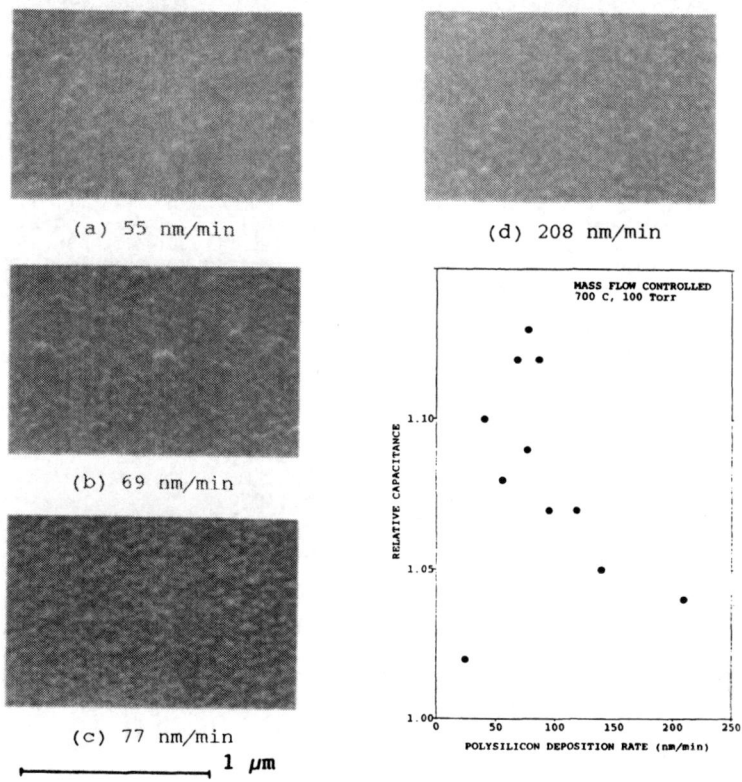

(a) 55 nm/min

(b) 69 nm/min

(c) 77 nm/min

(d) 208 nm/min

1 μm

Fig. 8. Effect of deposition rate   Fig. 9. Effect of deposition rate
on the surface texture.   on capacitance.

Electrical measurements [10,13,14] on rough polysilicon structures have indicated that the small increase in the leakage current observed with a rough capacitor plate is negligible compared to the typical increases of several orders of magnitude observed when $SiO2$ is used instead of reoxidized silicon nitride on a rough electrode. Reliability measurements on rough and smooth capacitors have indicated a lower lifetime for the rough capacitors at high fields (8-12 MV/cm), but they are expected to perform better than smooth capacitors at the normal operating field of 3 MV/cm because of their higher field acceleration coefficient.

CONCLUSIONS

The changes in the surface morphology and corresponding increase in cell capacitance that can be achieved under various polysilicon deposition conditions has been reported in this paper. A 70 % increase in capacitance has been observed for films deposited in the reaction rate controlled regime. This technique can provide the increase in capacitance required for high density DRAM's.

ACKNOWLEDGEMENTS

The author would like to thank A. Martin, M. Tuttle, P.C. Fazan and H.C. Chan for discussions. The assistance from the process development, manufacturing, and Y.E. SEM Lab groups at Micron is also gratefully acknowledged.

REFERENCES

1. N. C. C. Lu, IEEE Circ. Dev. Mag., Vol. 5, No. 1, 27(1989).
2. W. Wakamiya, Y. Tanaka, H. Kimura, H. Miyatake, and S. Satoh, Symp. on VLSI Tech., Digest of Technical papers, 69 (1989).
3. H. Shinriki, Y. Nishioka, Y. Ohji, and K. Mukai, IEDM Tech. Digest, 684 (1986).
4. P. C. Fazan, and R. R. Lee, IEEE Trans. Electron Dev., EDL-11, 279 (1990).
5. T. Mine, S. Iijima, J. Yugami, K. Ohga and T. Morimoto, Ext. Abs. 21st Conf. on SSDM, 137 (1989).
6. V. K. Mathews and C. Yu, Ext. Abs. 179th ECS Meeting, Vol. 91-1, No. 374, 565(1991).
7. R.Bisaro, J.Magarino, N.Proust, and K.Zellama, J. Appl. Phys., 59 (4), 1167 (1986)
8. M. Sakao, N. Kasai, T. Ishijima, E. Ikawa, K. Terada and T. Kikkawa, IEDM Tech. Digest., 655 (1990).
9. H. Watanabe, N. Aoto, S. Adachi, T. Ishijama, E. Ikawa and K. Terada, Appl. Phys. Lett., Vol. 58, No. 3, 251 (1991).
10. V. K. Mathews, P. C. Fazan, G. S. Sandhu, A. Ditali and H. R. Rhodes, Ext. Abs. 179th ECS Meeting, Vol. 91-1, No. 375, 567 (1991).
11. P. Joubert, B. Loisel, Y. Chouan, L.Haji, J. Elec. Soc., Vol. 134, No. 10, 2541 (1987).
12. T. I. Kamins and T. R. Cass, Thin Solid Films, Vol. 16, 147 (1973).
13. H. C. Chan, V. K. Mathews and P. C. Fazan, to be published in IEEE Electron Device Letters.
14. P. C. Fazan, V. K. Mathews, H. C. Chan and A. Ditali, to be published in Applied Physics Letters.

CONTROLLED INTERFACE ROUGHNESS IN GaAs/AlAs SUPERLATTICES

WILLIAM R. MILLER, JR.,* W. J. BOETTINGER,** W. F. TSENG,**
J. PELLEGRINO,** AND J. COMAS**

*Current Address: Guest Scientist at NIST, on leave from Penn State University, Middletown, PA 17057

**National Institute of Standards and Technology, Gaithersburg, MD 20899

ABSTRACT

We report the results of our study of controlled interface roughness in low-order GaAs/AlAs superlattices. Samples were prepared using either the interrupted growth or the migration-enhanced epitaxy (MEE) technique. The samples were prepared with m atomic planes of GaAs and m atomic planes of AlAs ($m \times m$) per modulation wavelength and repeated p times. For this study, $m = 1$ or 3. The samples were studied using X-ray diffraction. The interrupted growth samples both showed a split in one diffraction line indicating layers were not of integral order while the MEE samples showed no splitting, indicating integral order layers.

INTRODUCTION

There is considerable interest in the fabrication,[1] basic physics,[2,3] and device applications[4] of multilayer structures and superlattices. The physical properties of such entities can be strongly affected by the structure and composition of the interfaces between layers. Most theoretical work is based on the assumption of abrupt interfaces. Departures from ideal interfaces, i.e., interfacial disorder, may come from factors such as substrate roughness and variations in deposition rate.

Recently, Schuller et al.[5] introduced the concept of "controlled interface roughness" which fundamentally differs from random roughness. Controlled roughness comes from deposition of a noninteger number of atomic planes in each layer and results in an interfacial layer that consists of an AlAs/GaAs mixture that varies in a systematic manner in succeeding interfaces. The resulting diffraction pattern for controlled roughness is different than that for random roughness. In the case of random roughness, the line is broadened, but with no change in position. For controlled roughness, the line is sharp and its position altered. Using their notation, a diffraction spectrum consists of two series of peaks, denoted by $n^{\pm m}$ where n gives the order of the main reflection and $\pm m$ labels its satellites. For a 1×1 sample, a peak 0^{+1}, counting from the (000) line, should coincide with the 1^{-1} peak, counting from the (002) peak. Since there is a noninteger number of atomic layers, the 0^{+1} and 1^{-1} lines do not coincide, but the peak near $2\theta = 15°$C is split.

Schuller et al. were able to prepare samples which showed this behavior by carefully controlling the growth rates to yield a noninteger number of atomic planes. In our work, the objective was to see whether or not this behavior occurred in low-order GaAs/AlAs superlattices prepared by (1) the interrupted growth techniques and (2) by the migration-enhanced epitaxy technique. We prepared and studied by X-ray diffraction 1×1 and 3×3 samples using both deposition modes.

EXPERIMENTAL

The samples were grown in a molecular beam epitaxy system. The growth rates were measured in situ by reflection high-energy electron diffraction (RHEED). The semi-insulating GaAs (001) substrates were prepared using a (sulfuric acid/hydrogen peroxide/deionized water) etch followed by the formation of a thin protective oxide layer which was thermally desorbed in the growth chamber.

Two growth techniques were used: interrupted growth and migration-enhanced epitaxy. In the interrupted growth method, an AlAs layer is grown first. The Al shutter is then closed, and a pause follows, during which only the As shutter is open. The GaAs layer is then grown, followed by the same pause. The AlAs/pause/GaAs/pause sequence is then repeated until a 500-nm superlattice is completed.

In the MEE mode, only one shutter is open at any time. Conditions are maintained so that one monolayer is incident on the substrate while the shutter is open. The growth sequence consists of opening the Al shutter long enough to grow one monolayer, followed by a pause in which all shutters are closed. The As is opened to grow one monolayer, followed by a pause with all shutters closed. These four steps are repeated until the desired number of AlAs layers are grown. The GaAs layers are then grown in a similar four-step sequence. This is repeated until 500 nm is grown.

The samples were studied using a single-axis diffractometer with resolution of 0.01 deg. The Cu K_α line source provides a beam of $\lambda = 0.15406$-nm X-rays which are collimated by rectangular Soller slits and detected by a solid Ge detector.

RESULTS

The results are shown in figures 1 to 4. For the 1×1 MEE sample, figure 2(a,b) shows a single line at 15.58 deg, while for the 1×1 interrupted growth sample, figure 1(a,b) shows two peaks at 15.46 deg and 15.83 deg; i.e., the expected peak has split into two peaks. For the 3×3 interrupted growth sample, figure 3 shows two peaks at 59.00 deg and 59.55 deg, while figure 4 for the 3×3 MEE sample shows a single peak at 58.96 deg; again, the expected peak has split. Using these values of angles, calculations indicate that the interrupted growth samples are of a noninteger number of atomic planes, and the MEE samples are of integer values.

CONCLUSION

These results suggest a systematic difference between MEE samples and interrupted growth samples. It further suggests that an MEE sample will be of an integer number of layers, while an interrupted growth sample will more likely be – depending on factors such as control of deposition rate – of a noninteger number of layers and hence will show more controlled interface roughness.

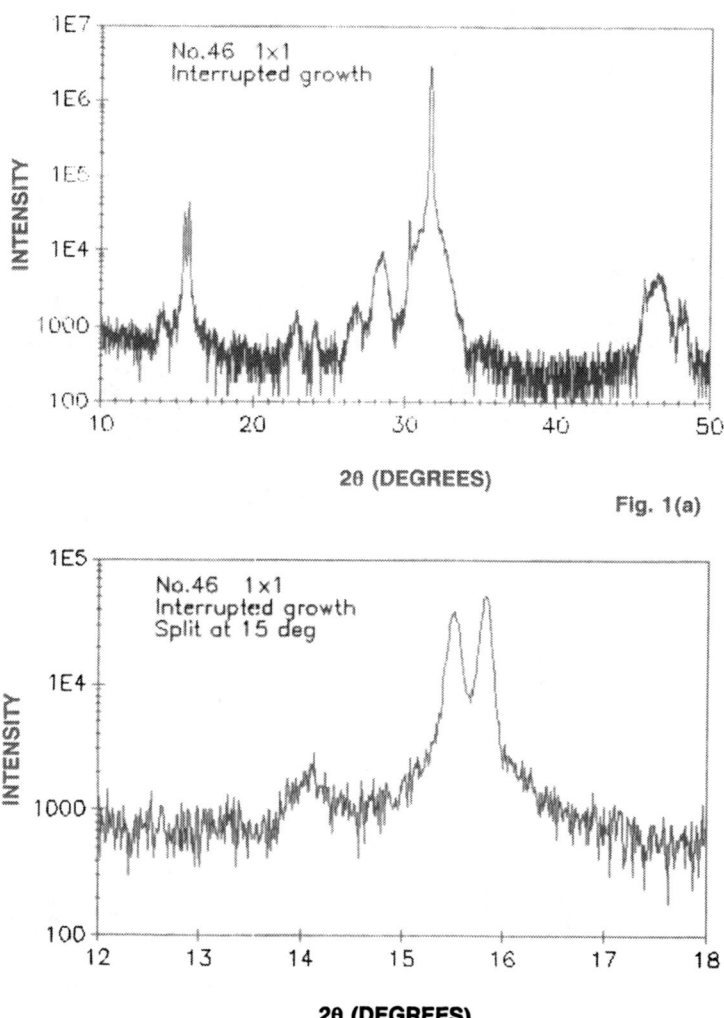

Fig. 1(a)

Fig. 1(b)

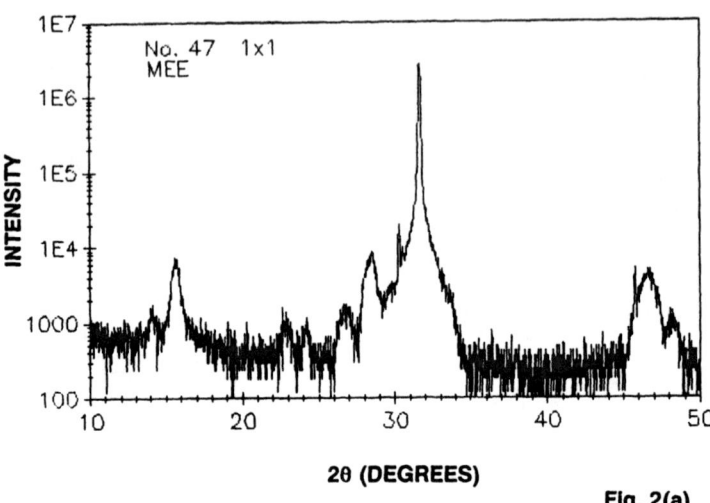

Fig. 2(a)

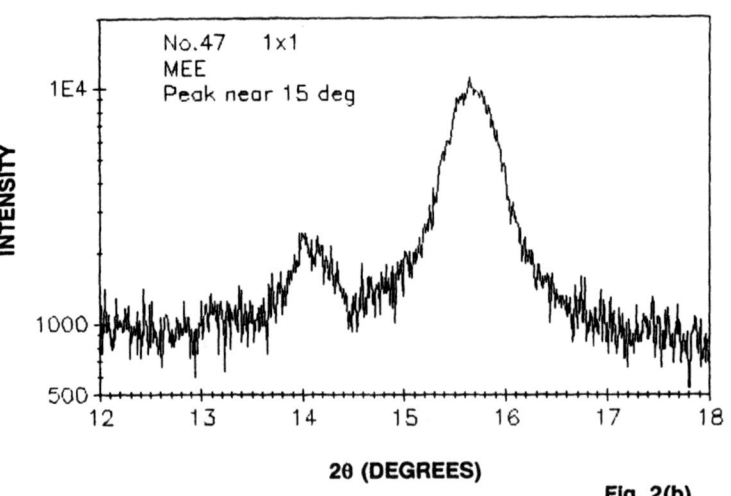

Fig. 2(b)

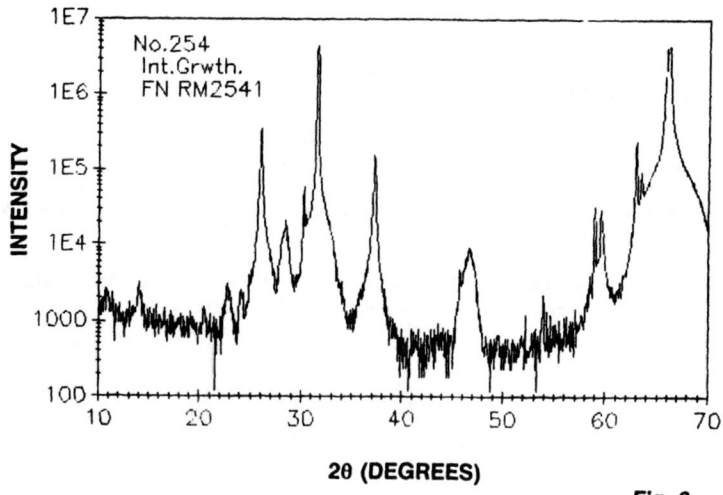

Fig. 3

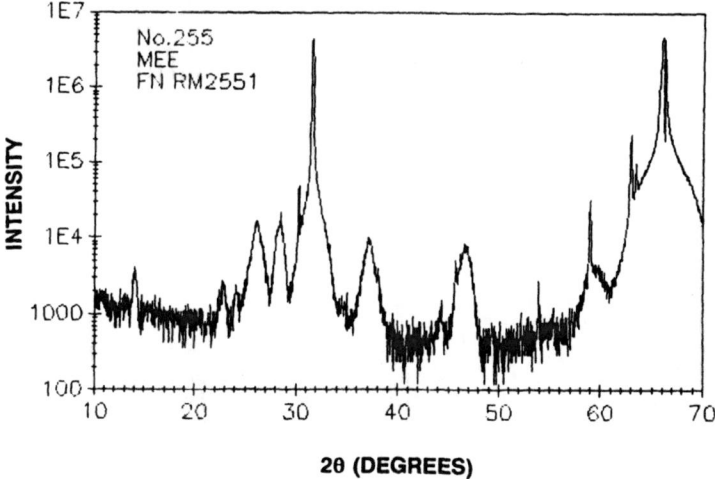

Fig. 4

REFERENCES

(1) M. A. Herman and H. Sitter, *Molecular Beam Epitazy* (Springer-Verlag, New York, 1989), chap. 6.

(2) M. G. Cottam and D. R. Tilley, *Introduction to Surface and Superlattice Excitations* (Cambridge University Press, New York, 1989), chap. 7.

(3) Gerald Bastard, *Wave Mechanics Applied to Semiconductor Heterostructures* (Halsted Press, New York, 1988), pp. 18-27.

(4) M. Shur, *Physics of Semiconductor Devices* (Prentice-Hall, Englewood Cliffs, New Jersey, 1990), pp. 597-612.

(5) I. K. Schuller, M. Grimsditch, F. Chambers, G. Devane, H. Vanderstraeten, D. Neerinck, J.-P. Locquet, and Y. Bruynseraede, Phys. Rev. Lett., 12 (10), 1235-1238 (1990).

PERIODICITIES IN THE X-RAY DIFFRACTION OF
LOW ORDER ALAS/GAAS SUPERLATTICES

JOSEPH PELLEGRINO, S.QADRI*, W.TSENG, W.R.MILLER**, AND J.COMAS,
NATIONAL INSTITUTE OF STANDARDS AND TECHNOLOGY, MD., *NAVAL
RESEARCH LAB WASH. D.C.,**ON SABBATICAL LEAVE FROM PENN. STATE
UNIV., MIDDLETOWN PA.

ABSTRACT

In this work we examine the physical properties for the superlattice system $(GaAs)n_1$ $(AlAs)n_2/GaAs(100)$ for low values of n_1 and n_2, i.e., $n_1 = n_2 = 3, 6, 12$. Normal, interrupted growth, and migration enhanced epitaxy (MEE) growth techniques were used to grow the superlattice structures in a molecular beam epitaxy system. X-ray diffraction spectra were obtained, and the major and satellite peak positions were analyzed to obtain the superlattice periodicity. An analysis of the major diffraction peaks and their associated satellites produced superlattice periodicity in good agreement with theory. Diffraction peaks were also observed in regions adjacent to the primary diffraction peaks which did not occur in the expected satellite positions. An analysis of these peaks relative to the primary peak indicate periodicities corresponding to layer thickness greater than the intended period. One possible cause for these periodicities is growth conditions that exist during the growth of the superlattice which result in the deposition of fractional monolayers. In this study we present results which suggest that an arsenic-deficient growth condition may be a contributing factor in the deposition of fractional monolayers.

INTRODUCTION

The molecular beam epitaxy (MBE) growth technique permits the fabrication of superlattices with monolayer periodicities. Because these structures involve individual layers several tenths of a nanometer (angstroms) thick, X-ray diffraction can be effectively used for structural examination. There is much interest in the structure of superlattices and compositionally modulated materials formed by imposing an atomic-scale periodicity during the growth of a film. One advantage of the MBE deposition technique in this regard is that the growth rate can be controlled to a considerable degree by regulating the temperature of the effusion cells and the substrate. The ability to control the growth rate, as well as the availability of in-situ reflection-high-energy-electron-diffraction (RHEED) capabilities to measure the growth rates, makes MBE well suited for fabricating structures requiring precise layer thicknesses. One category of material which demands a high degree of layer thickness control at the atomic level is superlattices. High-quality reproducible superlattice structures are required for optical and quantum confinement devices such as high electron mobility transitors, surface light-emitting lasers, self-electro-optic effect devices and quantum well lasers. In this work we attempt to correlate certain growth parameters with the structural periodicity of the superlattice.

EXPERIMENTAL

The superlattice materials used in this study were fabricated using the MBE growth technique. Semi-insulating GaAs(100) was used as the substrate material. Substrate preparation consisted of a standard (sulfuric acid/hydrogen peroxide/deionized water) etch followed by the formation of a thin protective surface oxide. This oxide is thermally desorbed from the substrate in the growth chamber prior to growth. Three different growth techniques were employed to grow the samples: conventional, interrupted growth, and migration-enhanced epitaxy (MEE). All three growth techniques discussed here are heavily dependent on substrate temperatures, the ratio of group V to group III fluxes, and the arsenic flux.

In the conventional (regular) growth mode, the arsenic shutter is kept open throughout the growth. The Al shutter is opened for the time needed to grow the necessary number of monolayers; then it closes the same time that the gallium shutter opens for the time to grow the appropriate number of GaAs monolayers. The growth sequence for the regular growth process can be represented as $n_1(AlAs)/n_2(GaAs)/n_1(AlAs)$ and so on until the active region thickness has been attained.

In the interrupted-growth process, the superlattice region is formed by first growing the AlAs layer. The aluminum shutter is then closed, and a pause interval follows in which only the arsenic shutter is kept opened. This pause interval allows the growth surface to equilibrate before the next layer is grown. This is then followed by growth of the GaAs layer, followed by the same pause interval. This basic sequence, AlAs/pause/GaAs/pause, is repeated for the superlattice active region. The pause intervals in the interrupted growth process are believed to contribute to a reduction in interface roughening by allowing complete layers of GaAs and AlAs layers to form.

For the MEE growth of the superlattice, only one shutter is opened at a time, and growth conditions are maintained such that one monolayer of constituent flux is incident upon the substrate during the time in which the shutter is opened. The theory of the MEE growth technique is based on the principle that when the first layer of atoms or molecules reaches a solid surface, there is a strong interaction (chemisorption), and subsequent layers tend to interact much less strongly (physisorption). For GaAs the presence of As is self-limited in that only a monolayer of As will remain on the surface, whereas this is not the case for gallium and aluminum. The group III elements can accumulate on the surface if growth conditions permit. When the substrate is heated properly, a condition exists such that only the first chemisorbed layer remains attached. In this way the film grows stepwise, i.e., a single monolayer per pulse, provided at least one complete monolayer coverage of a constituent element is formed before the next pulse is allowed to react with the surface. Sequence-wise, the MEE growth process is represented as: Al/pause/As/pause/Ga/pause/As/pause. Two requirements associated with the MEE technique are low growth temperatures and a V/III flux ratio which is lower than the conventional and interrupted-growth methods. The latter requirement arises because the migration of Ga and Al atoms is enhanced in an As-free atmosphere with respect to As-saturated surfaces.

The X-ray diffraction data were obtained from a single-axis rotating-anode diffractometer which uses a copper target source. Each sample was 2θ scanned between 2θ's of (20-40) degrees. The detector sampling rate was 1 degree per minute. This angular range was chosen to ensure detection of the (200) peak and its associated satellites.

THEORY

Extensive work has been done on the use of X-ray diffraction to characterize the structural parameters of superlattices.[1-3] In all these works Bragg's law is the primary equation used in the X-ray analysis of crystalline structure:

$$2d \sin \theta_B = n\lambda. \tag{1}$$

By using X-rays of known wavelength λ and measuring the Bragg angle θ_B, one can determine the spacing d of the planes in a crystal. The periodicity of multilayer structures results in an X-ray diffraction effect similar to the periodicity of the crystal lattice. Due to its periodic nature with multiple interfaces, the diffraction from a multilayer is therefore modulated and exhibits a well-defined satellite structure. The superlattice period Λ can be obtained from positions of the satellite peaks θ_{sl} of order n according to the following equation[4]:

$$\frac{2\sin\theta_n - 2\sin\theta_{sl}}{\lambda} = \pm\frac{n}{\Lambda}. \qquad (2)$$

A simulation program was used to produce superlattice X-ray diffraction spectra for comparison purposes. The program is based on a kinematical step model introduced by Segmuller and Blakeslee[5] to simulate X-ray scans from multilayers with sharp interfaces.

RESULTS AND DISCUSSION

Figure 1 contains data of the experimental and simulated X-ray spectra for the 12x12 AlAs/GaAs superlattice. The superlattice structure was grown using the regular growth method. The experimental plot shows that the compositionally modulated superlattice gives rise to satellite peaks on both sides of the (200) peak. These peaks exhibit reduced intensity as they move away from the (200) substrate plus superlattice peak. The regularity in angular position of the satellite peaks which are narrow with respect to the main (200) peak indicate that the superlattice is of high quality. Further evidence of this fact is revealed by noting that adjacent satellite peaks alternate in intensity. Figure 1b of the simulated X-ray spectrum for the 12x12 structure shows excellent agreement with the experimental data of Fig.1a.

Fig.1a X-ray diffraction spectrum

12x12 AlAs/GaAs superlattice

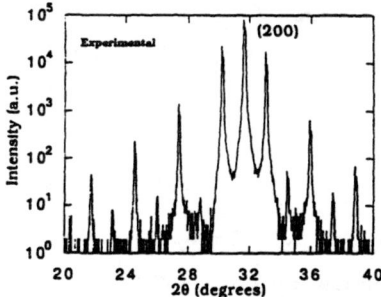

Fig.1b Simulated X-ray diffraction spectra

12x12 AlAs/GaAs superlattice

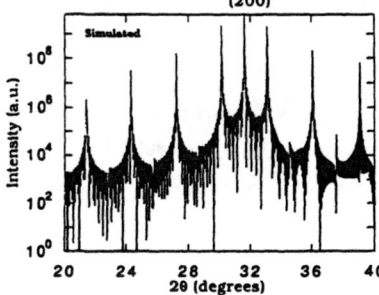

Figure 2 contains the X-ray spectra for the 6x6 AlAs/GaAs superlattice. The data of Fig. 2a are for a 6x6 superlattice grown at 620°C using the regular growth method. The As/Ga ratio is roughly equal to 20. The ±1 satellite peaks shown are broad and not very intense when compared with the satellite peaks of Fig. 1. Also, an additional set of satellite peaks, which are unintentional and undesirable, are observed between the ±1 and (200) peaks. Figure 2b is for the 6x6 structure grown at 620°C using the interrupted-growth method. The unintentional satellite peaks are still evident; however, the interrupted-growth method has resulted in a higher quality superlattice spectrum, i.e., the peaks are narrower and more intense. The improvement is probably due to a smoothing of the superlattice layers which occurs for the interrupted growth process. In Fig.2c the 6x6 superlattice was grown at 580°C using the same conditions as in Fig. 2b. One observes from the data of Fig. 2c that the unintentional satellite peaks have disappeared and that the ±1 satellite peaks are narrower and more intense. A lower growth temperature is consistent with less As re-evaporation from the growth surface.

Figure 3 shows the X-ray spectrum of the 3x3 AlAs/GaAs superlattices for various growth conditions. In Fig.3a the 3x3 superlattice is grown at 580°C using the interrupted-growth method, with an As/Ga flux ratio equal to 18. The labeled ±1 satellite peaks are broad and not very intense, and there is no evidence of unintentional satellite peaks. The spectrum of Fig. 3b for the 3x3 structure was obtained using a lower growth temperature, 540°C, and a reduced As flux, As/Ga = 13. Lowering the arsenic flux has resulted in a degraded superlattice structure as evidenced by the presence of unintentional satellite peaks. In addition, the ±1 peaks are broader and less well-defined than in Fig. 3a. The growth conditions for the spectrum of Fig.#3c are the same as in

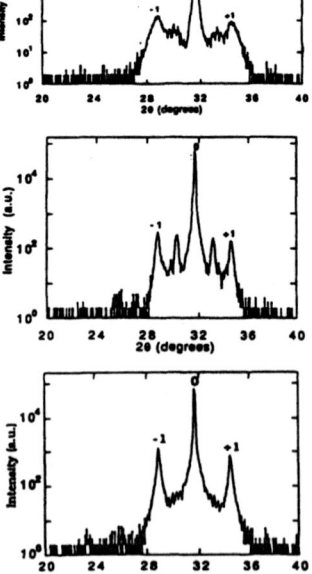

Fig.2a X-ray diffraction spectrum 6x6 AlAs/GaAs superlattice, T_{growth}=620°C,

As/Ga flux ratio equals 28, regular growth

Fig.2b X-ray diffraction spectrum 6x6 AlAs/GaAs superlattice, T_{growth}=620°C,

As/Ga flux ratio equals 28, Interrupted growth

Fig.2c X-ray diffraction spectrum 6x6 AlAs/GaAs superlattice, T_{growth}=580°C,

As/Ga flux ratio equals 20, Interrupted growth

Fig.#3b except the temperatures of the gallium and aluminum cells were lowered so as to establish higher As/Ga and As/Al flux ratios, thereby eliminating an As-deficient growth condition. This has removed the unintentional superlattice peaks observed in Fig. 3b. It is also observed that the ±1 peaks are more intense and much narrower than the corresponding spectra in Figs.3a and 3b.

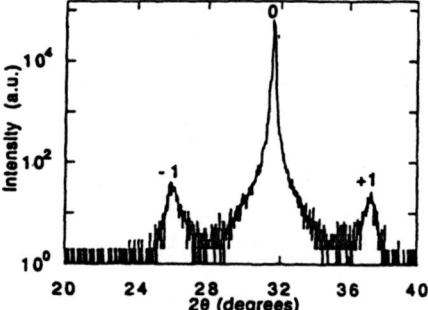

Fig.3a X-ray diffraction spectrum 3x3 AlAs/GaAs superlattice, $T_{growth}=580°C$, As/Ga flux ratio equals 17, Interrupted growth

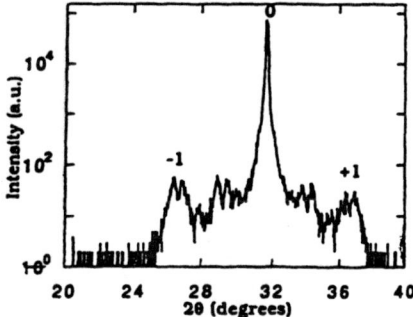

Fig.3b X-ray diffraction spectrum 3x3 AlAs/GaAs superlattice, $T_{growth}=580°C$, As/Ga flux ratio equals 13, Interrupted growth, Reduced arsenic flux

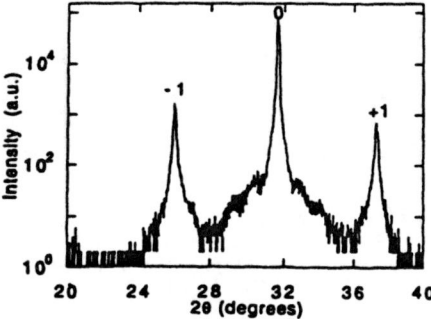

Fig.3c X-ray diffraction spectrum 3x3 AlAs/GaAs superlattice, $T_{growth}=540°C$, Al/Ga flux ratio equals 23, Interrupted Growth, Reduced arsenic, gallium, and aluminum fluxes

CONCLUSION

X-ray diffraction data have been obtained for 12x12, 6x6, and 3x3 AlAs/GaAs superlattices and compared with a simulated X-ray spectrum for the case of the 12x12 structure. Good agreement is observed between the experimental and simulated data. Unintentional periodicities have been observed in the 3x3 and 6x6 AlAs/GaAs superlattices. The observation of unintentional satellite peaks has been correlated with one or both of the following experimental conditions: 1) Elevated growth temperature and 2) Reduced As/Ga ratios. Both conditions are consistent with an arsenic-deficient growth condition.

REFERENCES

1. Vandenberg, J.M., Gershoni, D., Hamm, R.A., Panish, M.B., Temkin, H., J.Appl. Phys. 66(8), 3635, (1989)

2. Bartels, W.J., Hornsta, J., Lobeek, D.J.W., Acta Cryst. A42, 539-545, (1986)

3. Speriosu, V.S., Vreeland, Jr., T., J. Appl. Phys. 56(6), 1591, (1984)

4. Hermann, M.A., and Sitter, H., Molecular Beam Epitaxy, Springer Verlag (1989) p.182

5. Segmuller, A., and Blakeslee, A.E., J. Appl. Crystallogr. 6,413 (1973)

REAL TIME X-RAY STUDIES OF INTERFACE KINETICS IN EPITAXIAL STRAINED LAYERS

ROY CLARKE AND WALDEMAR DOS PASSOS
Department of Physics, University of Michigan, Ann Arbor, MI 48109

WALTER LOWE
AT&T Bell Laboratories, Murray Hill, NJ 07974

BRIAN RODRICKS AND CRISTINE BRIZARD
Advanced Photon Source, Argonne National Laboratory, Argonne, IL 60439

ABSTRACT

A new time-resolved x-ray method of probing the kinetics of interfacial strains in semiconductor heterostructures is presented. High-resolution synchrotron radiation measurements of the strain relaxation during rapid thermal annealing (RTA) show that the lattice strain of an as-grown strained layer structure $GaAs\text{-}In_xGa_{1-x}As\text{-}GaAs/GaAs$ is relieved cooperatively by a series of sluggish discontinuous transitions. We find that ion implantation enhances the annealing kinetics of InAlAs strained layers.

INTRODUCTION

High resolution x-ray diffraction is increasingly popular as a reliable non-destructive method for the structural characterization of epitaxial semiconductor layers [1]. Layer thicknesses, composition, strain and coherence, can be determined using this method with excellent precision [2]. With the development of new extremely bright synchrotron sources it is feasible to extend such structural measurements into the time domain. Using x-ray synchrotron radiation in conjunction with dispersive optics and fast x-ray area detectors we have been able to study the structure of heterointerfaces not only before and after but also *during* thermal processing.

EXPERIMENTAL METHODS

Prior to the time-resolved studies we employed conventional double crystal x-ray diffraction techniques [3] to obtain the structure's rocking curve both before and after processing. In this method the x-ray beam is first reflected by a reference crystal and then by the sample. The reference crystal monochromatizes the beam and is usually made from a virgin block of the sample substrate material. A measurement of the rocking curve is obtained by recording the integrated intensity as a function of the angle of incidence, ω. For single layer structures two principal peaks, one from the epilayer and the other from the substrate, are usually observed together with an interference fringe pattern if the sample is highly coherent. Rocking curves from more complex structures are more difficult to interpret and it is then necessary to employ some kind of computer modeling of the rocking curve using dynamical x-ray diffraction theory [4].

The structural kinetics measurements are carried out using a radiation-hardened charge coupled device (CCD) x-ray detector consisting of 340 x 580 pixels each 22.5 μm square (Fig. 1). The operation and characteristics of the detector have been described previously [5]. In order to collect real-time data, a horizontal slit is used to mask the surface of the CCD chip. After a predetermined exposure time ($\approx$ 100 msec in this experiment) the charge collected on the exposed area is transferred to the masked area of the chip using the parallel row-transfer register. In this mode the detector operates as a "streak camera" recording one-dimensional diffraction information with a time-resolution determined by the speed of the parallel transfer ($\approx$ 20 μsec/row in the present detector). The CCD is oriented such that the horizontal angular dispersion of the beam is utilized to record a moderately high-resolution ($\Delta\omega = 10$ arcsec) diffraction profile.

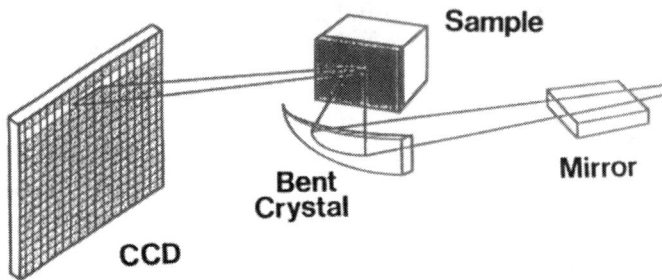

Fig.1 Experimental arrangement for angular dispersive real-time x-ray scattering studies. The curved crystal is asymmetric-cut Ge(111). The arrangement corresponds to the configuration of the X-16B AT&T Bell Labs beamline at the National Synchrotron Light Source, Brookhaven National Laboratory.

RESULTS AND DISCUSSION

We describe first some measurements on the annealing of a single quantum well (see inset of Fig. 2) by a quartz halogen lamp. The rapid thermal annealing cycle is shown in Fig. 2.

The sample was prepared without growth interruption [6]. It is clear that marked irreversible changes take place. For example, before annealing [Fig. 3(a)] the rocking profile is asymmetric and shows two small interference fringes to the low-angle side of the substrate peak. The latter is broadened by non-uniform strains in the as-grown sample. Modeling of the interface structure using dynamical scattering theory reveals that the fringes arise in the *as-grown* sample because part of the GaAs buffer layer (the upper $\approx$ 1000 Å) and the whole of the GaAs capping layer (2100 Å) are slightly dilated, in the direction perpendicular to the layers, by $\epsilon_\perp \approx 5.3 \times 10^{-4}$ with respect to bulk GaAs. The origin of this distortion is unclear.

After annealing [Fig. 3(b)], the rocking curve exactly matches the calculated curve for an *ideal coherent structure* with pseudomorphic $In_xGa_{1-x}As$ and completely relaxed GaAs buffer/cap layers. Note also the dramatic sharpening of the $\omega = 0$ peak after annealing and the appearance of pendellösung fringes which originate from a phase shift of the waves scattered from the substrate (buffer) relative to those coherently scattered from the GaAs cap layer. It is interesting to note that we obtain identical data [Fig. 3(b)] without annealing if growth interruption is employed. This confirms that kinetic effects are responsible for the imperfect structure corresponding to Fig. 3(a).

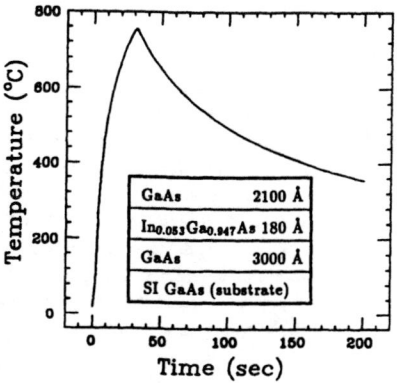

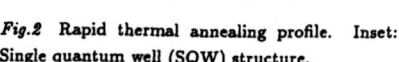

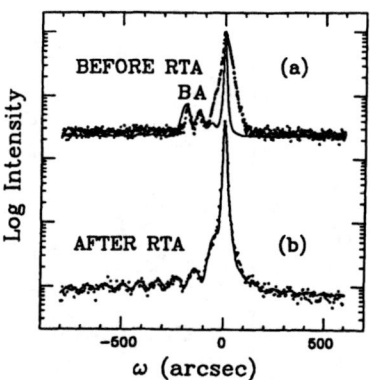

Fig.2 Rapid thermal annealing profile. Inset: Single quantum well (SQW) structure.

Fig.3 (400) rocking curves of the SQW, (a) as grown and (b) after RTA processing (see Fig. 2). The labels A and B refer to interference peaks (see text). Solid lines are a fit to dynamical scattering theory [4].

Now we turn to the time-resolved data. Figure 4 shows the variation in position of the peaks in the rocking curve, measured now by the real-time dispersive technique, as the sample is undergoing RTA. The time-resolved data, when analyzed each 100 msec using the same interference model as for the static results in Fig. 3, show the dynamic evolution of the sample from the as-grown inhomogeneous strain configuration to the ideal relaxed state.

In addition to the substrate peak shifts (upper curve S) caused by thermal expansion and contraction, there are many interesting and previously unobserved transitions during the RTA run shown in Fig. 4. We focus attention first on the behavior of the two interference fringes A and B in the *as grown* sample. On heating, the outer fringe (B) shifts to lower angle and soon after undergoes a sudden shift in the opposite direction. The shift to lower angles is caused by the differential thermal expansion between the In-GaAs layer and the GaAs layers, and the upward shift occurs as the resulting mismatch strains at the interfaces are suddenly relieved. On cooling, a reverse differential thermal expansion process is observed and subsequently (at t = 50 sec) a strongly discontinuous transition takes place where A and B are replaced by a single peak. The dynamical scattering analysis of this phase reveals that the dilation of the GaAs at the InGaAs buffer layer interface has relaxed at this transition but the cap layer is still strained. At later times, not shown in Fig. 4, there is a further discontinuous transition (at $\approx$ 300 °C) involving the remaining interference peak ($\omega \approx$ -170 arcsec) where the lattice spacing of the GaAs cap layer relaxes to its bulk value.

In this way the inhomogeneous strain relaxation of the thick GaAs layers of the SQW starts at the substrate, as one would expect, and proceeds upwards to the GaAs cap of the $In_xGa_{1-x}As$ layer. The important point is that the strain relaxes discontinuously and cooperatively in each of the GaAs segments of the SQW. Interestingly, it appears

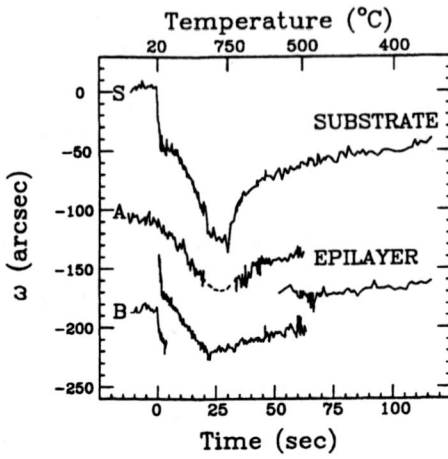

Fig.4 X-ray rocking angle measured by the CCD detector as a function of time during RTA. A quartz-halogen heating lamp is turned on at t = 0 and off at t = 30 sec. during which the sample temperature reaches 750 °C. Curves A and B refer to the peaks in Fig. 3(a). Curve S corresponds to scattering from the substrate and unstrained GaAs layers. The dashed portion of curve A is where it merges with the very intense substrate peak.

that macroscopic strained and unstrained GaAs regions of the sample can coexist in the same layer for relatively long intervals ($\approx$ 10 sec in the example shown here).

In a further set of experiments we looked at RTA effects on ion-implanted layers. Specifically, we investigated the effect of ion-implantation on the annealing kinetics of $In_xAl_{1-x}As$ /InP prototype Heterostructure Insulated Gate Field Effect Transistor (HIGFET) layers. Of particular interest is the quality of the layer structure, especially after a rapid thermal annealing treatment used for implant activation. We found that there are dramatic differences in the structural coherence after RTA depending on whether the layers are virgin or implanted.

For these studies an undoped 5250 Å-thick $In_xAl_{1-x}As$ (x = 0.54), layer was grown by Molecular Beam Epitaxy on a semi-insulating (SI) InP substrate. Si ions were implanted by a double energy (E) process under similar conditions that apply to HIGFET technology, namely $E_1 = 30$ keV at dose $D_1 = 2 \times 10^{13}$ cm^{-2} followed by $E_2 = 70$ keV, $D_2 = 4 \times 10^{13}$ cm^{-2}. RTA (capless annealing under Ar gas flow) at 750 °C for 30 sec was used for implant activation.

Figure 5 compares the rocking curves of the virgin (unimplanted) sample before and after RTA. There is essentially no difference. In Fig. 6 we again compare the x-ray profiles of the sample before and after RTA, this time in the implanted state. An interesting, and (as far as we are aware) previously unobserved, effect occurs: whereas there was virtually no change in the virgin sample, the implanted structure now becomes much more coherent and uniform as the result of an identical RTA treatment. Note the appearance of a well defined fringe pattern [Fig. 6(b)] which fits almost exactly the calculated profile for the ideal structure. Note that the value of $\epsilon_\perp$ after RTA in implanted samples is almost exactly what is expected ($\epsilon_\perp = 0.416\%$) for a perfectly pseudomorphic overlayer at this composition (including the tetragonal distortion) and so there is no dislocation formation during RTA.

What is the mechanism whereby long-range coherence is established in the annealing

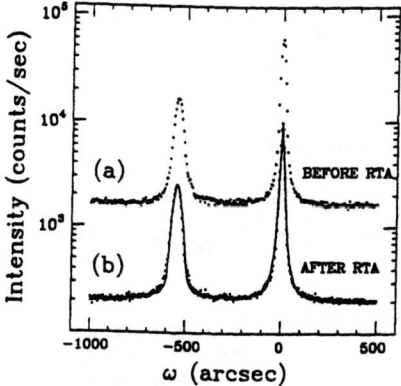

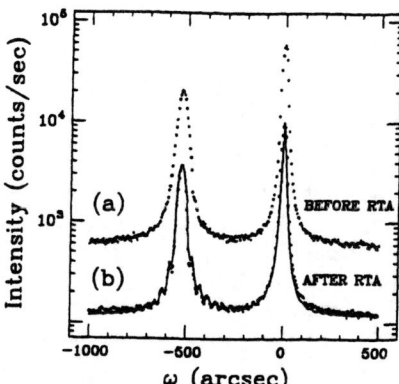

Fig.5 Typical (400) double crystal x-ray rocking curves (dots) of the unimplanted sample (a) before and (b) after RTA. The solid line represents a dynamical fit to the x-ray data.

Fig.6 Typical (400) rocking curves of the implanted sample (a) before and (b) after RTA. Note the distinct formation of the pendellösung fringes only after RTA.

process when implanted ions are present ? We are led to the conclusion that the kinetics of step migration, and concomitant interface smoothing, is influenced strongly by the implants. One possibility is that the efficiency of coupling the RTA light source into the overlayer may be enhanced by the presence of interband states created by implantation [7]. This would lead to enhanced absorption in the infra-red region together with increased heating rates and, consequently, more effective annealing of defects. In this way the interface steps could be smoothed without subjecting the sample to prolonged annealing thus avoiding deterioration of the overlayer stoichiometry.

Finally, in Fig. 7 we show the variation in position of the InAlAs peak in the rocking curve, measured now by the real-time dispersive technique, as the sample is undergoing RTA. The time-resolved data again shows that, on heating, the InAlAs layer first undergoes a differential thermal expansion. On cooling, a reverse differential thermal expansion process is observed before the InAlAs peak position settles into a seemingly stable equilibrium position. Surprisingly, however, a discontinuous transition takes place at $\approx$ 220 sec. The dynamical scattering analysis reveals that the InAlAs layer is relaxing into metastable states before reaching a final relaxed equilibrium position. Figure 8 shows another plot of the position of the InAlAs peak in the rocking curve as the sample is subjected to a second RTA process. This time we observe that no further lattice relaxation has taken place and that the InAlAs peak has attained its stable position characteristic of pseudomorphic growth.

In summary, the measurements obtained in a time-resolved mode by specially developed synchrotron radiation techniques reveal an unexpectedly complex pathway of interfacial strain kinetics by means of which an entire semiconductor heterostructure can attain its ideal, stable state. In particular, we have shown that interfacial strain relaxation is not, as previously believed, a gradual process occurring through local misfit dislocation migration but a discontinous, cooperative mechanism which proceeds via a

230

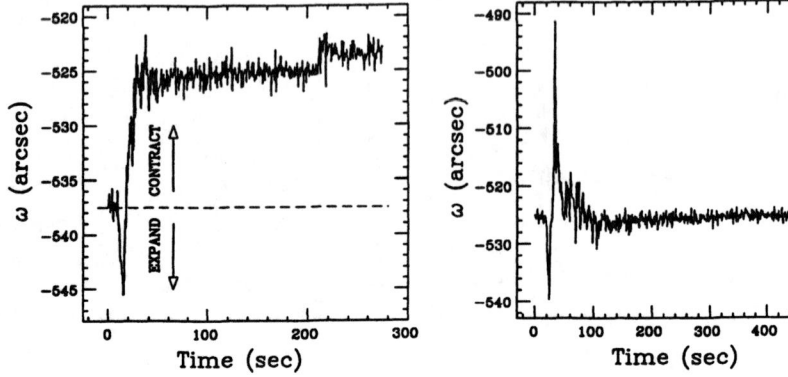

Fig.7 X-ray InAlAs peak position at the CCD detector as a function of time during RTA.

Fig.8 X-ray InAlAs peak position at the CCD detector as a function of time as the same sample undergoes a second RTA.

sequence of metastable states characterized by domains of differing strain. Finally, we have also identified a new mechanism whereby the RTA process appears to be more efficient in ion-implanted samples than in virgin samples. The high sensitivity of x-ray rocking curve analysis facilitates studying the subtle structure changes during thermal processing.

ACKNOWLEDGEMENTS

The authors wish to acknowledge the support of the Army Research Office (URI Program) under contract DAAL-03-87-K0007 and NSF grant DMR 8805156. We thank Yi-Jen Chan for growing the samples used in this study.

REFERENCES

1. *Thin Films Growth Techniques for Low-dimensional Structures* eds. R.F.C. Farrow, S.S.P. Parkin, P.J. Dobson, J.H. Neave and A.S. Arrott (Plenum, New York 1987);

2. M.H.Lyons and M.A.G.Halliwell, Inst. Phys. Conf. Ser. **76**, 445 (1985); L.Tapfer and K.Ploog, Phys. Rev. B **33**, 5565 (1986).

3. W.J.Bartels and W.Nijman, J. Crystal Growth 44, 518 (1978).

4. M.J.Hill and B.K.Tanner, J. Appl. Cryst. 18, 446 (1985).

5. B.Rodricks, R.Clarke, R.Smither and A.Fontaine, Rev. Sci. Instrum., 60, 2586 (1989).

6. K.H. Chang, P.R. Berger, R. Gibala, P.K. Bhattacharya, J. Singh, J.F. Mansfield and R. Clarke, in *Dislocations and Interfaces in Semiconductors* eds. K.Rajan, J.Narayan and D.Ast, (Metallurgical Soc. 1988) p.157.

7. W.O. Adekoya, M. Hage-Ali, J.C. Muller and P. Siffert, J. Appl. Phys. 64, 666 (1988).

EPITAXIAL GROWTH AND CHARGE DENSITY WAVE OF TaSe₂

Toshihiro Shimada, Fumio S. Ohuchi, and Bruce A. Parkinson
Central Research and Development, E.I. DuPont de Nemours & Co. Inc.
Experimental Station,Wilmington, DE19880

ABSTRACT

We report an epitaxial growth of $TaSe_2$, a family of transition metal dichalcogenides that exhibit Charge Density Waves (CDW). The films that have been characterized with RHEED, LEED, XPS and STM showed two different phases. Occurrence of CDW in the ultrathin films has been detected by XPS and LEED.

INTRODUCTION

Charge Density Waves (CDW) have been the subject of particular interest in solid state science[1]. They are phase transitions characterized by periodic lattice distortion and charge density modulation. Many layered materials exhibit CDW which are mainly induced by the instability of low dimensional free electrons. Among these materials, tantalum dichalcogenides (TaX_2; X=S or Se) have been studied extensively because they provide a series of strong CDW transitions with temperatures.

Although CDW are observed cumulatively from each layer of the materials, it is not determined *a priori* whether CDW can exist in one layer or not, and if exists, how such CDW are interacted with the substrate. To answer these questions, it is essential to fabricate such materials in a thin film form. We herein report the epitaxial growth and the behavior of CDW in ultrathin films of $TaSe_2$. A variation of molecular beam epitaxy (MBE) introduced by Koma et al[2], van der Waals epitaxy, has been used to grow $TaSe_2$ on other transition metal dichalcogenides (MX_2). A battery of techniques including Reflection High Energy Electron Diffraction (RHEED), Low Energy Electron Diffraction (LEED), X-ray Photoemission Spectroscopy (XPS) and Scanning Tunneling Microscopy (STM) has been utilized to characterize the films.

EXPERIMENTAL

Substrates used in the present experiments were single crystals of MoS_2, $MoTe_2$ and $SnSe_2$, of which lattice mismatch with $TaSe_2$ are -10%, +1% and +10%, respectively. They were grown by halogen vapor transport in quartz ampules with iodine as a transport agent. Prior to the substrate transfer to the growth chamber, they are cleaved under the ultrahigh vacuum (UHV) condition to obtain clean (0001) surfaces. Deposition of $TaSe_2$ was performed in a turbo-molecular pumped chamber , of which base pressure was 1x10^{-9} torr. Selenium was sublimed from an element source made of a quartz crucible. Ta was evaporated from a Joule heated wire, of which purity was 99.95%. The growth rate was 0.5 - 1 Å / min with Se-excess conditions (Se: Ta ~ 10 : 1). The rates of each elements were measured with a quartz oscillator. The growth was carried out at substrate temperatures (T$_{sub}$) ranging between 250°C and 650°C. The typical thickness of the films reported here is 2-6 nm (equivalent to 3-10 monolayers).

After the growth, the sample was characterized by XPS and LEED *in situ*. Since the films with fractional coverages are easily contaminated when exposed to air, this *in situ* characterization is essential for the present study. The XPS system was Surface Science Laboratory ESCA-100 with the energy resolution of 0.7eV in FWHM at the Au (4f$_{7/2}$) peak. STM was performed with Digital Instruments Nanoscope II under ambient atmosphere.

RESULTS AND DISCUSSION

Judging from RHEED patterns, T$_{sub}$ was the only important parameter for the crystallinity of the films. Epitaxial growth was achieved between 300°C and 650°C. Depending upon the growth conditions, we have identified two different peak shapes of Ta(4f$_{7/2}$) core level XPS as shown in Figure 1. A round shape (Fig.1a) and a skew triangle shape (Fig.1b) were observed from the films grown under the conditions of T$_{sub}$=300~340°C and T$_{sub}$=420~650°C, respectively. We refer these films as L-films (grown at low temperatures) and H-films (grown at high temperatures). When the L-films were heated above 420°C, they are converted to H-films irreversibly. XPS has shown little substrate and thickness dependence.

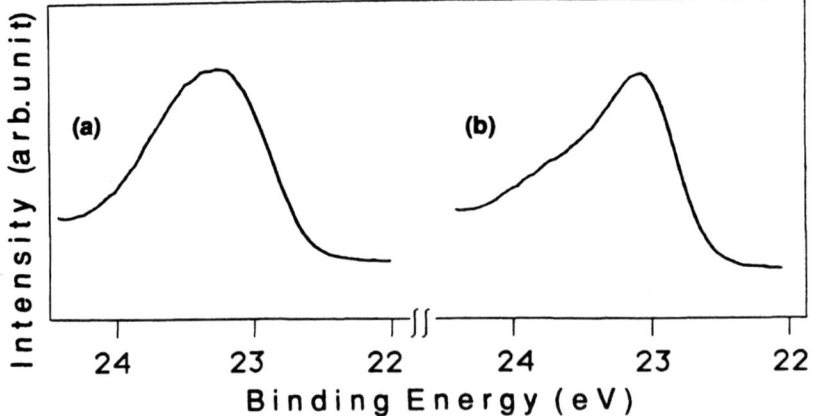

Fig.1. XPS Ta 4f $_{7/2}$ core levels of the films grown at
(a) 300~340°C (L-films) and (b)420~650°C(H-films).

First we describe characterization of the L-films. Since CDW is the temperature dependent phase transition, Ta (4f$_{7/2}$) core emission from the L-films were observed at different temperatures (Figure 2). Gradual broadening was identified as the temperature was decreased. A subtraction spectrum (d) clearly shows this difference. Figure 3 shows a LEED pattern of the L-film observed at 160K. It is not as clearly resolved as those from bulk single crystals, but there is an indication for incommensurate superstructure with a hexagonal lattice. The LEED pattern did not show significant temperature dependence. In contrast to the L-films, XPS of the H-films showed no temperature dependence. LEED patterns of the H-films are sharp hexagonal 1x1 structures and neither exhibited temperature dependence.

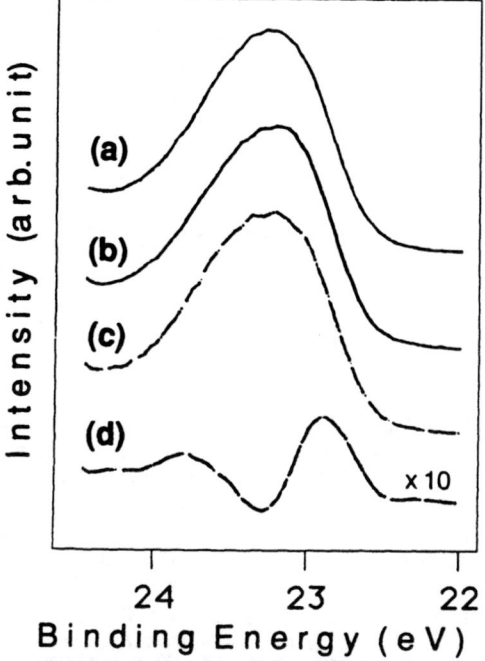

Fig.2.

Temperature dependence of XPS Ta 4f $_{7/2}$ core levels of L-films. The temperatures are (a) 480K; (b) 300K; (c) 160K; (d) a subtraction spectrum (c)-(a).

Fig.3.

LEED of L-films.

Bulk TaSe₂ is know to form two different structures of which coordination around Ta atoms is either octahedral (OC) or trigonal prismatic (TP) [1,3]. This is shown schematically in Figure 4. The OC-bulk TaSe₂ is metastable below 780°C and only obtained by quenching a growth vessel to water. It transforms to TP when heated to about 300°C.

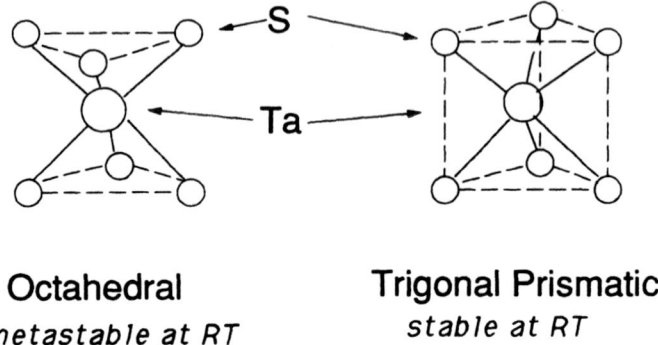

Octahedral

metastable at RT

Trigonal Prismatic

stable at RT

Fig.4. Structure of bulk TaSe₂.

CDW of OC is characterized by √13 x √13 superlattices and 1st-order phase transition at 473K (incommensurate and commensurate). Below this temperature is commensurate charge density modulation characterized by a splitting of Ta ($4f_{7/2}$) core level. The splitting becomes significant with further decreasing temperature. LEED patterns of OC shows √13 x √13 superstructure which rotates about 12° at the transition temperature. In TP, CDW is characterized by 3 x 3 structure and appears only below 90K. Reported Ta ($4f_{7/2}$) core level XPS from TP shows skewed triangle shape which has little temperature dependence even at the transition[4].

By comparing our films with these bulk samples, it is identified that our L- and H-films correspond to OC and TP of the bulk TaSe₂, respectively. In addition, we consider temperature independent LEED patterns of the L-films might be attributed to a "frozen incommensurate" CDW, which differs from the nature of OC-bulk TaS₂. XPS of Ta ($4f_{7/2}$) from L-films supports this hypothesis because the broadening at low temperature indicates some portion has similar temperature dependence as OC-bulk while some does not.

Since the incommensuration associated with the L-films must be due to the nature of the films, we examined following three possibilities:

    i) lattice distortion in the films arising from lattice mismatch with substrate,
    ii) pinning of CDW by defects in the films,
    iii) size effect induced by the domain structure of the films.

There is little substrate dependence found even with ±10% lattice mismatch, therefore we can eliminate i). Regarding ii) and iii), we have found XPS broadening of L-films is closely related to the stoichiometry. When the film is exposed to excessive selenium vapor after the growth, the stoichiometry changes slightly to become selenium rich. This excess selenium, however, disappear after annealing above 110°C under UHV. Although the binding energy of the additional component is similar to that of elemental selenium, it appears to suppress the temperature dependence of the $Ta(4f_{7/2})$ core level. We suspect this selenium is adsorbed at the domain boundary of the films and controls the electronic structure near the Fermi level. These domains probably also produce boundary conditions to CDW thus causing temperature independent incommensuration. We consider iii) is the most plausible.

Our results of L-films appear to be contradictory to the fact that the OC-bulk $TaSe_2$ is only grown at high temperatures above 880°C. We have attempted to determine the structure of the films but they were too thin to apply the diffraction method directly. Recently etching by tip-sample interaction in STM has been discovered [6], in which the etching patterns frequently show aligned triangular shapes. They are considered to be reflecting the orientation of the crystals with trigonal symmetry. We have applied this technique to the epitaxial films and obtained etching patterns of the L-film as shown in Figure 5. Three sequential layers are apparent and all of them have the same direction of the triangle etching patterns. Layers of the OC-bulk $TaSe_2$ stack in the same direction, while the alternating direction is expected from the most stable form of TP-$TaSe_2$. Observed etch pattern is, therefore, consistent to OC-type stacking.

As for the mechanism for the growth of OC type at such low temperatures, we suggest the non-equilibrium nature of MBE. Bulk $TaSe_2$ is usually grown by a two-temperature method named halogen vapor transport, which is much nearer to the equilibrium with the existence of a transport agent.

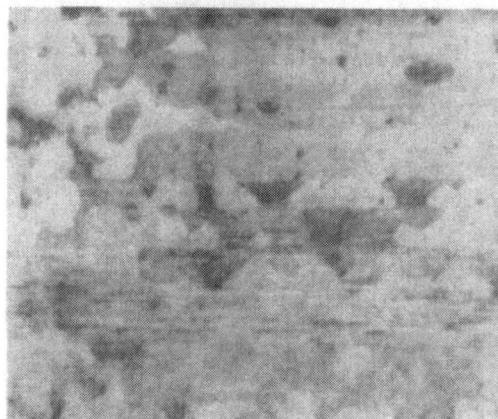

Fig.5.

STM etching pattern of the L-film.

CONCLUSION

Epitaxial growth of CDW material $TaSe_2$ was attempted and two phases corresponding to different polytypes were obtained. CDW was detected by XPS and LEED from the film in low temperature phase. The growth condition of CDW-showing octahedral coordinated $TaSe_2$ was far different from conventional halogen vapor transport, indicating non-equilibrium nature of molecular beam epitaxy.

ACKNOWLEDGEMENT

One of the author (T.S.) wish to thank Prof. Atsushi Koma of the University of Tokyo for his encouragement and discussion.

REFERENCES

1. R.H.Frindt and A.D.Yoffe,Advances in Physics 36,1 (1987)
2. A.Koma, K.Sunouchi, and T.Miyajima, Microelectron. Eng. 2,129 (1984); J. Vac. Sci. Technol. B3,724 (1985)
3. R.Husiman and F.Jellinek, J. less common Metals, 17, 111 (1969)
4. H.P.Hughes and R.A.Pollak, Philosophical Magazine 34,1025 (1976)
5. B.A.Parkinson, J. Am. Chem. Soc. 112,7498 (1990)

PART V

Laser Induced Transformations

MATERIAL REQUIREMENTS FOR REVERSIBLE PHASE CHANGE OPTICAL RECORDING

Kurt A. Rubin,
IBM Research Division, Almaden Research Center, 650 Harry Road, San Jose, California 95120

ABSTRACT:

The science and technology underlying phase change reversible optical storage, with an emphasis on the media, are reviewed. The transformation kinetics and their effect on writing (amorphization), erasing (crystallization) and long–term data stability are discussed. Phase separation is shown to affect cyclability and cause increased media noise. A readback CNR of 65.0 db was obtained by eliminating grooves as a source of media noise and using a recording layer which crystallizes into a cubic phase. Recording performance at short wavelengths is discussed.

Introduction

The purpose of this paper is threefold: (1) illustrate the general principals underlying phase change recording, (2) summarize the current state of the art, and (3) indicate where future opportunities are given improvements to the media and the optical head. A complimentary paper reviews the current status of some of the more drive related parameters of phase change[1].

Optical storage technology is based upon focusing laser light with a high quality objective lens to a small spot on the recording media (see Figure 1). The smallest size of the focused spot is limited by diffraction. The FWHM diameter of the light beam (ϕ) with a Gaussian intensity distribution focused on the disk is: $\phi = 0.56\lambda/NA$ where λ is the wavelength of light and NA is the numerical aperture of the lens. Using a laser operating at a wavelength of 780 nm with a lens NA of 0.55 results in a spot size of $\phi = 0.8$ μm. The spacing of the tracks and the areal density of recorded data are also limited by the size of the focused spot. For example, a track pitch and spot spacings of 1.6 μm corresponds to 40 Megaspots/cm^2. The use of an objective lens with a ~1 mm working distance results in a large separation between the optical head and the media which prevents the lens from hitting the disk. This separation helps ensure data integrity and together with the ability to actively servo on the track allows the media to be removable, resulting in a drive with essentially unlimited near–line data storage capacity.

Successfully implementing erasable phase change recording requires developing media which can be erased after writing. Usually materials which undergo a crystalline–to–amorphous (writing) and an amorphous–to–crystalline (erasing) transition are chosen. In addition, to read the data, the two states must have different optical properties. An example is indicated in Figure 2 which shows, in transmission, written amorphous spots as bright regions on a darker, crystalline background.

Among the greatest challenges present in phase change technology are understanding and improving the number of reversible cycles and reducing the media noise. Despite these challenges, there are several reasons why research and development are being pursued. One is that direct overwrite works well, which means that data already written on a track is erased at the same time new data is written. This eliminates the long disk rotation latency time required to first erase a track before writing it. Second, the signal from phase change media is very strong; it is about 1000 times that of magneto–optic media based on rare–earth transition metals. Unfortunately, the noise from phase change media is also larger so at present the signal–to–noise ratios are comparable for the two media. An opportunity exists to use this large signal for increased recorded data density using more sophisticated recording techniques such as pulse width modulation (PWM) if the media noise can be reduced and writing noise does not become a problem. Third, because there is no need to preserve the polarization state of the reflected light since the readback signal is based on intensity modulation, the optics in a phase change head are quite simple. Finally, a bias magnet is not needed so the optical drives can be more compact and use less electricity as is desirable for laptop computers.

Efforts are underway to develop standards for this media. This will allow media to be interchanged between drives from different manufacturers. There is also interest in using media for recordable CD's[2]. This may require modifying the disk standard since the reflectivity is much less than that of aluminized disks.

Since data is detected as intensity modulations, there is no need to preserve the polarization state of the reflected light. This suggests that novel and compact systems could be invented in the future by using integrated optics. One example of a simple head has already been implemented[3]. Optically, this is among the simplest of all heads. It consists of a laser and photodetector which are mounted on a slider which flies ~1 μm above the media. The small head/disk distance eliminates the need for an objective lens and focus actuator. This type of very lightweight head can have access times comparable to magnetic hard–disk drives. Of course, there are many issues such as removability and sensitivity to dust and debris which must be addressed. However, it does illustrate how simple the optics can be by using phase change media.

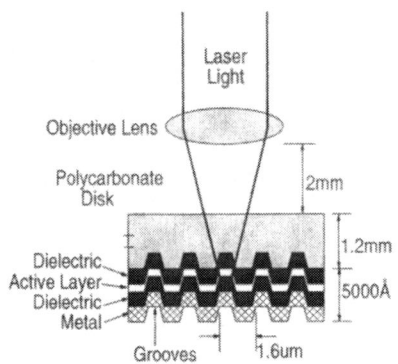

Figure 1. High areal data densities are achieved by focusing a semiconductor laser to a diffraction limited spot size. A typical thin film disk structure is illustrated schematically.

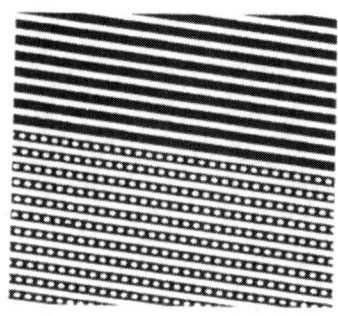

Figure 2. Transmission optical micrograph of spots recorded on an optical disk. The amorphous material is indicated by the brighter, more transparent, regions. In this example, the track pitch was 2.5 mm. The track pitch is typically 1.6 mm or less in commercial media.

Media

The substrate of an optical disk is typically 1.2 mm thick and is produced by injection molding of polycarbonate plastic. Several thin film layers are deposited on the disk by sputtering or evaporation. These are encapsulated with a UV cured resin to provide resistance to abrasion. In the thin film stack, a recording layer which is separated from the plastic disk substrate by a dielectric layer is used to store data. This dielectric layer thermally isolates the plastic, which has a glass transition temperature of ~130 °C, from the recording layer which can reach temperatures of >1000 °C. The recording layer is typically over–coated with a dielectric and a metal layer. The thickness of all these layers are chosen to yield optimal reflectivity of the amorphous and crystalline state while enhancing cyclability and minimizing the laser power required for writing. The disk has closely spaced grooves which are molded parallel to one another to provide a signal for tracking so the laser beam is prevented from wandering off the data.

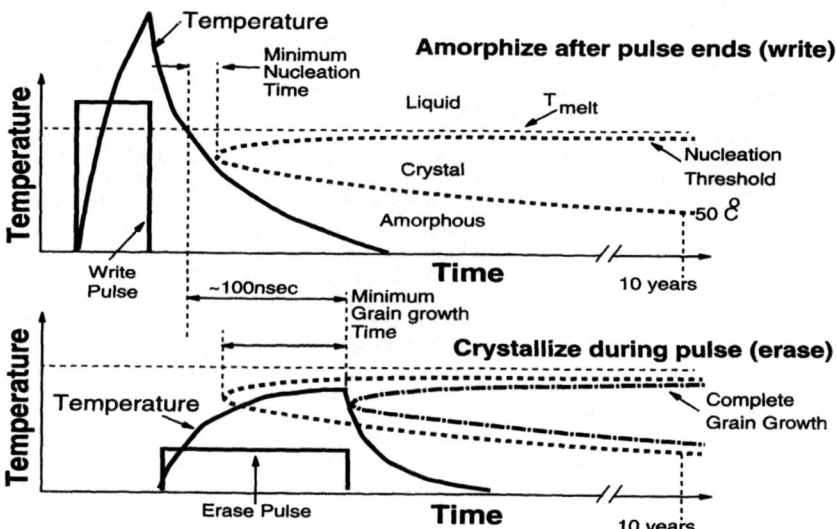

Figure 3. Time–temperature–transformation diagrams illustrating how amorphization or writing (top) and crystallization or erasing (bottom) occur. The laser pulses are indicated by the rectangular solid lines. The temperature rise and fall is indicated by the solid heavy line. The threshold for nucleation and complete grain growth are indicated by dashed and dot–dashed lines, respectively.

Phase Transformations – Writing and Erasing

The principles underlying reversible phase transformations are illustrated schematically in Figure 3 by the use of a time–temperature–transformation (TTT) diagram. Writing by amorphization involves the crystalline, liquid, and amorphous structural states[4]. A short, high powered laser pulse is used to heat a spot on the disk above the melting temperature of the recording layer. After the laser pulse ends, cooling occurs rapidly so that nucleation and growth of the crystalline phase is avoided, yielding an amorphous mark of written information. Erasure is accomplished by heating the amorphous marks to a high temperature for enough time that both nucleation and complete grain growth occurs. This can be either by solid state crystallization directly from the amorphous state or alternatively by cooling slowly from the liquid state so that solidification into the crystalline state occurs.

The time available for crystallizing an amorphous mark is quite limited. For example, with a 0.8 µm diameter circular spot and a disk media speed of 8 m/s the dwell time of the focused laser spot is 100 nsec. Both nucleation and complete grain growth must be completed in approximately this amount of time. At the higher media speeds desired for high performance drives, or with smaller sized written marks, the time for crystallization is correspondingly less. However, the fast crystallization time must not preclude the amorphous state from remaining amorphous for many years at somewhat near room temperature so that the recorded information is not lost. This means that the line on the TTT diagram corresponding to nucleation must remain above room temperature even after 10 years.

Much research has been performed during the past two decades to solve this problem. The majority of work involved alloys, for example, adding elements to Te[5][6][7]. However, it was not possible to find alloys which were fast and stable. This problem was solved when it was realized that material compositions which are stoichiometric phases can usually be crystallized quickly[8]. This is illustrated in Figure 4 for the GeTe binary alloy system. Although elemental tellurium crystallizes very rapidly, the amorphous state crystallizes at 10°C, so it is not stable at room temperature. When Ge is added, to make an alloy, the crystallization temperature increases. However, the minimum crystallization time increases even more quickly and rapidly reaches excessively long values. Significantly, as the stoichiometric composition GeTe is approached, the crystallization time decreases rapidly. In fact, GeTe crystallizes faster than elemental tellurium. The amorphous state of GeTe crystallizes at ~180°C (10°C/second heating rate) and is thus much more stable than Te. The reason for the fast crystallization can be understood by examination of the GeTe phase diagram, also shown in Figure 4. The composition GeTe is a stable compound. Crystallization from the amorphous to crystalline state can occur partitionlessly since no long range diffusion is required. In contrast, compositions on either side of the compound composition need to phase separate in order to crystallize. Since the phase separation requires diffusion, longer crystallization times result.

This idea of using materials which crystallize without phase separation to achieve fast erase times is quite general. Several different material systems have since been explored which exhibit fast crystallization and all crystallize into single phase compositions. Among the most studied are compositions in the Ge–Te–Sb ternary systems, particularly on or near the tie–line connecting GeTe and Sb2Te3[9]. Another system which exhibits fast crystallization is the In–Sb–Te system[10]. In some cases, the compounds crystallize too quickly, making it difficult to laser amorphize them. In these cases, the ability to amorphize can be enhanced by increasing the cooling rate of the structure. This can be accomplished by adding an aluminum layer next to the recording layer[8]. The disadvantage of a cooling layer is that it may desensitize the recording layer too much so that more laser power is required to heat a spot above the melting temperature. When higher laser power is used the cost of laser diodes increases and the laser lifetime decreases which is not desired. A thin dielectric layer is placed between the cooling and the recording layers to prevent interdiffusion and increase the sensitivity. This allows GeTe, which otherwise crystallizes too quickly, to be amorphized. An alternative is to solve the problem of materials which crystallize too quickly by purposely adding extra elements or moving away from stoichiometric compositions. However, this can result in undesired phase separation as will be discussed later. This is an area which needs more study.

Fast crystallizing materials are very important, since they enable single–pass direct overwrite[11][12][13]. A high power, short width laser pulse is used to form an amorphous spot at the desired location. In between writing pulses, the laser power is kept at a high, bias level. This causes erasure by crystallization of the previously written data. Usually that crystallization is not complete and residual amorphous regions remain. A measure of the degree of crystallization is provided by the erasability which is the difference, in db, of the carrier level at the new frequency and what remains of the carrier at the old frequency. An erasability of 25 db is adequate for most applications. However, in many cases, the erasability can be much less and yet the media will perform satisfactorily. This was illustrated by Ishida et al[12]. In their work, a bias power of 10 mW, which was one half the writing power, produced an erasability of 32 db. When the bias power was lowered to 7 mW, the crystallization was much less complete, resulting in an erasability of 12.5 db. TEM observations showed that the center of the old spots were crystallized. However, away from the center, where the temperature did not get as high, the old amorphous spot was not erased. Surprisingly, this residual amorphous data does not cause a problem. This was shown by measuring the readback jitter, which is the variation in time of the pulse widths of each spot. Even though the erasability was poor, the jitter did not increase. This is because, when the new amorphous spots were written, the high temperatures caused crystallization of the old amorphous data immediately surrounding the new mark. In a drive, the marks are detected within a certain window of time, so by crystallizing a ring around this new data, the residual old data fell outside the detection window and did not cause jitter. It is important to note that this effect, which is observed only with phase change media, increases the power margins for writing new data. Having wide margins is particularly important for removable media which can have property and performance variations from disk to disk and drive to drive.

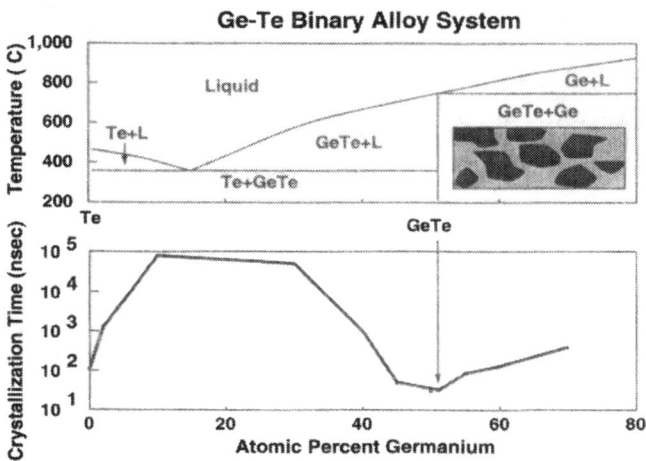

Figure 4. Phase diagram for the GeTe binary alloy system (top)[14]. Minimum time to crystallize a laser amorphized ~1 μm diameter spot (bottom).

Reversible Cycling

Cycling has probably been the most controversial issue associated with phase change media recently. The perceived lack of adequate number of reversible cycles has contributed to the slow acceptance of the technology. There has been uncertainty regarding how many cycles can be achieved and how many are needed for applications. However, much progress has been made in improving the number of cycles in the last several years. It is also becoming recognized that a large number of cycles is not needed for many applications.

Very good cycling, using direct overwrite can be obtained. For example, the structure NiCr alloy / 20 nm ZnS–SiO$_2$ / 20 nm Ge–Te–Sb / 150 nm ZnS–SiO$_2$ / polycarbonate could be cycled more than 2 million times[15]. While small variations of CNR and erasability were observed, the bit error rate did not significantly increase. Although these results are quite encouraging, many issues including the effects of process, composition and structure on performance remain. Even more importantly, the mechanisms associated with failure are not understood. This is of concern since it takes significant time to cycle even a single track on a disk one million times. It is important that the cycling failure mechanisms be elucidated so that accelerated testing can be performed.

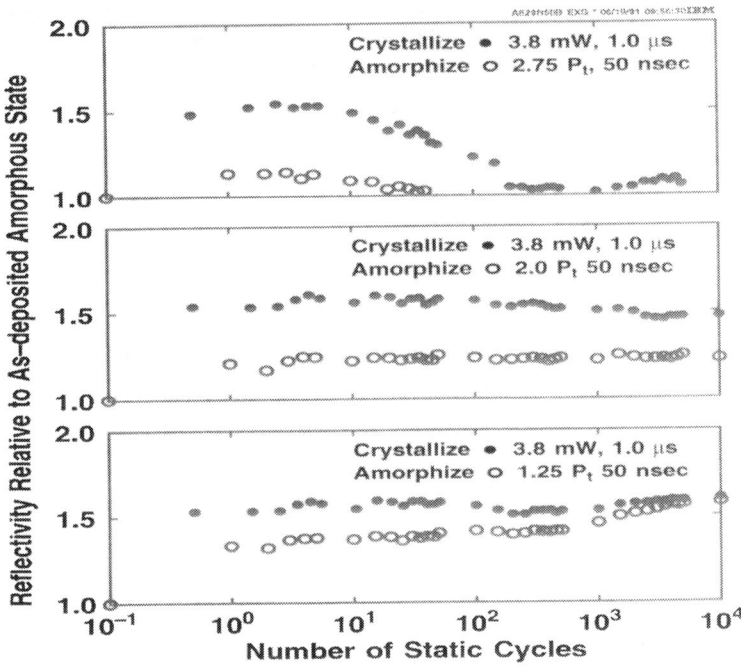

Figure 5. Static cycling of the structure 250 nm glass / 75 nm (GeTe)$_{85}$Sn$_{15}$/ 250 nm glass / polycarbonate. The erase pulses were a constant 1 μs. The amorphization pulse powers were 2.75 P$_t$ (top), 2.0 P$_t$ (middle) and 1.25 P$_t$ (bottom). The best cycling occurred with powers of 1.75–2.0 P$_t$. The initial reflectivity value of 1.0 is plotted at 0.1 cycles.

Information about the failure mechanisms can be gained by using static and dynamic cycling along with optical and electron microscopy. In static cycling, the laser beam is focused on a stationary spot and the sample is not moved until the cycling has been completed. As an example, consider the structure 250 nm glass / 75 nm (GeTe)$_{85}$Sn$_{15}$ / 250 nm glass formed by evaporation on polycarbonate substrates. Figure 5 shows static cycling where the laser pulse conditions were constant except for varying the amorphization power. Three amorphization powers, 1.25, 2.0 and 2.75 P_t were used, where P_t was the minimum power causing any amorphization at all. The relative reflectivity is plotted on the vertical axis. A value of 1.0 equals the reflectivity of the initial, amorphous as deposited state of the recording layer. At a medium amorphization power of 2.0 P_t reasonably good cycling was observed. There was good contrast between the low reflectivity amorphous and high reflectivity crystalline states. When the power was increased, rapid degradation of the reflectivity of both the amorphous and crystalline state occurred. This was due to the formation of pinholes in the recording layer. That damage was irreversible and is not desired.

At lower amorphization powers, such as 1.25 P_t, better cycling might be expected since the maximum temperatures experienced by the recorded mark was considerably less. However, this was not the case for this structure. Cycling failure occurred with the amorphous state reflectivity gradually increasing to that of the crystalline state until there was no difference of the two reflectivities. Further experiments showed that the low power failure was reversible. This is illustrated in Figure 6. First an amorphous pulse of 1.75 P_t was applied and the spot cycled well for 1000 times. Then the power was reduced to 1.25 P_t and cycling continued. As expected, smaller initial contrast was observed since a smaller region of the recording layer was heated above the melting temperature. The cycling ended with the reflectivity resulting from the high power pulse gradually increasing to the crystalline value. No contrast between an amorphous and crystalline state was observed at this point. Interestingly, when the power was increased to 1.75 P_t, the cycling recovered and good contrast was observed between the amorphous and crystalline states. Then when a low power pulse was applied, the cycling again failed by formation of the high reflectivity state.

Optical microscopy was performed on spots which had been cycled with the amorphizing power of 1.25 P_t. After 1 cycle, there was a fairly large amorphous spot. The diameter of the amorphous mark decreased progressively with increased cyclic number. The high reflectivity surrounding that amorphous mark indicated that it was crystalline. Since the amorphizing power was constant, melting occurred to a diameter larger than the final size of the amorphous mark. This suggested that the critical growth rate for amorphization of the recording layer changed with cycling. The increased crystallization speed could result from phase separation of the recording layer into a faster crystallizing composition. While the phase diagram of the TeGeSn system is not well studied, most likely the liquid state is fully miscible since the Ge–Te and Sn–Te binary systems are miscible in the liquid state. This suggests that phase separation occurs during solidification at the liquid–solid boundary during cooling. Each cycle would allow more segregation to occur. While molten, the liquid would not have time to completely remix since the viscosity is high just above the melting temperature. One reason the 1.75 P_t amorphizing pulses reversed the failure could be that at the higher temperatures the viscosity of the liquid was lower, allowing remixing of the segregated components. A second explanation for the cycling failure is that the number of nucleation sites increased at the edge of the molten spot after each cycle.

A similar experiment using a low and medium amorphization power was performed dynamically. In this case the media noise was found to increase with successive passes of the crystallizing beam. Associated with the increased media noise, TEM analysis by Suzuki and Rubin showed that precipitates were formed at the edge of the melt crystallized region[16]. These precipitates had a different crystal structure from the remaining GeTeSn cubic phase although, due to their small size, their identity was not determined. The media noise was found to be reversible, as shown in Figure 7. By raising the laser power to 16 mW, several erase pulses lowered the media noise by 3 db. The media noise was alternately increased and decreased by decreasing and increasing the laser power, respectively. We suspect that, similar to the static cycling case, the precipitates were pushed out further by using the higher power 16 mW laser pulses. At higher erase powers, they formed under the less intense portion of the laser beam, and thus contributed less to media noise. Further support for the idea that phase separation occurred was provided by a study of the solid state crystallization due to furnace annealing of GeTeSn[17]. There it was found that $(GeTe)_{85}Sn_{15}$ falls in a 2 phase region which results in precipitation into an fcc GeTeSn compound and elemental Ge after extended heating.

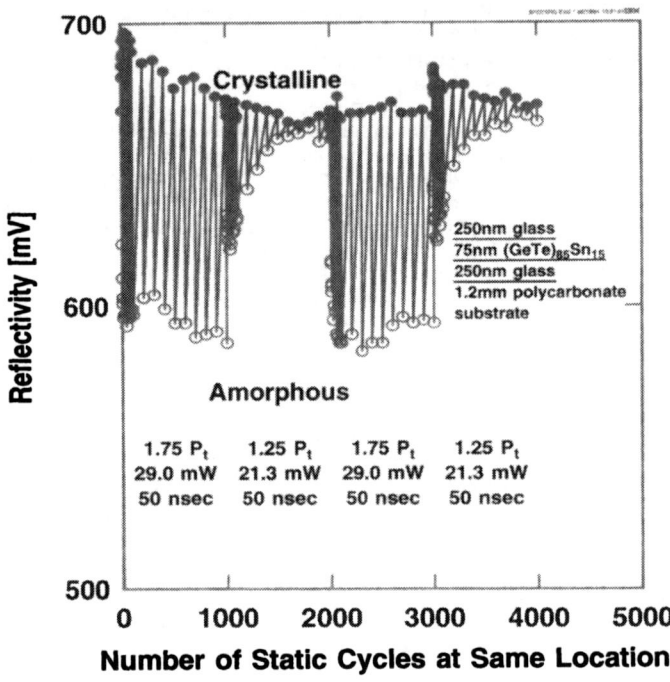

Figure 6. Static cycling of the structure 250 nm glass / 75 nm $(GeTe)_{85}Sn_{15}$ / 250 nm glass / polycarbonate. The erase pulses were a constant 1 μs. The amorphization pulse powers were alternated between 1.25 and 1.75 P_t. The contrast between the amorphous and crystalline state gradually reduced with each cycle using the lower amorphization power. Cycling recovered using the higher power.

The idea of metastable phase formation during laser crystallization appears quite generally among many material systems of interest for phase change recording. As indicated earlier, the material system which is undergoing the most study now is the Ge–Te–Sb ternary system. There are 3 compounds, $GeSb_4Te_7$, $GeSb_2Te_4$ and $GeSbTe$ which exist along the tie–line connecting Sb_2Te_3 and GeTe. However, all along that tie–line and even at compositions somewhat off from it, crystallization is quite fast. In particular, the minimum crystallization time decreases monotonically as the composition approaches Sb_2Te_3. The crystal structure is layered, and it is believed that the off–stoichiometric compositions are readily accommodated by the insertion of extra layers between the layers of the equilibrium phases[18].

Stability and Lifetime

Generally the long term stability of the amorphous phase against crystallization increases with increased melting temperature[19]. The stability of several materials is illustrated in Figure 8. The crystallization temperature (T_x) is found empirically to fall in the range of 1/3 and 2/3 the melting temperature (T_m). Thus Te, which has a low T_m, also has a low T_x. While increased data stability is desirable, higher T_m implies the need for more laser power which is not desired. Even more importantly, the ability to cycle well significantly degrades as higher melting temperature materials are used. Materials which crystallize at >150 °C (10 °C/minute heating rate) are preferable to ensure long term stability of the amorphous data.

Long term stability of the complete disk structure is also necessary. Results of accelerated aging studies of Ge–Te–Sb media incorporated in a complete disk structure have been published[20]. Using the very stringent failure criterion that doubling the defect rate indicated failure, a complete disk structure has a projected lifetime of >50 years at 32 °C based on aging at 80% humidity and 90 °C, 80 °C and 70 °C. While these results are quite encouraging, the mechanisms associated with failure due to defect generation are not completely understood.

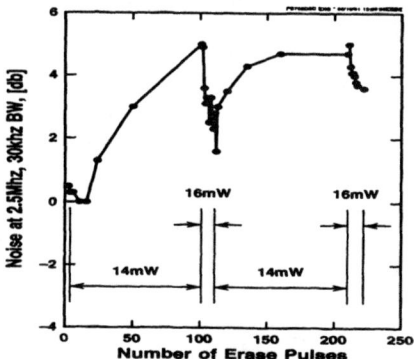

Figure 7. Media noise from the structure: 250 nm glass / 75 nm (GeTe)85Sn15 / 250 nm glass, at 5 m/s media speed was reversible by changing the crystallization power.

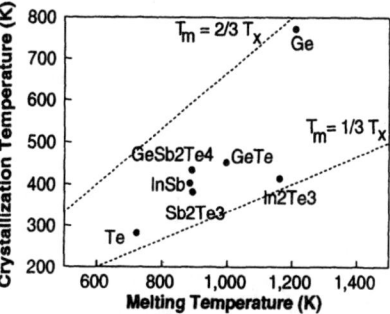

Figure 8. Stability against crystallization of amorphous state of several materials with heating rates of ~10 °C/minute. Increased stability is achieved by choosing higher melting temperature materials.

Readback

One way to assess readback performance is to measure the carrier–to–noise ratio (CNR). An example of the CNR from a spectrum analyzer showing log power versus frequency is given in Figure 9. It is obtained upon readback from a set of equally and closely spaced amorphous marks. The CNR is the difference of the carrier and the noise levels. In general, when media is compared using identical recording conditions, the sample with the higher CNR will support greater recorded data density. This is because data can be recorded by controlling the width of a domain, using a technique known as pulse width modulation (PWM). The carrier level measures the strength of the written signal. It is proportional to the reflectivity difference between the amorphous and crystalline state. The noise level is the average of the levels of frequencies immediately above and below the carrier. It is a measure of the local and undesired variation of the carrier level. There are many sources of noise in a phase change optical drive. Usually, because the signal is so strong, the predominant one is media noise. Media noise, which results from local variations in the reflected light, can arise from scattering or from local variations of the absorption due to surface roughness and inhomogeneities in the microstructure and optical properties of the thin film layers or the grooves. It is desirable to lower the media noise as much as possible. This will allow greater recorded data density.

One way to reduce the media noise is to choose a recording layer composition which crystallizes into a cubic structure[11]. In this way, noise due to crystalline grains is eliminated. If the crystal structure is not cubic, the optical properties depend upon the orientation of the grains resulting in a high media noise. The reflectivity of a cubic grain is independent of orientation. One might suspect then that when using phase diagrams for choosing recording layer compositions for phase change recording that only compounds which crystallize into cubic phases should be considered. However, in the Ge–Te–Sb system, compositions which in their equilibrium phase are hexagonal, will often crystallize into a metastable cubic phase. Similarly, the addition of Sn to GeTe helps stabilize it into an fcc structure rather than a rhombohedral structure. Thus, there is a much wider range of compositions from which to select the recording layer than might be suggested by equilibrium phase diagrams.

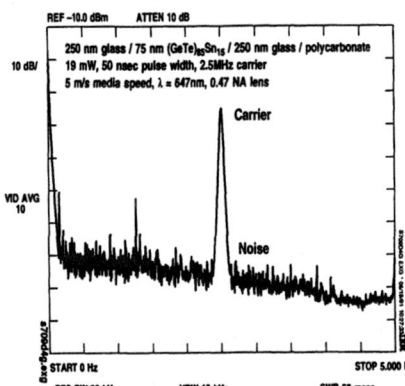 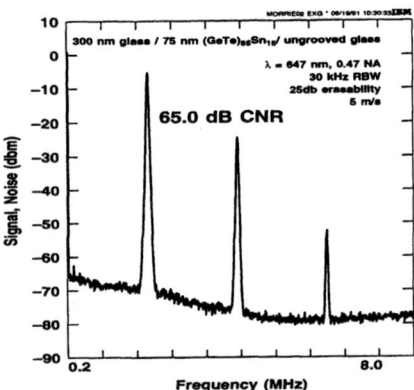

Figure 9. A CNR of 55 db was obtained from the structure: 250 nm glass / 75 nm $(GeTe)_{85}Sn_{15}$ / 250 nm glass / grooved polycarbonate disk.

Figure 10. High CNR from non grooved disk with the structure: 300 nm glass / 75 nm $(GeTe)_{85}Sn_{15}$.

Very high CNR can be achieved with cubic materials. When the composition, $(GeTe)_{85}Sn_{15}$, which crystallizes into a cubic structure, was deposited on a disk which did not have any grooves, the CNR was 66.4 db as shown in Figure 10. The media noise from the grooves of a grooved disk with the same composition recording layer limited the CNR to only 55 db as shown in Figure 9. If better quality grooves can be made which have less noise, then higher CNR will result. Alternatively, sampled servo techniques could be used where tracking pits are alternated with sections of data and no tracks are needed.

Improved Data Density at Short Wavelengths

Optical storage is targeting recording at shorter lengths since the areal density of recorded information increases inversely with the square of wavelength. Chalcogenide films are a strong candidate for recording at blue wavelengths since there is a large difference between the reflectivities of the amorphous and crystalline phases at short wavelengths. This is illustrated in Figure 11 for GeTe. $(GeTe)_{85}Sn_{15}$ also has a strong signal at short wavelength. To take advantage of that signal, it is necessary to write amorphous marks of smaller size. It was verified by TEM that marks approximately 0.25 μm in diameter were formed in the thin film structure: 250 nm glass / 75 nm $(GeTe)_{85}Sn_{15}$ / 250 nm glass / polycarbonate by reducing the laser power until just the center of focused beam had enough intensity to melt the recording layer[16]. This method of writing with the center does not produce well–defined mark location or size since it is so sensitive to slight reflectivity differences or small variations of the incident power. In contrast, when the FWHM diameter of a beam is used to write, the spot size is well defined and reproducible.

The strong signal and ability to write small marks shows that recording at short wavelengths is possible. This was further confirmed by experiments at 488 nm performed by Rugar and Rubin[21]. The same disk as that used to obtain the data in Figure 10 was used. In order to push the limit of recorded density, 3 high frequency carriers were written with 3 different frequencies. The CNR was 50 db. The three tracks were spaced closer and closer until the crosstalk signals from the adjacent tracks were 30 db below the signal measured while reading data from the center track. This was accomplished with a track spacing of 0.78 μm. The readback mark size and data spacing were 0.42 μm and 0.84 μm, respectively. This resulted in a raw data density of 1.5×10^8 marks/cm^2.

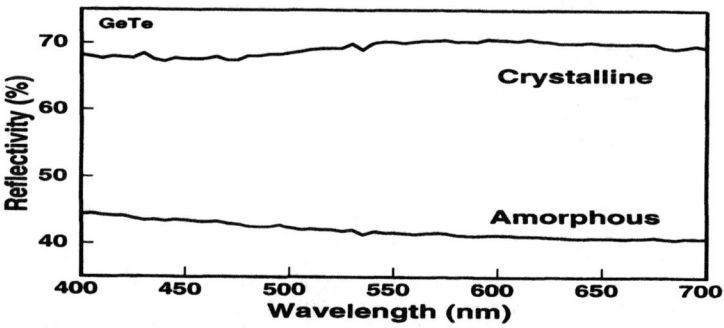

Figure 11. Wavelength dependence of the reflectivity of evaporated films of GeTe. The crystallized state was achieved by furnace annealing.

Conclusions

Fast erasing phase change optical storage media is achieved using compound compositions which crystallize partitionlessly. Although both stable and metastable compositions can be used, phase separation may be a problem with metastable compositions. Writing by amorphization requires structures with high quench rates of ~10 oC/nsec. Materials which undergo a semiconductor to semimetal transition have a strong signal. Media noise is minimized by use of materials which crystallize into a cubic crystal structure. Single beam direct overwrite works very well and erasabilities in excess of 25 db are readily achieved. Recording of blue wavelengths is quite promising, since the signal is strong and small stable spots can be formed. Although much progress has been made in improving the number of cycles, more work must be done to understand the failure mechanisms and the sensitivity of the cycling to process conditions, media structure and recording conditions. Areas requiring more intensive study include the origin of microstructural defects during cycling, the relationship of mechanical properties of the thin film overcoat layers to cycling, cause of media noise, and the thermal and optical properties of the liquid and amorphous states. Long term, there are good opportunities for increased recorded data density if the media noise is reduced.

Acknowledgments

The author thanks M. Chen and T. Suzuki for many stimulating discussions, and W. McChessney for sample preparation. In addition, he is grateful to D. Rugar for providing performance measurements at 488 nm and T. Suzuki for TEM analysis and to J. Kaufman for reflectivity measurements.

References

[1] K. A. Rubin, to appear in J. Mag. Soc. Jap., September 1991.
[2] E. Ohno, K. Nishiuchi, N. Yamada and N. Akahira, in Optical Data Storage, 1991, Tech. Dig. Ser., 5 (Opt. Soc. Amer.) 92.
[3] H. Ukita, H. Nakada and Y. Katagiri, in Optical Data Storage, 1991, Tech. Dig Ser., 5 (Opt. Soc. Amer) 130.
[4] K. A. Rubin, R. W. Barton, M. Chen,V. B. Jipson and D. Rugar, App. Phys. Lett., 50 (1987) 1488.
[5] S. R. Ovshinsky, J. Non–Cryst. Solids 2, (1970) 99.
[6] D. Strand and D. Adler, Proc. Soc. Photo–Opt. Inst. Eng. 420 (1983) 200.
[7] P. C. Clemens, Appl. Opt. Lett. 22 (1983) 3165.
[8] M. Chen, K. A. Rubin and R. Barton, App. Phys. Lett., 49 (1986) 1255.
[9] K. Nishimura, M. Suzuki, I. Morimoto and K. Mori, Jap. J. App. Phys, 28 Supp. 28–3 (1989) 135.
[10] Y. Maeda, H. Andoh, I. Ikuta, and H. Minemura, J. Appl. Phys. 64 (4), (1988) 1715.
[11] M. Chen and K. A.Rubin, Proc. Soc. Photo–Opt. Inst. Eng., 1078 (1989), 150.
[12] T. Ishida, S. Ohara, N. Akahira, T. Ohta and T. Yoshida, Jap. J. App. Phys, 28 Supp. 28–3 (1989) 129.
[13] Y. Maeda, H. Andoh, I. Ikuta, M. Nagai, Y. Katoh, H. Minemura, N. Tsuboi, Y. Satoh, N. Gotoh, and M. Ishigaiki, App. Phys. Lett., 54 (1989) 893.
[14] After Hansen, ed., Constitution of Binary Alloys (McGraw–Hill, New York, 1958).
[15] T. Ohta, M. Uchida, K. Yoshioka, S. Furukawa and K. Kotera, Proc. Soc. Photo–Opt. Inst. Eng., 1078 (1989), 27.
[16] T. Suzuki and K. A. Rubin, IBM Corporation, unpublished data.
[17] M. Libera, M. Chen and K. A. Rubin, to appear in J. Mat. Res.
[18] N. Yamada, E. Ohno, K. Nishiuchi, N. Akahira and M. Takao, J. App. Phys., 69 (1991) 2849.
[19] K. A. Rubin and M. Chen, Thin Solid Films, 181 (1989) 129.
[20] T. Ohta, S. Furukawa, K. Yoshioka, M. Uchida, K. Inoue, T. Akiyama, K. Nagata and S. Nakamura, Proc. Soc. Photo–Opt. Inst. Eng., 1316 (1990), 367.
[21] D. Rugar and K. A. Rubin, IBM Corporation, unpublished data.

THE RELATIONSHIP BETWEEN CRYSTAL STRUCTURE AND PERFORMANCE
AS OPTICAL RECORDING MEDIA IN Te-Ge-Sb THIN FILMS

D. STRAND, J. GONZALEZ-HERNANDEZ, B.S. CHAO, S.R. OVSHINSKY, P. GASIOROWSKI
AND D.A. PAWLIK

Energy Conversion Devices, Inc., 1675 West Maple Road, Troy, MI 48084

ABSTRACT

The crystallization properties of Te-Ge and Te-Ge-Sb alloys prepared by
thermal evaporation were analyzed using various characterization techniques.
Similar to previous results, our data for Te-Ge shows that alloys that
deviate slightly from $Te_{50}Ge_{50}$ stoichiometry show drastically slower
crystallization kinetics. Raman spectroscopy and x-ray diffraction show that
alloys having non-stoichiometric atomic ratios phase separate during
crystallization into a $Te_{50}Ge_{50}$ phase plus pure crystalline tellurium or
germanium. It is this relatively slow process of phase segregation which
limits the crystallization rate. Phase segregation during crystallization of
non-stoichiometric Te-Ge can be eliminated by adding antimony to samples
having a tellurium concentration of from 45 to 55 atomic percent over a wide
range of Ge:Sb ratios. These alloys can have laser induced crystallization
times of less than 50 nsec. The thermal crystallization temperature is
reduced only slightly when antimony is substituted for germanium.

INTRODUCTION

The idea of phase change technology, which was originated by S.R.
Ovshinsky more than twenty years ago [1,2], is now being successfully applied
in commercial optical disk based data storage systems. Although phase change
technology has long been recognized for the simple record and read processes
it uses, its commercial success was predicated on its direct overwrite
capability. Direct overwrite is the replacement of pre-existing recorded
data with new data in a single action. Compared to currently available
magneto-optical data storage systems, which require a first revolution of the
disk to erase existing recordings, followed by a second revolution for
recording new data, information can be recorded up to twice as fast using
direct overwrite phase change media.

In order to qualify for use as a direct overwrite media, phase change
materials had to be developed which could be crystallized in the same duration of laser exposure as was available for making them amorphous. This
means that the crystallization process has to be very rapid. Stoichiometric
compounds have been shown to exhibit the necessary crystallization speeds,
but often show the disadvantage of rapid loss of crystallization speed when
the composition deviates even slightly from stoichiometry [3,4]. In this
work, we show that adding Sb to the Te-Ge binary system greatly increases the
compositional range over which rapid crystallization is achieved. Further,
we correlate the crystal structure to the crystallization kinetics to explain
these and more subtle effects in the relationship between composition and
crystallization temperature.

EXPERIMENTAL

Te-Ge-Sb alloy films were prepared by thermal evaporation on unheated
(100) crystalline silicon, glass and polymethylmethacrylate substrates.
Multiple element alloy sources and single element sources were used singly
and in combination for the evaporation of the films used in this study. The

Mat. Res. Soc. Symp. Proc. Vol. 230. ©1992 Materials Research Society

compositions of the films were determined using energy dispersive x-ray spectrometry (EDS). Auger compositional depth profiles were measured to determine compositional uniformity and light element impurities. The sudden decrease in the optical transmission (measured at 630 nm) which accompanies crystallization during heating at a constant rate of 60 °C/min was used to indicate the crystallization temperature (T_x) of the as-deposited amorphous films. X-ray diffraction (XRD) was used to determine the crystalline structure in annealed films. The films were annealed by heating under argon atmosphere to a pre-determined temperature at a heating rate similar to that used in the T_x measurements. For the Raman measurements the 488 nm line from an argon ion laser was focused to a power density of about 3 W/cm^2 onto the sample using a cylindrical lens. The phase transformation kinetics data used for determining crystallization rate was measured using a static tester we constructed which focuses substrate-incident 830 nm semiconductor laser radiation to a spot approximately 1 µm diameter on an as-deposited 80 nm thick amorphous film on a polymethylmethacrylate substrate.

RESULTS

Figure 1 shows the Raman spectra obtained from two Te_xGe_{100-x} films with x = 43 (curves (a) and (b)) and x = 53 (curve (c)). Spectra (a), (b) and (c) correspond to the as-prepared, 400 °C and 300 °C annealed samples respectively. In the as-prepared amorphous sample, the broad Raman lines centered at about 125, 163, 225 and 275 cm^{-1} are characteristic of the scattering from Te-Te (125 and 163 cm^{-1}), Te-Ge (225 cm^{-1}) and Ge-Ge (275 cm^{-1}) bonds in a covalently bonded random network material. The stronger intensity of lines related to the Te-Te bonds results from the high Raman cross-section of tellurium. After annealing, the broad amorphous Raman features in the spectrum of both compositions vanish and sharper peaks associated with identifiable crystalline phases appear. For the x = 43 sample the peak at 297 cm^{-1} indicates the presence of crystalline germanium. The two peaks at 124 and 142 cm^{-1} in the film with x = 53 indicate the presence of crystalline tellurium. Although during the crystallization process the major volume fraction of each of these films crystallized into a TeGe rhombohedral phase, as indicated by XRD, no obvious Raman signal from that phase is observed. This indicates that this semi-metallic compound has a low Raman cross-section.

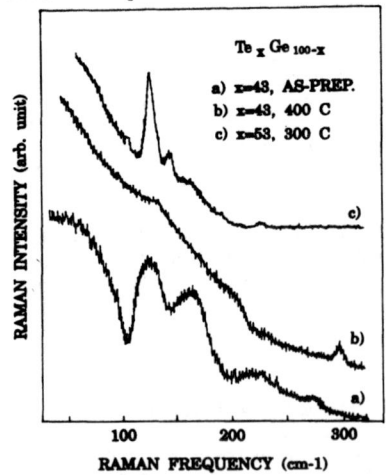

Figure 1. Raman intensity vs. Raman frequency for two Te_xGe_{100-x} films with x = 43, as-prepared (a) and annealed at 400 °C (b), and x = 53, annealed at 300 °C (c). The broad peaks in curve (a) are related with Te-Te, Te-Ge, and Ge-Ge bonds in the amorphous structure. The peak at about 300 and 100-150 cm^{-1} in curves (b) and (c) corresponds to c-Ge and c-Te, respectively.

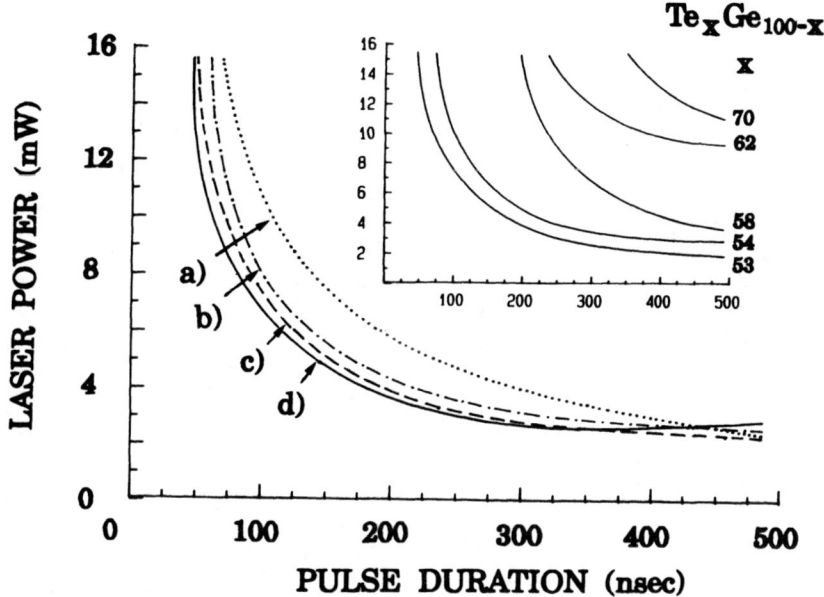

Figure 2. Phase transformation kinetics curves for the Te rich
Te_xGe_{100-x} films (insert) and the Te-Ge-Sb alloy films:
(a) $Te_{46}Ge_{50.5}Sb_{3.5}$, (b) $Te_{48}Ge_{50.5}Sb_{1.5}$, (c) $Te_{51}Ge_{32}Sb_{12}$,
and (d) $Te_{55}Ge_{24}Sb_{21}$ and $Te_{54}Ge_{41}Sb_5$.

The curves in figure 2 indicate where the onset of crystallization
occurs in a test matrix of incident laser pulses having various pulse
amplitudes and durations (phase transformation kinetics, or PTK plots) [5]
for Te-Ge-Sb alloy films having a wide range of compositions. In the figure
insert, results of similar measurements are shown for Te rich Te_xGe_{100-x}
alloys. Similar to previous results [3], we find that the crystallization
speed of Te_xGe_{100-x} alloys strongly depends on the value of x. Films
having a composition close to stoichoimetric TeGe can be crystallized using
short laser pulses, however, much longer laser pulses are needed to
crystallize films whose compositions are slightly away from stoichiometry. A
different behavior is observed in the Te-Ge-Sb alloy films. The phase
transformation kinetics of these films are similar to those of stoichiometric
TeGe films. Further, the phase transformation kinetics of various compositions in the Te-Ge-Sb alloy system are similar regardless of the Ge:Sb ratios
as long as the Te level is 45 to 55 atomic percent. Crystallization times of
less than 50 nsec were observed in compositions having Sb levels from 1.3 up
to 35 atomic percent.

The crystallization temperature (T_x) of the Te-Ge-Sb alloy films is
summarized in figure 3(a). Our measurements show that the T_x of

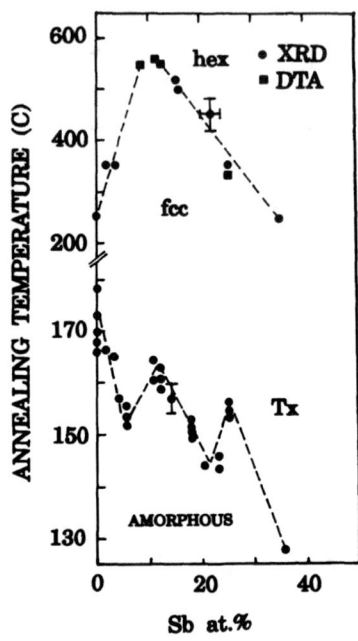

Figure 3. Dependence of the crystallization temperature, T_x, (a) and the crystalline-to-crystalline transition (fcc to hexagonal) temperature (b) as a function of the Sb content in the Te-Ge-Sb alloy films.

Te_xGe_{100-x} alloy thin films, in the vicinity of a stoichiometric TeGe, is in the range of 165-178 °C. Similar to what we observed in the crystallization speed, we found that T_x is also very sensitive to the ratio of Te:Ge [3]. Any slight deviation from stoichiometric TeGe results in a higher T_x. The T_x decreases slightly when antimony is added to Te-Ge alloy films. However, the T_x does not follow a monotonically decreasing trend with an increase of the amount of Sb in the alloys, as seen in figure 3(a). As antimony is added to the alloy, the T_x first drops from 165 °C at zero atomic percent antimony to 153 °C at 5 atomic percent antimony. The T_x then increases to about 162 °C at 11 atomic percent antimony. Adding more antimony to the composition gradually decreases the T_x again, followed by a second relative maximum at 25 atomic percent antimony.

All of the Te-Ge-Sb alloy films having up to 35 atomic percent antimony were amorphous as-prepared. Films having a composition of $Te_{50}Sb_{50}$ are crystalline (hexagonal phase) as-prepared. When heated to a temperature slightly above their T_x, the amorphous Te-Ge-Sb films crystallize into a fcc structure (see figure 3(a)). The crystalline fcc phase transforms to a more stable crystalline structure having rhombohedral symmetry (hexagonal structure) upon annealing at higher temperatures. This crystalline-to-crystalline phase transition temperature depends on the film composition, as shown in figure 3(b).

The data in figure 3(b) was collected using two separate experimental techniques: differential thermal analysis (DTA) and thermal annealing as performed in conjunction with XRD measurements. A stoichiometric TeGe film has a transition temperature at about 250 °C. The transition temperature increases when antimony is added, to 560 °C at 11 atomic percent, and then gradually decreases with further increases in the Sb concentration.

Figure 4 shows XRD data measured on a $Te_{52}Ge_{23}Sb_{25}$ film in the as-prepared state (a) and after annealing at temperatures of 300 °C (b) and 400 °C (c). The broad features shown in curve 4(a) centered at an angle of about 28° and 47° in the 2θ scale are typical of the amorphous structure of the as-deposited alloy film. In film annealed at or below 300 °C (curve (b)), all of the diffraction peaks are related to the crystalline fcc structure. Additional diffraction peaks, marked by arrows in curve 4(c), clearly indicate that the crystalline-to-crystalline phase transition occurred in the film at annealing temperatures between 300 and 400 °C.

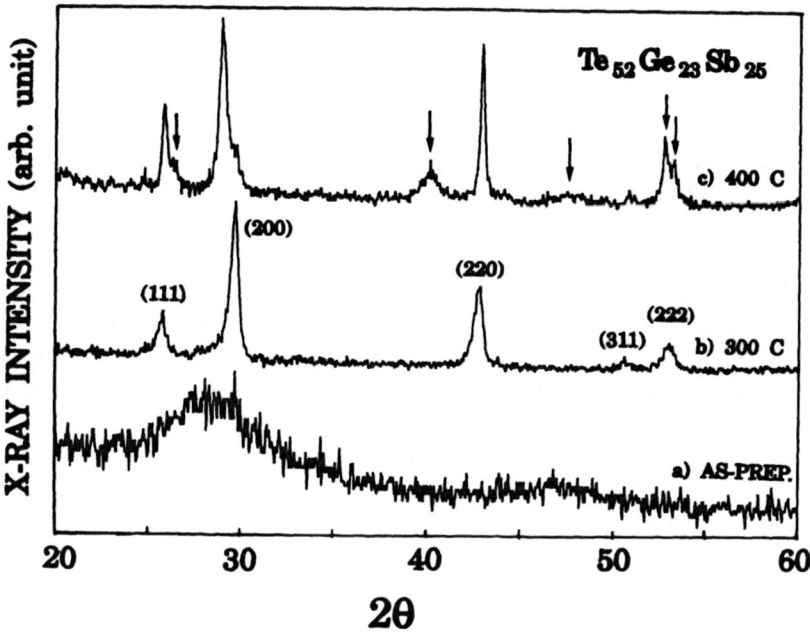

Figure 4. X-ray diffraction patterns for $Te_{52}Ge_{23}Sb_{25}$ film as-deposited (a), annealed at 300 °C and 400 °C. The arrows indicate the lines that are only associated with the hexagonal structure.

DISCUSSION AND CONCLUSION

It has been previously reported that the crystallization kinetics of $Te_x Ge_{100-x}$ strongly depend on the value of x (figure 2, insert) [3]. Deviations from stoichiometry result in a much slower crystallization rate compared to the $Te_{50}Ge_{50}$ alloy. Phase segregation with slow atomic diffusion during crystallization has been suggested as the limiting mechanism for crystallization in non-stoichiometric compounds. In the present study, we have used several characterization techniques to analyze the structure of crystalline Te-Ge and Te-Ge-Sb alloys.

Using Raman sepctroscopy we have been able to detect phase segregated crystalline tellurium in fully crystallized Te-Ge films with tellurium concentrations of 53 atomic percent or higher. Similarly, our Raman and XRD measurements clearly show the presence of crystalline germanium in alloys which have Ge concentrations of 57 atomic percent or higher and which have been annealed at 400 °C or higher (figure 1). Higher sensitivity for the detection of crystalline tellurium by Raman scattering results from a higher Raman cross-section at the energy used for excitation. Further, due to resonance effects, crystalline tellurium incorporated in an absorbing matrix can be detected when present at less than one percent of the volume. Based upon the Raman data shown in figure 1(a), we expect to be able to detect the presence of crystalline germanium at a volume fraction of about three percent in an absorbing matrix.

A similar Raman analysis has been carried out in the Te-Ge-Sb alloy system with Te concentrations ranging from 45 to 56 atomic percent and Sb concentrations from 1.3 to 50 atomic percent. In that compositional range no phase segregation has been observed, which means that if there is any of crystalline Te or Ge present, the volume fraction must be below our detection limits. The EDS data indicates that there is enough Te in some of our alloys to readily be detected by Raman, if present, and enough Ge to be detectable by Raman, if present, in other alloys. The PTK curves in figure 2, showing only slight differences in performance for various compositions, support our premise that antimony averts phase segregation.

As illustrated in figure 3(a), the Te-Ge-Sb alloys have lower T_x than stoichiometric TeGe. However, it is clear that the T_x reached local maximum values in films with Sb concentrations of about 11 and 25 atomic percent. It is expected that ordering in the amorphous alloys varies smoothly with antimony concentration. If this is so, the slight increase in T_x ($\sim$ 2%) for these two compositions is probably related to slightly different crystallization kinetics. At a composition of about 11 atomic percent antimony, crystallite size obtained from the XRD data is largest [6] and the fcc crystalline structure is most stable. This suggests that at this composition, the contribution to the growth process by growth, relative to nucleation, becomes more important.

REFERENCE

[1]. J. Feinleib and S.R. Ovshinsky, J. Non-Cryst. Sol. 4, 564 (1970).
[2]. J. Feinleib, J. deNeufville, S.C. Moss, and S.R. Ovshinsky, Appl. Phys. Lett. 18, 254 (1971).
[3]. M. Chen, K.A. Rubin, and R.W. Barton, Appl. Phys. Lett. 49, 502 (1986).
[4]. N. Yamada, E. Ohno, K. Nishiuchi, N. Akahira and M. Takao, J. Appl. Phys. 69, 2489 (1991).
[5]. K.A. Rubin, R.W. Barton, M. Chen, V.B. Jipson, and D. Rugar, Appl. Phys. Lett. 50, 1488 (1987).
[6]. J. Gonzalez-Hernandez, B.S. Chao, D. Strand, S.R. Ovshinsky, P. Gasiorowsky and D.A. Pawlik, unpublished.

DIRECT EVIDENCE OF THE Si INTERSTITIALICY INJECTION AND FAST DIFFUSION EFFECT IN Si DURING PULSED LASER MELT PROCESS

YIH CHANG*, J. CHEN**, S. TALWAR*, E. Y. SHU*, and THOMAS. W. SIGMON*
*Solid State Laboratory and Department of Electrical Engineering, Stanford University, Stanford, CA 94305.
**Technology and Manufacturing Group, Intel Corp., Santa Clara, CA 95051.

ABSTRACT

High temperature gradient induced fast diffusion effect during pulsed laser melt processes is reported for the first time in the Si semiconductor. We use both the oxidation-induced stacking faults and dislocation loops as markers to determine the degree of enhanced diffusion in the unmelted Si substrate by tracing their movement after laser melt. The cross-sectional transmission electron microscope is employed to investigate the defect structures before and after laser melt. The expanded dislocation loops exhibit a significant diffusion and climb occurring during the nanosecond-scale processing duration. We also formulate and solve the governed diffusion equation, based on a high temperature gradient induced electric field, to simulate the dislocation movement. The agreement between the experimental and simulated results verifies that the fast diffusion effect indeed occurs in the Si semiconductor during the pulsed laser melt period.

INTRODUCTION

An anomalous point defect injection and fast diffusion effect is reported during the characterization of the redistribution behaviors of each atomic species in the heteroepitaxial $In_xGa_{1-x}As/n+$-GaAs(100) layers fabricated using pulsed laser induced epitaxy (PLIE) [1]. A hypothesis, based upon the combined effects of concentration impulse and huge temperature gradient across the liquid/solid (l/s) interface, is proposed to explain the significant solid phase diffusion observed during the pulsed laser melting process. In other words, the impulse of Ga, As, In, and Si exerted by the pulsed laser is believed to pile up at the l/s interface during the nanosecond-scale melting period. As the melt front reaches and regrows from the position of the maximum melt depth, these atoms are then injected into the unmelted GaAs substrate by a high temperature gradient induced fast diffusion effect.

In order to understand and confirm this fast diffusion effect, in this presentation, we describe an experiment designed to examine this phenomenon in detail. For this purpose we choose Si substrate material for this experiment. If the proposed high temperature gradient induced fast diffusion effect in fact exists, this phenomenon will also occur in Si semiconductor because of the high thermoelectric power [2]. The advantage of choosing Si is that the diffusion parameters of Si interstitialicy (Si_i) are established for certain temperature ranges, whereas we are far from understanding the diffusion of Ga and As interstitialicies in GaAs substrate. Therefore, we can formulate the diffusion equation and simulate the fast diffusion effect by using Si as the medium.

This study employs the modified processing steps for the formation of the oxidation-induced stacking fault (hereafter called OISF) to produce well-shaped dislocation loops and stacking faults with size less than 5000Å. The cross-sectional transmission electron microscope (TEM) is used to examine the defect structures before and after pulsed laser melting process. In the unmelted substrate, the curved and expanded dislocation loops exhibit a significant diffusion and climb process occurs. The observed dislocation motion is caused by the temperature gradient enhanced diffusion of Si_i into the Si substrate. These Si_i, in the neighborhood of the existing dislocations, then jump and condense into the dislocation core to cause the movement, with moving distance longer than 200Å in this nanosecond scale processing period. Since almost all of the diffusion parameters are available for Si semiconductor, we formulate and solve the governed diffusion equation to simulate the movement of the dislocation based on the proposed hypothesis.

Mat. Res. Soc. Symp. Proc. Vol. 230. ᶜ 1992 Materials Research Society

DESIGN CONCEPT AND PROCESSING STEPS

It is well known that dopant diffusion in Si is assisted by point defects such as Si interstitialicies and vacancies [3-7]. Thus, for accurate profile prediction, it is essential to develop an understanding of the kinetics of point-defect generation and diffusion, as well as their interaction with dopant impurities. Processes–such as thermal oxidation of Si, thermal nitridation of Si and SiO_2, oxygen precipitation in Czochralski grown Si, and low-pressure chemical-vapor-deposited (LPCVD) silicon nitride on Si–provide valuable opportunities for studying point defect structures. That is because these processes can generate sources of excess interstitialicies or vacancies. For example, thermal oxidation of Si can create stacking faults and dislocations beneath the Si/SiO_2 interface if nucleation sites are available for the condensation of the Si interstitialicies injected from the Si/SiO_2 interface [3]. Since we propose that the host atoms (Si atoms in this study) are injected into the unmelted substrate by a high temperature gradient induced fast diffusion effect, we expect the growth or expansion of the stacking faults and dislocation loops due to this injection of Si atoms. In other words, if stacking faults and dislocations exist beneath the surface, we can investigate the change of size and shape before and after the laser melt process in order to determine the degree of the enhanced diffusion.

Previous results, using stacking fault size on the order of a few tens of microns, allow the employment of an optical microscope to easily examine the Wright-etched configurations appearing on the surface of the Si wafers [5]. However, in this study, we must reduce the diameter of the stacking fault down to less than 5000Å. This is because the laser induced melt depths are less than 3000Å, with the energy fluence of the laser beam between 1.0~1.4 J/cm². The choice of the energy fluence also depends on several conditions, such as preventing surface damage but still allowing significant diffusion to be observed. Figure 1 shows the selected energy fluence ranges and the corresponding melt depths, seen to lie between 600~2200Å, shown by the cross-hatched region. For this circumstance, we must use TEM to observe the desired results. For this case, not only stacking faults but dislocation loops are also visible under the appropriate TEM imaging conditions. Although the sample preparation and crystallographic interpretation become more complicated, the advantage of the presence of dislocations is that it supplies one more piece of information with which to study this fast diffusion effect. Thus, the design concept of this experiment allows use of both stacking fault and dislocation loops as markers to determine the degree of enhanced diffusion effect in the unmelted region by simply tracing their movement after laser melting.

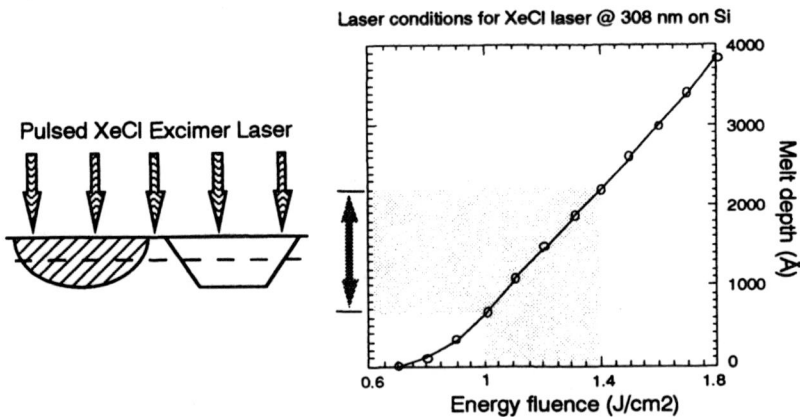

Fig.1 Selected laser energy fluence ranges required to melt the defect structures with melt depths within the cross-hatched range.

Details of the processing are as follows: the substrates are float-zone (FZ) grown, n-type, (001) Si wafer with resistivities of 5~10 Ω-cm. A dose of 10^{15} cm⁻² Si+ is implanted into the Si at an energy of 100 keV to nucleate defect sites near the surface. The projection range of this implantation is 1337 Å, shallow enough to restrict the growth of the stacking faults to less than 1μm. We increased the dose from 5×10^{13} cm^{-2}, normally used for optical microscopic investigation [6], to 10^{15} cm^{-2} in order to nucleate more defect sites for easier TEM examination. The implanted wafer is then oxidized at 950 °C in a dry oxygen ambient for 12 hours growing a SiO$_2$ film 1700 Å thick. During the SiO$_2$ growth period, the implant-damaged layer is recrystallized via a solid phase epitaxial regrowth. It is reported that the growth of a SiO$_2$ film of 700 Å is enough to grow embryonic OISF of less than 1 μm in length [5]. However, this growth period is still not long enough to fully develop extrinsic-type dislocations. We extended the oxidation duration by increasing the SiO$_2$ film from 700Å to 1700Å in order to annihilate some of the OISF and generate the dislocation structures which are believed to forme under the unfaulting reactions [7]. After oxidation, the wafers are HF-etched, to remove the SiO$_2$ layer, and processed with the pulsed XeCl excimer laser, in which the laser system is similar to that described in reference 1. Since the Burgers vector identification and the dislocation movement are easily determined in TEM cross-sectional imaging mode, we used cross-sectional TEM to examine these defect structures both before and after laser melt.

RESULTS AND DISCUSSIONS

Figures 2(a) and (b) show the bright-field TEM images of typical as-grown stacking faults and dislocation loops before laser treatment. After being characterized under different diffraction contrast conditions, these defects are identified to be extrinsic. The detailed formation mechanism is quite complicated, but to explain in a simpler way, these defects are generated by the condensation of the injected Si$_i$ from the Si/SiO$_2$ interface, as if an extra half plane of Si atoms is inserted into the perfect lattice. The most important feature of these defects is that both the stacking faults and AB portion of the dislocation loop, described in Fig. 2, are pure edge components. If they show any movement in the direction toward the unmelted substrate, they have to undergo a diffusion and climb process. This unique characteristic is critical in distinguishing the effects between diffusion and complicated stress-state-induced defect motion, and helps in the study of the fast diffusion process occurring during the PLIE process.

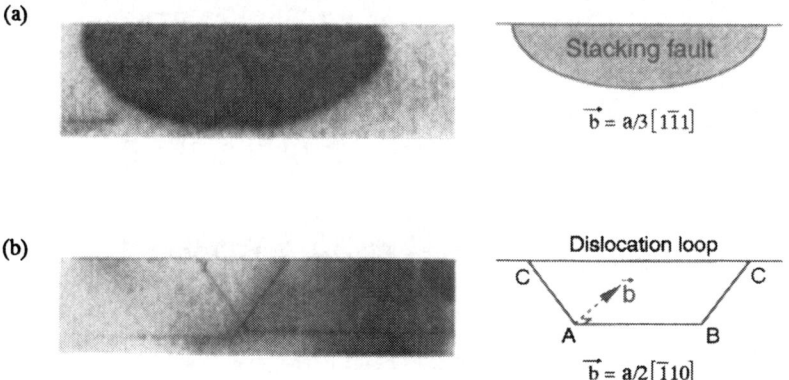

(a)

Stacking fault

$\vec{b} = a/3\,[1\bar{1}1]$

(b)

Dislocation loop

$\vec{b} = a/2\,[\bar{1}10]$

Fig. 2 Defect structures before laser melt. (a) Extrinsic stacking fault lying on $(1\bar{1}1)$ and bound by Frank partial with $\vec{b} = a/3\,[1\bar{1}1]$. (b) Extrinsic dislocation loop lying on $(1\bar{1}1)$ with $\vec{b} = a/2\,[\bar{1}10]$. AC and BC segments are 60° dislocations while the AB portion is a pure edge component.

Figures 3(a) and (b) show bright-field images of the defect structures following melting by the pulsed laser, using 100 pulses at two different energy fluences, 1.14 and 1.4 J/cm^2, respectively. We can easily trace the interface between the melted and unmelted layer by drawing a line from the position of the birds-beak of the faults to the discontinuity of the dislocation segment. By comparing the change of the shape of the dislocation loops with that shown in Fig. 2(b), we clearly see that a large amount of dislocation motion has occurred in the underlying unmelted substrate. However, the differences, if any, in the amount of dislocation movement for these two energy fluences are too small to be easily resolved using the TEM images. If we laser melt the defect structures at the same energy fluence but with a different number of pulses, we can clearly see the differences, shown in Figures 4. This figure illustrates that greater dislocation loop movement occurs with increasing number of pulses. This result is consistent with the proposed hypothesis since the temperature gradient induced fast diffusion occurs for each pulse.

(a)

(b)

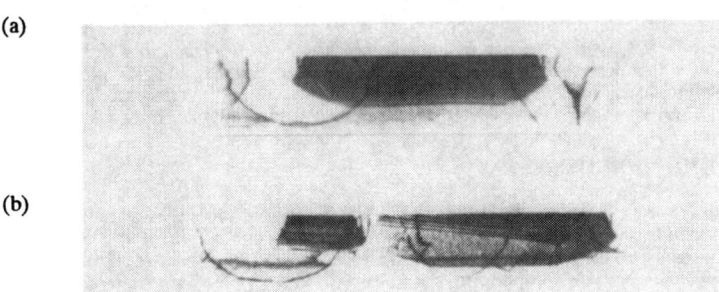

Fig. 3 Defect structures after laser melted using 100 pulses at an energy fluence of (a) 1.14 and (b) 1.4 J/cm². The melt depth of sample (a) is 2200Å while sample (b) is 1320Å.

(a)

(b)

(c)

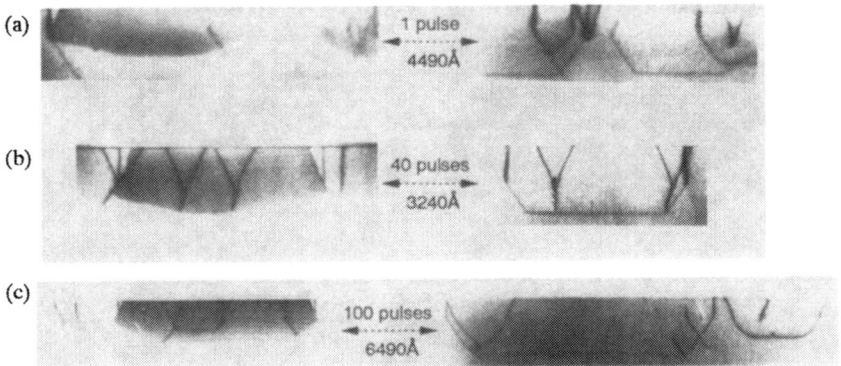

Fig. 4 Defect structures after laser melting at energy fluence of 1.14 J/cm² with (a) 1 pulse, (b) 40 pulses, and (c) 100 pulses. The melt depth of these samples is 1320Å.

In order to quantitatively correlate the dislocation motion with the temperature gradient induced fast diffusion effect during the pulsed laser melting process, we formulate a diffusion equation. The flux of Si_i, J_{Si_i} is written as

$$J_{Si_i} = -D_{Si_i} \frac{\partial C_{Si_i}}{\partial x} + \frac{q D_{Si_i} C_{Si_i}}{kT} \frac{\partial \phi}{\partial x} \qquad (1)$$

where D_{Si_i} is the diffusion constant of Si_i, C_{Si_i} is the instantaneous concentration of Si_i, k is the Boltzmann constant, q is the charge, T is the transient temperature, and ϕ is the electrochemical potential. Based upon our proposed hypothesis, the diffusion behavior of Si_i is assumed to be governed mainly by the temperature gradient induced electric field during the PLIE process. In this case, the ε-field term, $\partial\phi/\partial x$ in equation (1), can be written solely in terms of the thermoelectric power, s, times the temperature gradient, $\partial T/\partial x$. Using the mass conservation law, we obtain the following diffusion equation:

$$\frac{\partial C_{Si_i}}{\partial t} = D_{Si_i}\frac{\partial^2 C_{Si_i}}{\partial x^2} + f(x,t)\frac{\partial C_{Si_i}}{\partial x} + g(x,t)C_{Si_i} \tag{2}$$

where

$$f(x,t) = \frac{\partial D_{Si_i}}{\partial x} - \frac{qsD_{Si_i}}{kT}\frac{\partial T}{\partial x} \tag{3}$$

$$-g(x,t) = \frac{\partial D_{Si_i}}{\partial x}\frac{qs}{kT}\frac{\partial T}{\partial x} + \frac{qD_{Si_i}}{kT}\frac{\partial s}{\partial x}\frac{\partial T}{\partial x} + \frac{qsD_{Si_i}}{k}\frac{\partial^2 lnT}{\partial x^2} \tag{4}$$

In equation (2), the overall Si_i concentration profiles are mainly determined by $f(x,t)$ and $g(x,t)$, since $f(x,t)$ controls the shift while $g(x,t)$ effects the upward movement. After inputting all of the physical parameters and the initial Si_i profiles, assuming the concentration impulse at t = 0.0, we solve out equation (2) and obtain the Si_i (x,t) profiles. For further details the reader is encouraged to read Reference 8.

Once we obtain the transient $Si_i(x,t)$ profiles, we can simulate the dislocation movement and compare with the results, shown in Fig. 4(c). This simulation depends on the assumption that the Si_i diffuses via the temperature gradient enhanced diffusion mechanism. The Si_i then condensed into the dislocation core which results in the dislocation motion. The time dependent dislocation movement perpendicular to the (001) plane, $dr(x,t)/dt$, can be expressed as [3],

$$\frac{dr(x,t)}{dt} = \pi a_0^2 D_{Si_i}^{effective}\left[C_{Si_i} - C_{Si_i}^{equilibrium}\right]\cos 35.3° \tag{5}$$

where a_0 is the effective capture cross-section, $D_{Si_i}^{effective}$ is the effective diffusion constant,

$$D_{Si_i}^{effective} = D_{Si_i} \times C_{Si_i} / C_{Si_i}^{equilibrium} \tag{6}$$

$C_{Si_i}^{equilibrium}$ is the equilibrium concentration of Si_i [9],

$$C_{Si_i}^{equilibrium} = 1e27\ exp\left[-\frac{3.8}{kT}\right] \tag{7}$$

and $\cos 35.3°$ is the projection of $(1\bar{1}1)$ on $(\bar{1}10)$ in the cross-sectional TEM imaging. The a_0 value we assume here is $\sim 10 \times \vec{b}$, which is $\sim 50\text{Å}$ in this simulation since the interaction between dislocation core and Si_i, we believe, is in this range. The dependence of $D_{Si_i}^{effective}$ on the ratio between C_{Si_i} and $C_{Si_i}^{equilibrium}$ results from the assumption that, for the higher temperature gradient, the larger the ratio.

Figure 5 shows both the simulated and experimental results on the same dimensional scale. Comparison of Fig. 5(a) with (b) shows that the overall shape of the dislocation loops after laser melt are quite similar. Even more important is that the $\sim 250\text{Å}$ movement of the edge portion of the dislocation loop at a depth of 1500Å is of the same order of magnitude for both cases. The agreement between these two results confirms that our proposed fast diffusion effect indeed occurs during PLIE process in Si substrates. Although the 60° dislocation has a screw component which largely results from the complicated state of thermal stress, we still see that the magnitude of the movement for both the simulation and experimental results are quite close. This result conflicts with the work of Baumgart et. al., who found expansion of dislocation loops in *cw* laser-annealed implanted and virgin silicon wafers [10]. They proposed that both the lateral and in-depth temperature gradients were responsible for the glide of the 60° dislocations. We cannot rule out thermal stress effects since no quantitative information is available to describe the anisotropic 3–D thermal stress state occurring during the pulsed laser melting process. We still, however, believe that the diffusion and climb process is a dominant effect in the explanation of the dislocation motion, based on the agreement seen in Fig. 5.

(a)   (b)

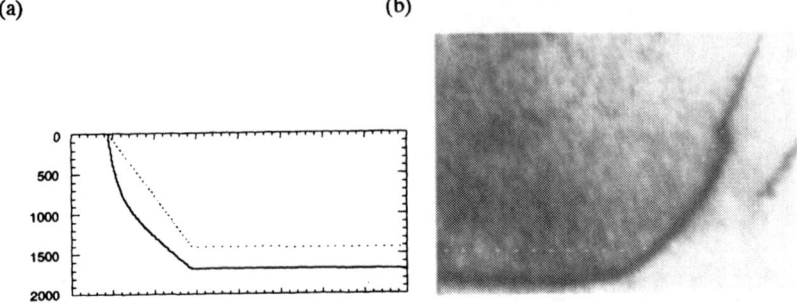

Fig. 5 Dislocation loop movement from (a) simulated results and (b) magnified TEM photograph from Fig. 4(c). The scale shown in (a) is in Å units.

To summarize, we have developed an experiment using oxidation-induced stacking faults and dislocation loops which serve as markers in the study of the effect of temperature gradient induced fast diffusion resulting from the pulsed laser melting of Si. The observed dislocation loop movement, confirmed by both TEM imaging and simulation results, indicates that this fast diffusion effect indeed occurs for Si.

REFERENCES

[1]   Y. Chang and T. W. Sigmon, presented at MRS Spring Meeting, 1991.
[2]   R. H. Bube, "Electrons in solids", 2nd edition, Academic press, 1988.
[3]   S. M. Hu, J. Appl. Phys., **45**(4), 1567 (1974).
[4]   C. L. Claeys, G. J. Declerck, and R. J. V. Overstraeten, Appl. Phys. Lett., **35**(10), 797 (1979).
[5]   Y. Hayafuji, K. Kajiwara, and S. Usui, J. Appl. Phys., **53**(12), 8639 (1982).
[6]   S. T. Ahn, H. W. Kennel, W. A. Tiller, and J. D. Plummer, J. Appl. Phys., **65**(8), 2957 (1989).
[7]   K. V. Ravi, Phil. Mag., **45**, 1081 (1974).
[8]   Y. Chang, Ph.D. Dissertation, Stanford University, Stanford, CA, 1991.
[9]   P. M. Fahey, P. B. Griffin, and J. D. Plummer, Rev. Modern Phys., **61**, 289 (1989).
[10] H. Baumgart, F. Phillipp, G. A. Rozgonyi, and U. Gosele, Appl. Phys. Lett., **38**(2), 95 (1981).

ANOMALOUS POINT DEFECT INJECTION DURING PULSED LASER MELTING PROCESSES: DIRECT EVIDENCE OF Ga_i and As_i PROFILES IN GaAs

YIH CHANG AND THOMAS W. SIGMON
Solid State Laboratory and Department of Electrical Engineering, Stanford University, Stanford, CA 94305

ABSTRACT

Significant point defect injection during a pulsed laser melt process is reported for the first time. Heteroepitaxial $In_xGa_{1-x}As$/GaAs layers fabricated by a pulsed laser induced epitaxy technique are used in this study. Transmission electron microscopy (TEM), energy-dispersive X-ray (EDX) and secondary ion mass spectrometry (SIMS) are employed to study the redistribution behavior of each species on the atomic scale. It is found that both the Si dopant species and the Ga, As, and In host atoms are injected into the underlying GaAs substrate. These species are then significantly redistributed, forming near spherical As-rich regions. Direct evidence of As_i and Ga_i (Ga and As interstitialcies) profiles in the GaAs substrate are also obtained for the first time. A hypothesis, based upon the combined effects of concentration impulse and large temperature gradients across the liquid-solid interface, is proposed to explain the significant solid phase diffusion observed during the pulsed laser melting process. We estimate the temperature gradient induced electric field during the process to be on the order of $10^4 V/cm$.

INTRODUCTION

The interaction of pulsed UV-lasers with materials is receiving extensive study due to potential applications. These include, for example, gas immersion laser doping, laser assisted redistribution of implanted dopant atoms, and the fabrication of superconductors by UV-laser ablation [1,2]. A new growth technique, known as pulsed laser induced epitaxy (PLIE), has been developed to fabricate heteroepitaxial layers using a pulsed UV-laser melt-recrystallization process. This growth mechanism, based on planar epitaxial regrowth from the underlying substrate following pulsed UV-laser melting of the overlayers, has successfully fabricated heteroepitaxial $In_xGa_{1-x}As$/GaAs and patterned Ge_xSi_{1-x}/Si layers [3-6].

Since the electrical properties of the pulsed UV-laser melted layers depend strongly on the concentration profiles of various species in the melted/unmelted region, we must fully understand the redistribution behavior of each species during the rapid melt-solidification processes. For example, Si, In, Ga and As profiles control the properties in the heteroepitaxial $In_xGa_{1-x}As$/n^+-GaAs(100) layers since the In content determines the energy bandgap and the Si dopant controls the Fermi level in the GaAs matrix. We have characterized the Si and In profiles in the melted region for the $In_xGa_{1-x}As$/GaAs layers fabricated by PLIE [5]. This presentation especially focuses on a detailed study of the redistribution profiles of each atomic species (Si, In, Ga, As) in the unmelted GaAs substrate. We find that both the dopant and host atoms are redistributed significantly in the underlying GaAs substrate. This leads us to discover the point defect injection driven by a fast diffusion effect during the PLIE process. Although there are several possible effects existing during the melt duration, a hypothesis based upon high temperature gradient induced fast diffusion is proposed to explain the point defect injection.

EXPERIMENT AND RESULTS

Fig. 1 schematically shows the laser processing steps used to grow $In_xGa_{1-x}As$/GaAs layers. The $In_xGa_{1-x}As$/GaAs layers used for this work are formed by laser melting a deposited

200Å In film on GaAs(100) substrates uniformly doped with Si (10^{18} cm^{-3}). The laser energy fluence is 0.61 J/cm^2 and 40 pulses are used. The formed In$_x$Ga$_{1-x}$As layer thickness is ~940Å with a composition x=0.21. For further details of the process flow and the characterization of the compositions and layer thicknesses of the In$_x$Ga$_{1-x}$As formed, the reader is encouraged to read Ref.5. Fig. 2 clearly shows the formation of a heteroepitaxial In$_{0.21}$Ga$_{0.79}$As layer which is coherently matched to the GaAs substrate. A misfit dislocation is also labelled in this [110] cross-sectional lattice image. Misfit dislocations are formed at the interface between In$_x$Ga$_{1-x}$As and GaAs since the In$_{0.21}$Ga$_{0.79}$As layer thickness is significantly larger than the critical thickness for commensurate growth, believed to be about 200Å for x=0.21 [7]. An island ~100Å in diameter is also seen in this micrograph and a misfit dislocation lies at the right top corner of the island.

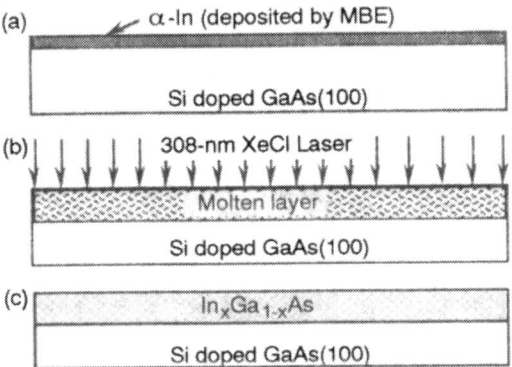

Fig. 1 Schematic diagram of heteroepitaxial In$_x$Ga$_{1-x}$As/GaAs layers fabricated by the pulsed laser induced epitaxy technique. (a) A thin layer of amorphous In is deposited onto GaAs substrate by MBE. (b) The deposited amorphous layer and part of the underlying GaAs are melted by the XeCl laser. (c) A heteroepitaxial In$_x$Ga$_{1-x}$As/GaAs layer is formed.

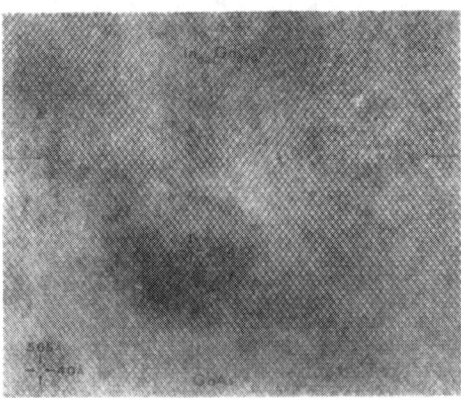

Fig. 2 (a) Cross-sectional lattice image of the heteroepitaxial In$_x$Ga$_{1-x}$As/GaAs layer near the interface. A misfit dislocation and one As-rich island are also labelled.

One important feature to be observed in Fig. 2 is that the island, which we assume lies exactly at the position of the misfit dislocation, is almost coherently matched to the matrix and located ~100Å below the $In_{0.21}Ga_{0.79}As/GaAs$ interface. Since the position of the islands is a key in determining whether the formation mechanism is due to liquid or solid phase diffusion, we have to identify precisely the relative location of the islands by means of the misfit dislocations. We argue that the misfit dislocation defines the $In_{0.21}Ga_{0.79}As/GaAs$ interface position because the misfit is induced by the release of the strain energy created by introducing the larger lattice parameter $In_xGa_{1-x}As$ matched onto the GaAs substrate. We also believe that the dislocation starts forming right at the $In_xGa_{1-x}As/GaAs$ interface instead of generating from the free surface, since the liquid/solid melt-front solidifies exactly from this position. Although thermal stress will be induced after the molten region solidifies, this stress is compressive since the density of the molten GaAs is 11% higher than that of solid GaAs at the melting temperature [8]. This will cause the misfit dislocation to form even further down into the GaAs substrate. Since the location of the dislocation is indeed the lowest boundary of the $In_xGa_{1-x}As/GaAs$ interface, we conclude that the islands lie in the unmelted region.

In order to study the composition of the island, we employ *in-situ* EDX analysis to determine the As/Ga ratios when performing the TEM measurements. Fig. 3(a) shows EDX spectra for four different locations. These measurements clearly show that the As/Ga ratios are different for the different positions. Since In replaces the position of Ga atoms when forming the $In_xGa_{1-x}As$ compound, we must extract the exact As/Ga ratio from the redistribution of the In atoms. We use the GaAs substrate as standard and then quantitatively calculate the compositions of the island and the strip area (the software used in this study is "Thin" in PV9900 developed by EDAX Corp.,Philips). Since SIMS measurement takes the average signals from ~100μm spot size, we show the EDX spectra of the strip area in order to compare with the profile obtained from SIMS measurement. We show the EDX spectra of the island in order to demonstrate the As-rich characteristics. However, we also take the EDX spectra of the region inbetween the islands. It showed a slightly lower As/Ga ratio. These results indicate that both the island and the strip area are As-rich. The formation of the As-rich clusters can offer important information of the point defect transport history since the As-rich clusters also have been reported elsewhere with different processing technique under various excess As conditions [9-11]. These results show that the similar As-rich clusters which are hcp-like are formed under the proposed interdiffusion of both Ga vacancies and As interstitials [9].

Fig. 3 (a) EDX spectra of the corresponding position. The energy range, from 2 to 14 KeV, is the same for these four spectra. The full scale and calculated As/Ga ratio for each spectrum are also labelled. These spectra are not yet normalized to eliminate the environmental background. (b) SIMS concentration profiles of In and Si plotted on a logarithm scale. (c) SIMS profiles of Ga and As plotted on a linear scale.

Since the Si concentration in the substrate is on the order of 10^{18} cm^{-3}–well below the EDX detection limit–we must perform SIMS measurements to determine its profile. A 14.5 keV Cs$^+$ primary ion beam is used by Charles Evans & Associates, to obtain the redistribution profiles of each (Si, In, Ga, As) species. Figs. 3(b) and 3(c) plot the Si, In, Ga and As profiles. We can clearly see that there is a good agreement between the SIMS and EDX measurements because the As-rich behavior shown in Fig. 3(c) is consistent with the results of the EDX. This specific consistency is a good way to clarify some of the artifacts involved in the SIMS measurement, and enables us to combine the TEM lattice images with SIMS profiles to study the redistribution behavior of each species on the atomic scale.

As seen in Fig. 3(b), the In is slightly distributed across the entire In$_x$Ga$_{1-x}$As layer, with a tail region extending deep into the GaAs substrate. This slightly redistributed behavior results from the liquid-phase segregation effect during the pulsed laser induced melt-solidification processes with equilibrium segregation coefficient equal to 0.1 [5]. However, the overall In concentration in this In$_x$Ga$_{1-x}$As layer is determined to be stoichiometrically x=0.21 using both RBS measurement and selected area diffraction in the TEM imaging [4]. The In tail, shown in Fig. 3(b), is not just an artifact from the SIMS measurement since EDX spectra also showed a significant amount of In present below the In$_x$Ga$_{1-x}$As/GaAs interface, seen in Fig. 3(a). However, the Si profile, uniform in the GaAs before the pulsed laser processing, is significantly altered after the rapid melt-solidification process. Three distinct regions, shown as pile-up, dip and bump regions in Fig. 3(b), lie between the free surface and the GaAs substrate. The pile-up and dip associated with the redistribution are characteristics of liquid phase segregation with an equilibrium segregation coefficient less than one [5,8]. Since the In$_x$Ga$_{1-x}$As/GaAs interface lies in between the bump and dip regions, both the Si bump and In tail regions are believed to be formed by the fast injection of In and Si into the unmelted GaAs substrate.

DISCUSSION

Fig. 3(c) shows direct evidence of As and Ga interstitialcy (designated as As$_i$ and Ga$_i$) profiles. Here the As$_i$ and Ga$_i$ are clearly present below the In$_x$Ga$_{1-x}$As/GaAs interface. The FWHM of the As peak is approximately 100Å, slightly larger than the size of the islands (~80Å) shown in Fig. 2. Since the depth resolution of the SIMS measurement is also about 100Å, we believe the position of the As peak corresponds to the island location. The Ga$_i$, however, has a much more diffused profile and a lower peak concentration. Once both As and Ga atoms are injected into GaAs substrate, Ga atoms diffuse faster since the Ga$_i$ diffusion coefficient is believed to be higher than that of As$_i$ [12], although no quantitative diffusion constants for Ga$_i$ and As$_i$ are available at present. In the melted region, however, the Ga and As are distributed quite uniformly, except for a small dip (shown in the figure) believed to be due to species partition driven by an interfacial potential difference across the liquid and solid GaAs phases [12]. Using all of the information drawn from Figs. 2 and 3, we conclude that both the Si dopant and Ga, As, and In host atoms are injected into the GaAs substrate from the molten region during the process. These atoms are then significantly redistributed or diffused via a nucleation and growth mechanism with a diffusion length longer than 100Å to form spherical As-rich islands in the unmelted region. We have also observed a similar fast diffusion behavior occurring in Si substrates [13]. Detailed understanding of this effect must be fully understood for pulsed laser processing of both GaAs and Si materials.

In order to estimate the degree of the pulsed laser processing induced fast diffusion effect, we can compare the diffusion length deduced from Fig. 3 with that calculated from thermal equilibrium conditions. However, we only can estimate the enhanced Si dopant diffusion length because no quantitative diffusion coefficients are available for Ga$_i$ and As$_i$ and In species in GaAs substrate. D_{Si}, the temperature and Fermi-level dependent diffusion coefficient of grown-in Si dopant in GaAs, is given as [13],

$$D_{Si}(T) = 78 \left[\frac{n}{n_i}\right]^2 exp \left[-\frac{4.0}{kT}\right]$$

where n is the Si concentration, k is Boltzmann constant, T is the processing temperature and

n_i (T) is the intrinsic free carrier concentration in GaAs which is given as [14],

$$n_i(T) = 3.82 \, e14 \, T^{3/2} \, exp\left[-\frac{8806}{T} + \frac{3.13 \, T}{T + 204}\right]$$

We extend the temperature ranges from the reported data to the melting point of the GaAs semiconductor in the calculation by assuming the reported diffusion mechanism still hold in such high temperature regime. The maximum diffusion length $\sqrt{Dt}$ is estimated to be 10^{-3} Å by assuming the processing duration t =100 nsec and temperature 1513°K, the melting point of GaAs. However, as shown in Fig. 3(c), the depth of the Si bump is ~100Å, which is 9 orders of magnitude higher than that for the grown-in case. In order to account for this huge enhanced diffusion, there must be a large driving force which drives the point defect injection into the unmelted GaAs substrate during the nanosecond-scale processing period.

In order to investigate the source of the driving force, we perform a calculation of the generated electric field in two different ways in order to estimate the magnitude of this characteristic driving force. The most direct way, labelled as ε_e, is to estimate the required internal electrochemical potential in order to keep the Si bump and As peak without smearing out, as shown in Fig. 3(b). The second method, labelled as ε_T, is to multiply the average temperature gradient simulated from the one dimensional heat and diffusion equations by the thermoelectric power of GaAs [7]. Here, we assume the temperature gradient is the same in GaAs and Si because the simulated result is presently only available for Si [1]. These two estimated values, ε_T and ε_e, are 10^4, and 5×10^4, respectively. The fact that ε_T and ε_e are of the same order of magnitude implies that the enhanced diffusion effect induced point defect injection is caused by the high temperature gradient which occurs during the PLIE process.

A hypothesis based on a high temperature gradient induced electric field which drives point defect injection from the molten region into the unmelted substrate is proposed to explain the significant diffusion occurring during pulsed laser processing. It is well known that application of a temperature gradient to a semiconductor establishes a concentration gradient of free carriers. Diffusion of these carriers will occur and be counteracted by the buildup of an electrical field. This is known as the Seebeck effect. Since there is a temperature gradient existing in the GaAs substrate during the pulsed laser melting period, we believe the concentration of Ga_i and As_i will follow the temperature profiles. These atoms then diffuse into the substrate with diffusion length depends strongly on the temperature gradient. That is, the diffusion of these atoms is driven by the Seebeck effect. We can confirm this enhanced diffusion effect by using Si as the substrate since currently there is no information available to quantitatively describe the temperature dependence of the concentration of the Ga_i and As_i in the GaAs semiconductor [15]. The details of the diffusion behavior of the host atoms and point defects in the Si substrate will be quantitatively described by the governed diffusion equation in ref. [16] in terms of the well-established temperature dependence of the Si_i concentration during the pulsed laser melting process.

However, in order to explain the formation of the As-rich island as shown in Fig. 2, an alternative concentration gradient of host atoms has to exist near the l/s interface since the Ga_i and As_i concentrations, solely driven by the temperature profile, are not high enough to cause the As-rich behavior appeared in both SIMS and EDX measurements. We believe this second concentration gradient is generated by the impulse of the pulsed laser exerting on the moving molten layer, and is piled-up at the l/s interface. Injection of these host atoms is then driven from the molten region into the unmelted GaAs substrate by the Seebeck effect. In this case, these host atoms then condense and form the As-rich cluster with diffusion length ~100Å. That is, impulse of the host atoms at the l/s interface is driven by the temperature gradient to form the As-rich island.

In conclusion, we have found that a significant redistribution behavior of Ga, As, In and Si atoms occurs in the unmelted GaAs substrate during pulsed laser induced epitaxial $In_xGa_{1-x}As$/GaAs layer formation. This point defect injection is explained by a hypothesis based upon the combined effect of concentration impulse and high temperature gradient across the liquid and solid interface. The electric field generated is estimated to be in the range of 10^4 V/cm.

REFERENCES

1   P. G. Carey, Ph.D. Dissertation, Stanford University, Stanford, CA, 1988.

2   S. Y. Chou, Y. Chang, K. H. Weiner, T. W. Sigmon, and J. D. Parsons, Appl. Phys. Lett., **56**, 530 (1990).

3   J. R. Abelson, T. W. Sigmon, K. B. Kim, and K. H. Weiner, Appl. Phys. Lett., **52**, 230 (1988).

4   Y. Chang, S. Y. Chou, T. W. Sigmon, A. F. Marshall, and K. H. Weiner, Appl. Phys. Lett., **56**, 1844 (1990).

5   Y. Chang, T. W. Sigmon, A. F. Marshall, and K. H. Weiner, "Thin Film Structures and Phase Stability", in *MRS Symposium Proceedings*, edited by B. M. Clemens and W. L. Johnson (Materials Research Society, Pittsburgh, PA, 1990), Vol.187.

6   Y. Chang, S. Y. Chou, J. Kramer, T. W. Sigmon, A. F. Marshall, and K. H. Weiner, Appl. Phys. Lett., **58**(19), 2150 (1991).

7   E. A. Fitzgerald, Jr., Ph.D. Dissertation, Cornell University, Ithaca, NY, 1988.

8   Landolt-Bornstein new series, III/17d and III/17a (1984).

9   B.-T. Lee, R. Gronsky, and E. D. Bourret, J. Appl. Phys., **64**, 114 (1988).

10 T. Sands, J. Washburn, and R. Gronsky, Mat. Lett., **3**, 247 (1985).

11 M. R. Melloch, N. Otsuka, J. M. Woodall, A. C. Warren, and J. L. Freeouf, Appl. Phys. Lett., **57**, 1531 (1990).

12 M. B. Panish, J. Electrochemical Soc., **113**, 1226 (1966).

13 W. A. Tiller, Department of Materials Science and Engineering, Stanford University, private communication.

14 J. S. Blakemore, J. Appl. Phys., **53**(1), 520 (1982).

15 M. D. Deal, Center for Integrated System, Stanford University, private communication.

16 Y. Chang, J. Chen, S. Talwar, E. Y. Shu, and T. W. Sigmon, *to be published in MRS Symposium Proceedings,* edited by M. Chen, *et. al.* (Materials Research Society, Pittsburgh, PA, 1991), Vol. 230.

CRYSTALLIZATION AND AMORPHIZATION OF SiC-CERAMIC PVD COATINGS AFTER LASER TREATMENT

O. Knotek, F. Löffler
Institut für Werkstoffkunde, Technical University Aachen,
Templergraben 55, D-5100 Aachen, Federal Republic of Germany

Abstract

Over the last ten years, surface coating engineering has become a key technology throughout the world, especially due to the successes achieved in the optics and microelectronics sectors. First advances in the wear protection field were attained with coated carbides, indicating a substantial technical and economic potential in these areas.

The development of wear protective coatings often reaches the limits of what is technically possible. For this reason, processes incorporating two different methods and combining the specific advantages of each are increasingly being developed for the deposition of certain coatings. In the study presented here, PVD (Physical Vapour Deposition) technology was combined with laser surface treatment technology to produce amorphous and crystalline SiC coating phases. The objectives of the study were to establish the respective advantages of crystalline and amorphous phases. The amorphous ceramic coating exhibits excellent corrosion resistance, whereas the crystalline phases generally attain higher adhesion and superior stability at high temperatures. There are therefore technical applications for both phases.

Method

The combined process should be seen against the background of the high reactivity of silicon with the metal substrates, which makes it impossible to induce phase transformations in SiC coatings by heat treatment in a vacuum furnace. A silicate-substrate reaction can be avoided only by means of surface heating in a depth range only slightly less than the coating thickness. Since temperatures of about 1 000 °C are required, the only feasible heat source for treating specimens in this way is a laser beam.

A suitable laser type is selected according to its attainable laser power and operating mode. For extremely high energy densities over extremely short periods, the choice is between the Nd-YAG and the CO_2 laser. Both types of laser are used for surface treatment of workpieces in production engineering applications, the Nd-YAG being a solid-state and the CO_2 a gas laser.

Input power for both types of laser was selected in the 1 kW range. The Nd-YAG laser was operated in the pulsed and the CO_2 laser in the continuous mode, so that results in the different modes could be compared.

There were three variation parameters for the Nd-YAG laser, as indicated in *Fig. 1*. Track shift is determined by the movements of the laser and the timing of the pulses; the focal position (def) indicates the distance from the focal point of the laser and the feed rate describes the relative motion between the specimen and the laser. An increase in any one of these three parameters entails a decrease in the energy input per unit area.

Mat. Res. Soc. Symp. Proc. Vol. 230. ⸰ 1992 Materials Research Society

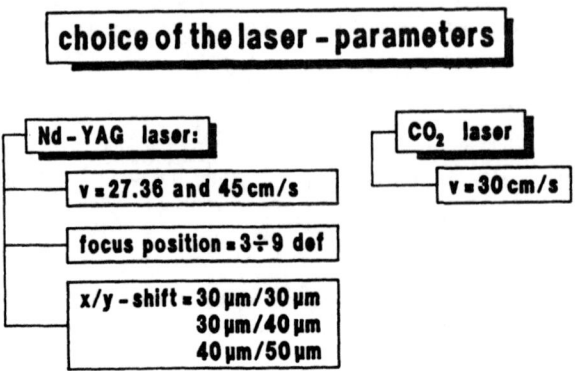

Fig. 1: Laser Parameters Used in the Tests

In continuous CO_2 laser treatment, the energy input is influenced chiefly by the feed rate, so that high feed rates are necessary to achieve low heat penetration depths.

The parameters shown in *Fig. 1* were selected to attain surface heating at 1 000 °C in the 1 to 3 μm range. Austenitic steels and carbides were coated with silicon carbide. The PVD processes used to produce the SiC coatings were magnetron sputter ion plating and arc ion plating.

Results

In addition to the laser type and settings, the coating parameters time and power were varied. All the silicon carbide coatings were present in the amorphous state after deposition, with increases in hardness, critical load and coating rate as the coating power rose. These correlations have already been discussed elsewhere /3,4/. The coating thicknesses of the samples for the Nd-YAG laser ranged from 1 to 7 μm, with increments of roughly 1 μm. *Fig. 2* shows the results. If the coating is thicker than the laser-affected zone, it will contain crystalline and amorphous layers. There is no influence on the interface. If the coating is thinner than the laser-affected zone, it crystallizes in a mixed phase with the substrate. The hardness decreases down to the substrate hardness, with no significant change in adhesion. Coating thickness should consequently be greater than the depth of the laser-affected zone, to ensure a good tribological coating.

Following laser surface heating, all the SiC coatings exhibited the desired crystalline state. X-ray microstructural analysis permitted precise investigation of phases, orientations, particle sizes and lattice parameters. The next section discusses silicon carbide coatings treated with the Nd-YAG laser.

After surface heating, the amorphous silicon carbide coating was transformed to the cubic ß-SiC phase, a so-called low temperature modification which represents the only stable phase in the Si-C system /5/. All Nd-YAG treated coatings exhibited the (111) and the (220) orientations with an intensity ratio of I(111) : I(220) ≈ 2 : 1. Irrespective of the coating

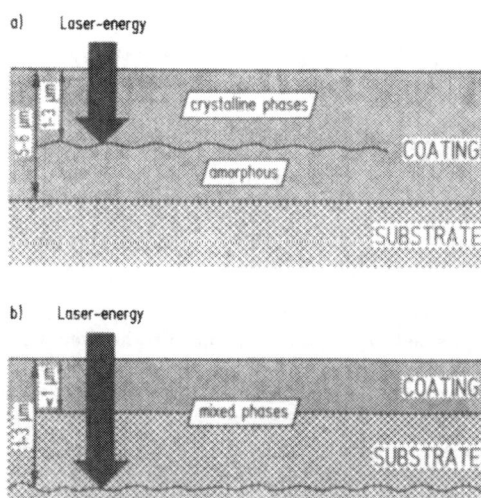

Fig. 2: Depth of the Laser Affected Zone
a) coating thickness > affected zone
b) coating thickness < affected zone

with a 30/30 shift and a feed rate of 27 cm/s. The associated high coating temperatures resulted in formation of this oxide as the specimens cooled in the ambient air.

The lattice parameters were within the range indicated in the ASTM catalogue, which lists a = 0.43474 nm for ß-SiC in powder form with an (111) orientation. *Fig. 3* shows the deviations from this value, clearly indicating the primary dependence on deposition power. The lowest deviations occur, for example, with the silicon carbide coatings deposited at a

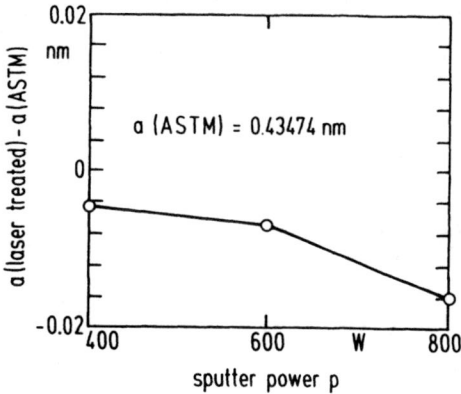

Fig. 3: Lattice Parameter Deviations from the ASTM Value for the (111)-Oriented β-SiC Phase

coating power of 400 W. This may be attributed to the higher film growth rates at high powers, causing high growth-induced stresses which the brief laser treatment cannot relieve.

The particle size determined for all SiC coatings ranged from 6 to 15 nm and showed no dependence on the laser parameters. It is apparent that particle sizes increase with rising coating voltage. At a power of 600 to 800 W during the coating process, particle sizes of $8-9$ nm are achieved in the laser-treated silicon carbide coatings, resulting in a nanocrystalline structure. This structure has a high proportion of interfaces due to the high grain density, and hence represents a separate material grade /6/.

Since a high feed rate was required for surface treatment with the CO_2 laser, it was difficult to determine the introduced energy densities and depths. The affected zone was consequently deeper than the coating zone, so that, in addition to the ß-SiC phases, Si-Co phases were formed, indicating reactions with the carbide substrate (approximately $6\,\mu$m SiC coating thickness). As the high energy also promoted a higher growth rate, particle size was some 31 nm. The comparatively coarse structure is revealed by SEM structural analysis of rupture structures (*Fig. 4*). The Nd-YAG treated specimens exhibit a smooth surface and a finely-structured coating, while the CO_2 treated coating has a columnar structure with a rough

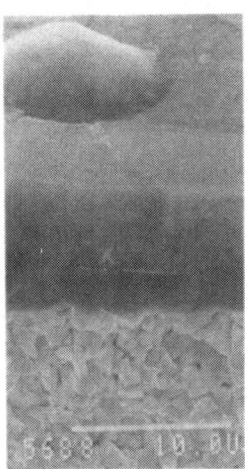

Fig. 4: Rupture Surface of Crystalline Silicon Carbide Coatings
a) Nd-YAG laser with 30/40 shift, b) Nd-YAG laser with 30/30 shift, c) CO_2 laser

surface. Although fine cracks are visible in the first rupture, these cease at a depth of approximately $1\,\mu$m. At a lower track width (e.g. 30/30 shift), scarcely any surface cracks are observable.

Changes in microhardness after laser treatment were not dependent on the laser parameters. Systematic changes are, however, observable in the mean hardness values for identical original coatings and different laser parameters (*Fig. 5*).

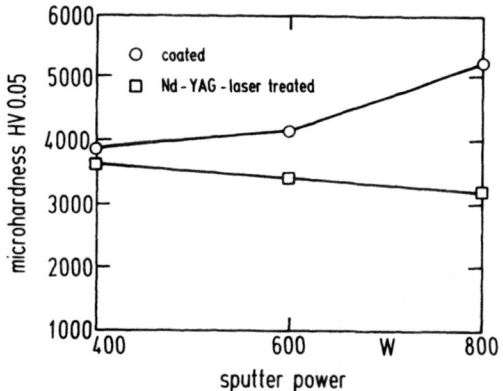

Fig. 5: Changes in Hardness Following Laser Treatment of SiC Coatings as a Function of Coating Power

The familiar increase in hardness with increasing power is reversed after surface treatment. The smallest decrease in hardness values occurred with the silicon carbide coatings deposited at a power of 400 W. This may be ascribed to the higher stresses in coatings with a higher film growth rate, i.e. a higher coating power, since a greater strength decrease is then possible after heat treatment. The resulting film hardness remains extremely high even after laser treatment, at approximately 3,500 HV 0.05.

In line with the hardness values, and also influenced by the particle size, the critical adhesive load for heat-treated silicon carbide coatings on austenitic steel was improved to $F_{crit} = 15$ N (800 W power). F_{crit} represents the diamond-cone load at which the coating spalls.

Summary

The pulsed Nd-YAG laser proved suitable for surface heating of SiC coated substrates in the μm range. The ceramic layers were reproducibly transformed from the amorphous to the nanocrystalline state, without influencing the substrate material. A ß-SiC phase was formed in all cases after Nd-YAG laser treatment. Improvements in coating adhesion accompanied by decreased hardness and a slight increase in roughness were observed.

Acknowledgements

The authors would like to thank Dr. Wissenbach of the Fraunhofer Institute for Laser Technology in Aachen for carrying out the laser treatments.

Literature

1.   N.N. Laserforschung und Technik — Förderungskonzept — Published by the Federal Ministry of Research and Technology.

2.   Wissenbach, K.: Umwandlungshärten mit CO_2-Laserstrahlen, doctoral thesis at the TH Darmstadt, 1985.

3.   Knotek, O.; F. Löffler: On Magnetron Sputtered Silicon and Aluminium Based Ceramic Coatings, International Tribology Conference, Nagoya (1990).

4.   Löffler, F.: Eigenschaften von keramischen Hartstoffschichten auf Silizium- und Aluminiumbasis, Research Reports Series 5, No. 180, VDI Verlag (1990).

5.   Olesinski, R. W.; G. J. Abbaschian: Bull. Alloy Phase Diagrams, 5(5), Oct. 1984, in: Binary Alloy Phase Diagrams, ed. by T. B. Massalski, Am. Soc. f. Metals, Ohio, 589–590, 1986.

6.   Gleiter, H.; P. Marquardt: Nanokristalline Strukturen — ein Weg zu neuen Materialien?, Z. Metallkunde, 75, 263–267, 1984.

PART VI

Ferroelectrics, Oxides and Ceramics

PROCESSING OF FERROELECTRIC MEMORIES

CARLOS A. PAZ DE ARAUJO, L. D. McMILLAN, AND J. F. SCOTT

Symetrix Corporation, Colorado Springs, CO 80918 and
The University of Colorado, Depts. of Physics and of
Electrical Engineering and Computer Engineering, Boulder,
CO 80309-0390 and Colorado Springs, CO 80933-7150

ABSTRACT

This paper describes the classification of existing integration schemes for ferroelectric memories, including both non-volatile and volatile devices, and the status of device development for CMOS (both SRAM and DRAM architecture), GaAs JFET structures, bipolar, and true ferroelectric FETs (in which the ferroelectric is deposited in the gate region to modify the source-to-drain current when the polarization is reversed). Emphasis in the paper is on electroding, film deposition, drying and baking, and annealing steps.

INTRODUCTION

Within the past four years there has been a renaissance in ferroelectric memory devices. Such memory systems were investigated in considerable detail in many large US electronics corporations from 1955-1975, with notable progress made at IBM, Bell Telephone Laboratories, Westinghouse, and RCA. Generally speaking they were not competitive with silicon integrated circuit devices (DRAMs or SRAMs) because of high operating voltages (thick films), endurance/fatigue problems, and cross-talk (disturb pulse problems); and after 1975 they became a low-priority research project in most industrial R&D laboratories. A review of this early technology has been given recently by the present authors.[1] Since 1986, however, there has been a resurrection of interest in these ferroelectric systems, driven by improvements in thin-film technology (both sol-gel and sputtering). It is now possible to produce rather dense memories (cell size as small as 2 microns on a side) of ferroelectric thin films (250 nm thick) from a variety of materials, to integrate them fully into standard Si or GaAs ICs, and to achieve very fast READ/ERASE/REWRITE speeds (2 - 10 ns), 3V or 5V operation without internal charge pumps, high endurance (10^{12} destructive READ/WRITE cycles), and extreme radiation hardness (100 MRAD Si equivalent). These attractive properties make these devices suitable for replacement of several existing nonvolatile memory systems in present use, including plated wire, magnetic bubble memory systems, core, and most EEPROMs. As a result, at least four US corporations are engaged in prototype production of low-density (1k - 16k) random access memories (McDonnell-Douglas, Raytheon, National Semiconductor/Krysalis, Ramtron), in addition to our company (Symetrix). Additionally, the high dielectric constant of ferroelectric films, which can exceed 4000 over a fairly wide temperature range, offers an improvement of two orders of magnitude over Ta_2O_5 for high-capacity DRAMs, which should minimize the need for trenching or exotic geometries in the 64 Mbit DRAM of the near future. This aspect of ferroelectric thin films is being exploited in Japan by companies such as

Mat. Res. Soc. Symp. Proc. Vol. 230. ©1992 Materials Research Society

Fujitsu, Toshiba, NMB, and NEC, and in the USA by Texas
Instruments, National Semiconductor, Ramtron, and Symetrix.
Many recent publications and three conferences have
addressed the important issues in this emerging technology [2-4].
However, the processing of the devices, which is apt to be the
subtopic of greatest interest to members of the Materials
Research Society, has received the least attention in publications in the open literature, because it involves issues most
often viewed as proprietary. The present paper attempts to
rectify that short-coming, to the degree that is possible at
present.

INTEGRATION SCHEMES

Table I below outlines the status of device development for
ferroelectric memories since 1986, including the anticipated
completion date of several commercial projects. Both volatile
and non-volatile devices are included.

Table I

FERROELECTRIC MEMORY DEVELOPMENT

NON-VOLATILE FERROELECTRIC MEMORIES	DESTRUCTIVE READ-OUT	SRAM + FE CAP -- CMOS 256 bit (Ramtron 1988) 5 micron rules
		DRAM ARCHITECTURE + FE CAP -- CMOS -- 1T and 2T Krysalis 512 bit; Ramtron 4K; National 4K (1987) (1991) (1991)
		SRAM, 2T DRAM -- JFET-based GaAs McDonnell-Douglas: 4K (1991) 16K and 64K (in progress)
		SRAM, 1T DRAM -- Bipolar/CMOS (Raytheon, 1992)
	NDRO	FE-FET -- CMOS (SNOS substitute) Westinghouse $BaMgF_4$ and/or $Bi_4Ti_3O_{12}$ (1991)

VOLATILE FERROELECTRIC MEMORIES	TRUE VOLATILE DRAMs	High-Dielectric (Non-switching) FE films:
		      *PZT (Fujitsu? Toshiba? NMB?)
		      *$SrTiO_3$ (NEC?)

	LONG-REFRESH DRAMs	PZT -- Ramtron (1991?) 4 kbit

PROCESS TECHNOLOGIES

   The process technology for ferroelectric thin films typically involves thirteen process steps, as diagrammed in Fig. 1,
even when the photolithography steps are not included. These
begin with the processing of Si or GaAs wafers, followed by
deposition of an insulating buffer layer. For most of the
ferroelectric memories the third step is deposition of the bottom
electrode, although in a true FE-FET this step is replaced by a
thin layer of Si/SiO_2. The fourth step is the deposition of the
ferroelectric film (usually PZT in most devices at present),
followed by drying/baking, deposition of the top electrode, and
annealing. It is these five steps, from bottom electrode
deposition to top electrode and annealing, that we will emphasize
in the present paper. The remaining steps consist of the deposition of an interlayer dielectric, opening of vias (etching),
preparation of interconnects (standard metal; e.g., Al), final
glass passivation (typically a borosilicate), reopening of vias
(etch), and attachment of bonding pads. These latter steps are
rather standard in the microelectronics industry; therefore we
concentrate on the electroding, film deposition, drying, and
annealing stages of greatest interest to materials scientists.

Figure 1.

PROCESS TECHNOLOGIES FOR FERROELECTRIC THIN FILMS

Si/GaAs Wafers

Buffer Layer (Insulator)

First Electrode

Ferroelectric Film Deposition

Drying and/or Baking

Top Electrode

Annealing

Interlayer Dielectric

Open Vias (Etching)

Interconnects (Al)

Glass Passivation (BSG)

Reopen Vias

Bonding Pads

ELECTRODING/CONTACTS

Our PZT/CMOS devices normally utilize Pt [111] electrodes that are deposited by sputtering. A typical device has 200 nm thick Pt electrodes, usually accompanied by 20 nm of Ti to improve mechanical adhesion to the Si IC. Depth profiling of these devices is accomplished by Auger analysis. The profiles, illustrated in Figs. 2-4 below, show well-defined Pt edges and highly uniform oxygen, lead, zirconium, and titanium levels throughout the interior of the film. Note the flatness of the oxygen curve in the middle of Fig.2.

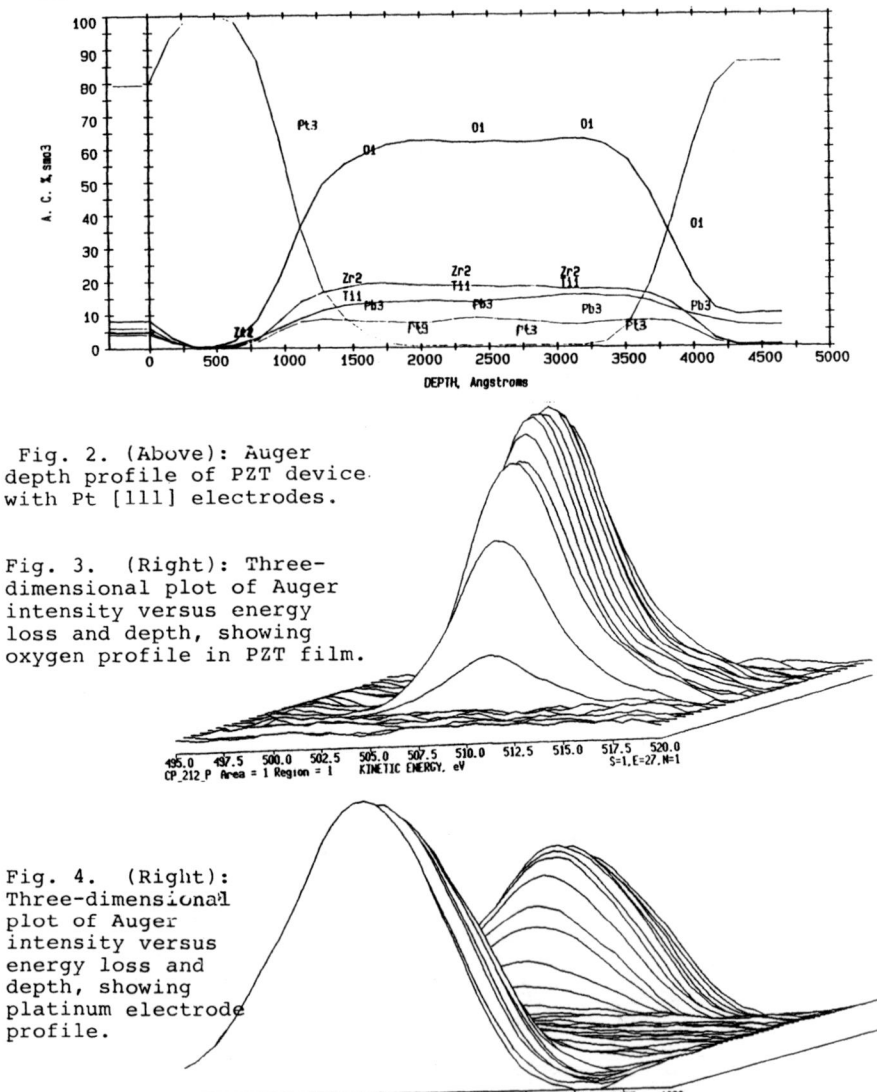

Fig. 2. (Above): Auger depth profile of PZT device with Pt [111] electrodes.

Fig. 3. (Right): Three-dimensional plot of Auger intensity versus energy loss and depth, showing oxygen profile in PZT film.

Fig. 4. (Right): Three-dimensional plot of Auger intensity versus energy loss and depth, showing platinum electrode profile.

In general electroding does not present difficulties for this technology, but several matters require care. Ti is often used to prevent adhesion failure for the Pt electrode. (A sketch of an adhesion failure map is shown in Fig. 5 below.) Ti can act as an electrochemical sink, however, if the Pt electrode is slightly porous. This leads to oxygen diffusion through the Pt electrode to the Ti, where it reacts to form rutile. This can greatly accelerate failure mechanisms in the PZT/Pt memory that involve oxygen vacancy production. Such mechanisms deteriorate retention (by creating space charge near the electrodes) and promote d.c. breakdown [5] through semi-conducting pathways of oxygen-depleted PZT.

Occasionally more bizarre failure mechanisms or defects arise in this electroding step. Figs.6-7 illustrate the appearance of one of our devices. This SEM photo was analyzed after 17 minutes of sputtering. It is from the middle of the PZT layer. The spectrum of the "bump" shows a relatively high concentration of Pt. The appearance of such a Pt peak during depth profiling is thought to be due to Pt diffusion through the PZT layer.

Fig. 5 (Right): Peeling pattern of Pt electrode on Pt/PZT Silicon device.

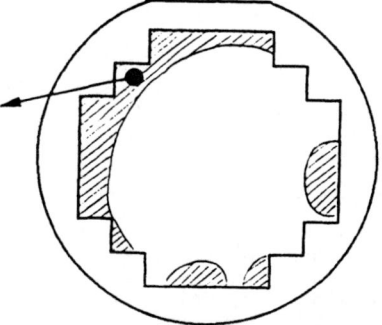

Fig. 6 (Below): 5-micron diameter "bump" in SEM micrograph thought to arise from Pt diffusing through the PZT.

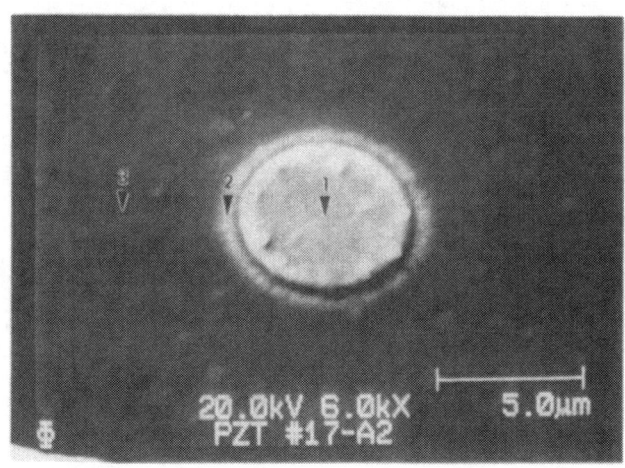

Fig. 7. Distribution of "bumps" thought to be Pt diffusion
through the PZT film. This SEM micrograph is shown with a
magnification of 1 cm = 10 microns. A higher magnification
photo (Fig.6, preceding page) illustrates the structure of a
single "bump" of elliptical shape and 4 - 5 microns diameter.

FILM DEPOSITION

The usual ferroelectric capacitor devices consist of PZT
(sol-gel is perhaps the favored deposition technique at present)
in a sandwich structure sketched below in Fig. 8. The sol-gel in
our materials is prepared from lead acetate, titanium isopropoxide, and zirconium n-propoxide in a common solvent.

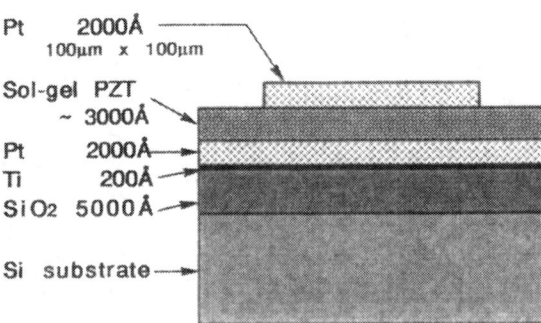

Fig. 8. Pt/PZT cell structure for silicon IC memory production.

Although the structure diagrammed in Fig. 8 appears quite
simple, it is structurally and electrically complex, due to both
the electrode/ferroelectric interface and the grain boundaries
within the ceramic ferroelectric. Fig. 9a below sketches somewhat schematically the typical microsctructure; and Fig.9b
gives an equivalent circuit model for this device, from which
one can predict P(E) hysteresis curves, I(V) current data, etc.

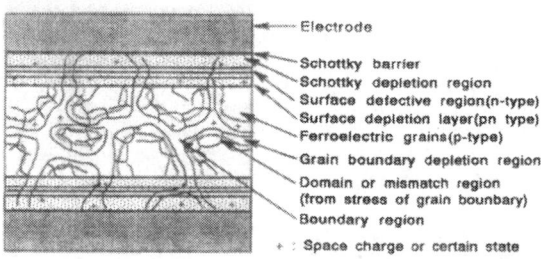

(a)   Micro structure in ferroelectric thin capacitors.

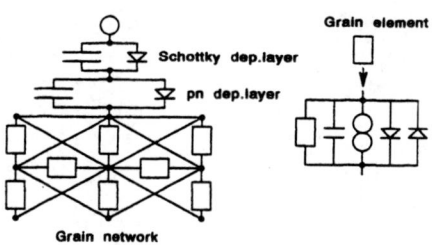

(b) The full network equivalent circuit model.

**Fig- 9.   The microstructure and circuit model of ferroelectric thin film
capacitors.**

A fuller discussion of equivalent circuit modeling for
these ferroelectric memory devices has been given elsewhere [6-8].

STEP-COVERAGE AND CRACKING

A systematic series of experiments has been carried out in our
laboratory to ascertain the nature of problems that might be
encountered in depositing PZT onto patterned electrodes. We have
found that it is optimal to deposit the PZT film onto a uniform
platinum surface and then pattern the circuit by etching the
bottom electrode after annealing. That is, we do not deposit the
PZT onto a previously patterned electrode. If the latter
procedure is followed, cracks result in the PZT surface. Fig. 10
below shows a test pattern with a gap height of 4700A and step
widths from 2 microns up. Figs. 11-12 illustrate crack-free PZT
films deposited onto uniform Pt; and Fig. 13 illustrates the
hysteresis curve measured for these films at applied voltages of
1, 2, 3, 4, and 5V.

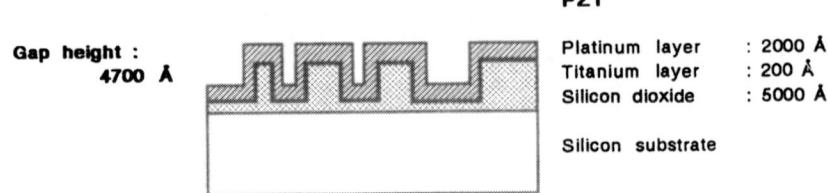

PZT

Gap height :
4700 Å

Platinum layer : 2000 Å
Titanium layer : 200 Å
Silicon dioxide : 5000 Å

Silicon substrate

Fig. 10. Test structure for Pt/PZT memory on Si to examine step coverage and cracking. The step width minimum is 2 microns.

Figs. 11 and 12. Crack-free PZT memories deposited on uniform Pt.

Of course, cracking may also occur in the later processing steps. This is discussed in a section that follows below.

UNIFORMITY AND REPEATABILITY

We have achieved a remarkably good uniformity of microstructure and electrical properties for devices made across a full 5" wafer. As shown in Fig. 13 below, the electrical hysteresis curves are virtually identical from the edges and center of the wafer. In addition, we have extremely high reproducibility from wafer to wafer (we can produce 20 wafers per week in our prototype line at Symetrix/Univ. of Colorado).

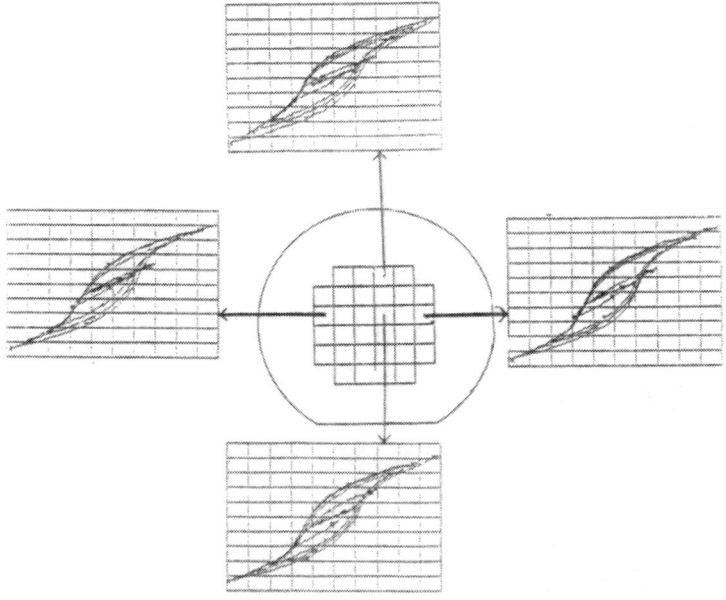

Fig. 13. Uniformity of electrical hysteresis properties of four ferroelectric memories taken from center and edges of a 5" wafer. The system was sol-gel PZT (40/60) on uniform Pt [111] electrodes, patterned by subsequent etching. The film thickness was approximately 300 nm. The absolute values of coercive field E_c and remnant polarization P_r are not shown. They typically range from E_c = 30 - 70 kV/cm and P_r = 20 - 30 uC/cm^2, depending upon doping and other proprietary processing parameters.

DRYING AND BAKING

These are the most sensitive steps. Most of the parameters are proprietary. It is in these steps that pore-elimination techniques are employed. Most of this information will appear in the form of patent applications from various corporations or will simply be held as trade secrets. It is not expected to reach the open literature.

TOP ELECTRODE

The deposition of the top electrode is usually the same as for the first electrode: It is sputtered platinum. Some groups use a different metal for the top electrode (e.g., gold); the latter can be flash-evaporated.

ANNEALING

This is the critical microstructure-forming step. It is extremely sensitive to the ambient conditions. Rapid-thermal-processing and furnace annealing are both commonly used; it is

also possible to use a hot plate.
   Figs. 14 and 15 below show microstructures of PZT films
annealed with four (proprietary) different techniques at two
different temperatures. In each case the microstructure
correlates extremely well with the electrical hysteresis curve
qualities. Figs. 16 and 17 illustrate the sharp Pt/SiO_2
boundary after a proper anneal, and the kind of cracking
problem that is occasionally seen after baking -- namely, long
(1000 micron) cracks that originate and terminate on device
edges. Both straight and curved cracks are observed.

Drying/Baking Process "A"

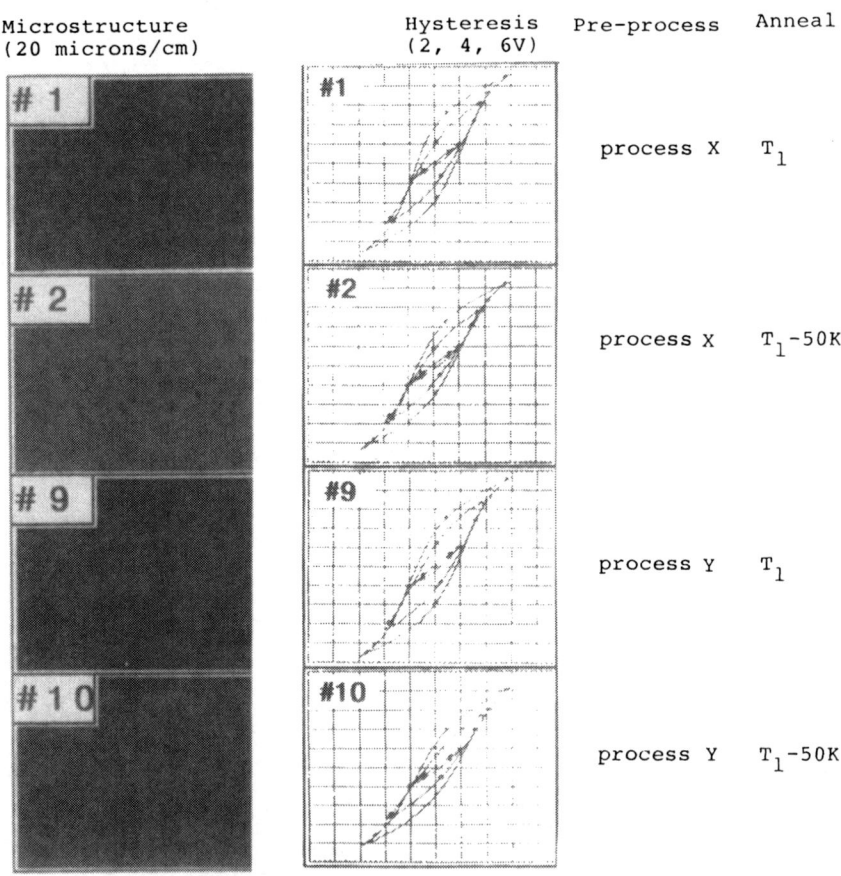

Microstructure (20 microns/cm)	Hysteresis (2, 4, 6V)	Pre-process	Anneal
#1	#1	process X	T_1
#2	#2	process X	$T_1 - 50K$
#9	#9	process Y	T_1
#10	#10	process Y	$T_1 - 50K$

Fig. 14 Correlation of microstructure and electrical
properties for PZT/Pt memory on Si.

Drying/Baking Process "B"

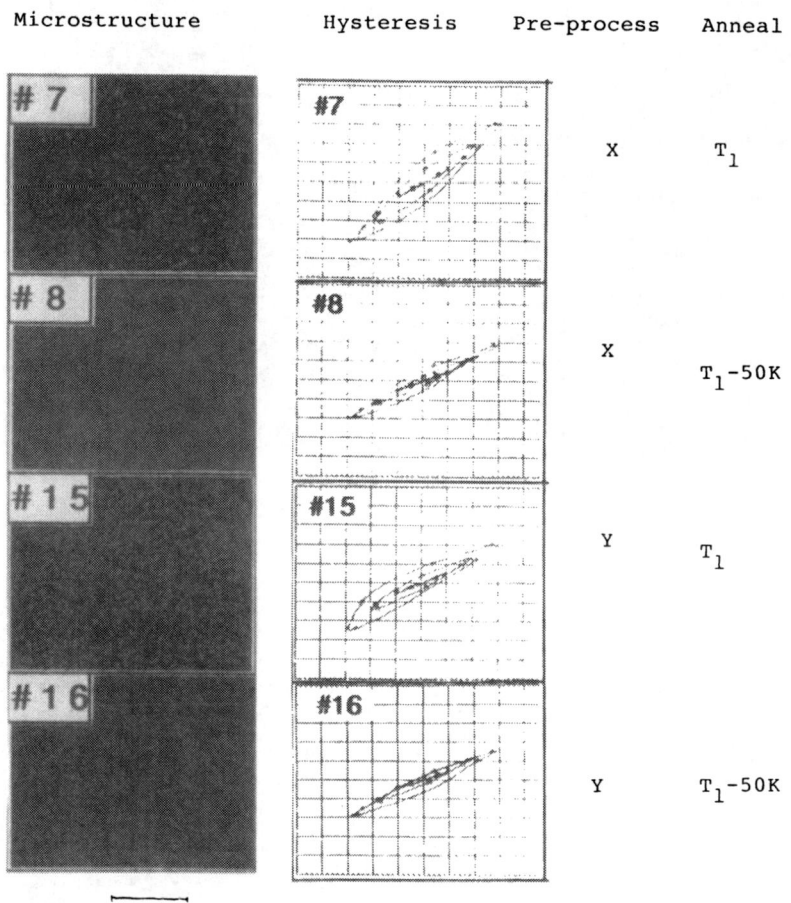

Fig. 15.   Correlation of microstructure and electrical
properties of a PZT/Pt memory on Si that uses a different
drying/baking process from that in Fig. 14.

The import of Figs. 14 and 15 is that they show large changes
in both microstructure and electrical hysteresis; a change in
annealing temperature of only 50°C (50K) makes a great difference.
And variation in drying or baking shows that increasing grain size
worsens electrical hysteresis parameters (coercive field and
remnant polarization).

 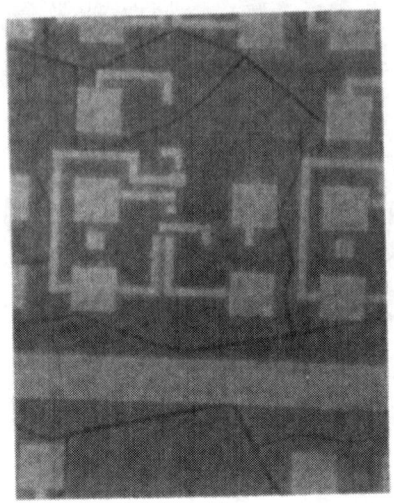

Pt-SiO₂ boudary   20 µm   Cracks in center   200 µm

Fig. 16. Pt/SiO$_2$ boundary.   Fig. 17. Cracking.

TESTING AND ELECTRICAL CHARACTERIZATION

The primary test methods are electrical hysteresis, current-
voltage I(V) measurements, capacitance-voltage C(V) measurements,
time-dependent d.c. breakdown measurements, current transients
during switching I(t), and fatigue/endurance. We give examples
of each of these below for sol-gel PZT (40/60) films ca. 250 nm
thick below, together with some observations concerning their
implications for device production.
Fig. 18 shows a typical I(V) curve. On this log-log plot
the current is ohmic up to about 1.5 V, above which a space-charge
limited current of form I = AV^2 is manifest (Child's Law). This
shows that at 5V operation the device behaves as fully depleted,
and the characteristic trapping energy is about 1.5 V. A similar
pinning energy of 1.44 eV was deduced by Dey and Zuleeg [9]
independently from the thermal activation energy of I(T) curves
for sol-gel PZT.

Fig. 18. (Right): I(V)
d.c. leakage current in
PZT (40/60), showing
space-charge limited
current above ca. 2V.

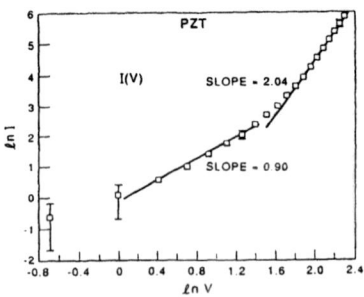

I(t) CURRENT TRANSIENTS

The displacement current from the ferroelectric switched capacitor in an FE RAM is the most important characteristic for the IC designer, for it gives the output voltage and time, and the dependence of those quantities on temperature, age, etc., that the sense amplifiers must discriminate for READ operations. Typical linear (nonswitching) and nonlinear (switching) responses for a 100 micron^2 (10 x 10 micron) cell at 20°C are shown in Fig. 19 below, together with its C(V) characteristics (Fig. 20). Some researchers assume that C(V) and hysteresis data are redundant; that is not generally so and depends upon the frequency of each. Hysteresis data at 60 Hz or 1 kHz sense very little of the interfacial charge (space charge) at the electrode-ferroelectric interface. C(V) data, by comparison, are quite sensitive to the ramp speed of the measurement, which is typically a saw-tooth with a primary Fourier component at about 0.1 Hz. The latter is quite sensitive to space charge. Hence C(V) data and P(E) hysteresis data should be regarded as complementary, not redundant.

An equivalent circuit model for electrical hysteresis is given in Fig. 21b, and it is compared with experimental PZT data in Fig. 20a. The nonlinear capacitances CS1 and CS2 represent interfacial space charge at electrodes. Neglecting these terms produces inferior fits to the data.

FATIGUE/ENDURANCE

Recently our group has shown that the endurance of PZT memories (measured as the number of destructive READ/ERASE/WRITE cycles required to reduce the remnant polarization measured to half its original value) is inversely proportional to the drive voltage. Our calculation is an extension of Ma's original statistical mechanical theory for magnetic memories [10] and is reproduced [6] in Fig. 22. This simple graph reconciles a lot of diverse fatigue data in the literature, since some groups make their measurements at 2 or 3V, rather than 5V.

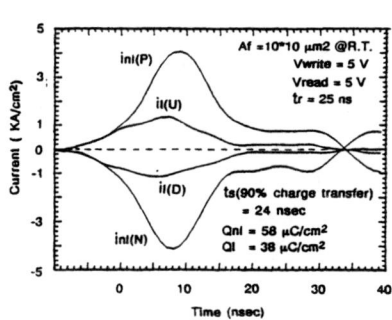

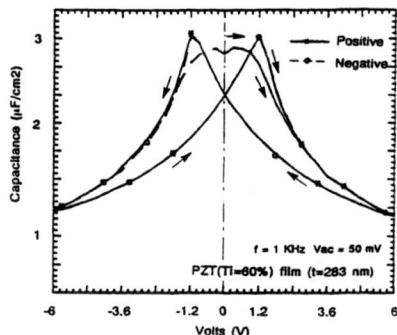

Fig. 19. I(t) displacement current data in a PZT memory.

Fig. 20. C(V) capacitance data showing switching in PZT.

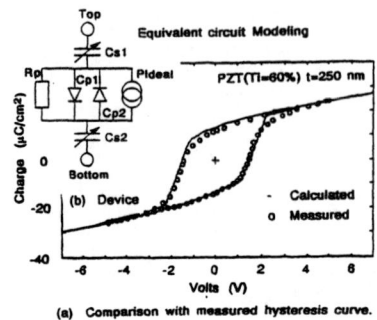

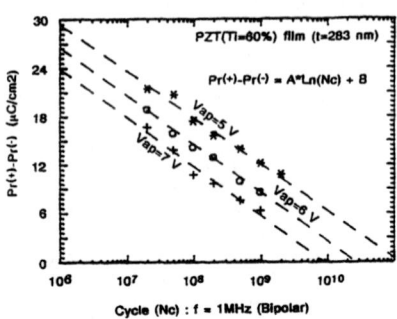

Fig. 21. Hysteresis data and circuit model.

Fig. 22. Fatigue in PZT (40/60) at different voltages (5 - 7V).

CONCLUSIONS

Ferroelectric memories can already be fabricated in low-density embodiments with extremely uniform characteristics within a wafer and excellent wafer-to-wafer reproducibility, using sol-gel techniques. Materials problems include optimization of electrodes and electrode-ferroelectric interfaces, systematic studies of baking/drying procedures, correlation of microscopic structures and topography with electrical character-istics, and exploration of annealing techniques (RTA, furnace, etc.). PZT sol-gel is the system of choice at present, but other oxides (e.g., bismuth titanate, barium bismuth titanate, strontium titanate-based compounds) and fluorides ($BaMgF_4$) hold considerable promise for FE FETs and high-capacity DRAMs. Replacement of much of the nonvolatile memory market now consisting of EEPROMs, magnetic bubbles, plated wire and core is anticipated within five years; this is a $1 billion/year market.

ACKNOWLEDGMENTS

All this work was done with H. Watanabe and T. Mihara. This work was supported at Symetrix by contracts with the Defense Nuclear Agency (DNA), DARPA, and the US Navy (White Oak, MD and Crane, IN) and at the University of Colorado by NSF grant DMR89-18900 and Army Research Office grant DAAL-03-90-G0002.

REFERENCES

[1] J. F. Scott and C. A. Araujo, Science 146, 1450 (1989).
[2] Proceedings of the First International Symposium on Integrated Ferroelectrics (ISIF), Ferroelectrics (special issue) (1990).
[3] Proceedings of the Second International Symposium on Integrated Ferroelectrics (ISIF), Ferroelectrics (1991).
[4] Proceedings of the Third International Symposium on Integrated Ferroelectrics (ISIF), Ferroelectrics (in press, 1992).
[5] J. F. Scott, B. M. Melnick, C. A. Araujo, and L. D. McMillan, J. Appl. Phys. [June 1, 1991, in press].
[6] C. A. Araujo et al., in Ref. 4.
[7] S. L. Miller et al., J. Appl. Phys. (in press, 1991).
[8] J. Evans, in Ref. 4.
[9] S. Dey and R. Zuleeg, Ferroelectrics 109, 1643 (1990).
[10] S. K. Ma, Statistical Mechanics (World Scientific, Singapore, 1985).

STRUCTURE AND CHARACTERIZATION OF SPUTTERED THIN FILMS BASED ON LEAD TITANATE.

A. PIGNOLET, P.E. SCHMID, L. WANG and F. LÉVY
Institute of Applied Physics, Swiss Federal Institute of Technology, CH-1015 Lausanne, Switzerland

ABSTRACT

Pure and doped lead-titanate (PT) and lead-zirconate-titanate (PZT) thin films have been deposited on platinum-coated silicon by rf-magnetron sputtering from pressed powder targets. The films have been deposited without substrate heating. The amorphous films were then annealed in an oxygen flow. The structure of the films is tetragonal or rhombohedral depending on composition. The electrical resistivity, dielectric permittivity, ferroelectric hysteresis and pyroelectric coefficient are reported.

INTRODUCTION

The macroscopic polarization of ferroelectric oxides has strong dependences on thermodynamic variables such as temperature, pressure or electric field giving rise to pyro-, piezo- and ferroelectricity. A great number of electronic applications based on these materials have been proposed. Thin films of ferroelectric oxides are very promising materials for a wide range of applications such as high-value capacitors [1], infrared sensors [2-6], ultrasonic sensors [7], non-volatile ferroelectric memories with low switching voltage [8-10] and various electro-optical devices [11, 12].

Recently the synthesis of thin films of ferroelectric oxides has been the subject of intense study [13-16]. Important materials under consideration include pure and doped lead titanate and lead zirconate-titanate (PZT). These materials exhibit a unique blend of electrical properties such as pyroelectricity, piezoelectricity, elasto-optic effect, linear and quadratic electro-optic effect. Many attempts have been carried out to prepare ferroelectric thin films by various methods such as ion plating, rf sputtering, ion beam sputtering, electron beam evaporation and chemical vapor deposition. Most of these films have been deposited on single-crystal substrates including $SrTiO_3$, MgO or Al_2O_3 [17-19]. However a better integration of ferroelectric films to microelectronic devices requires substrates compatible with silicon technology [20, 21].

We have studied the structural and electrical properties of pure and doped $PbTiO_3$ and PZT deposited on platinum-coated silicon by rf-magnetron sputtering. Polycrystalline films with a tetragonal or rhombohedral structure depending on composition were obtained by deposition without substrate heating followed by a post-deposition thermal treatment. The dielectric, ferroelectric and pyroelectric properties of these films have been investigated and are reported below.

EXPERIMENTAL PROCEDURE

Thin films of $PbTiO_3$ and PZT were deposited by rf-magnetron sputtering (NORDIKO system) onto platinum-covered (100) silicon and platinum-covered, thermally oxidized silicon (about 100 nm of SiO_2).

The vacuum chamber was pumped down to 10^{-4} Pa prior to film deposition and the optimized sputtering conditions are summarized in Table 1. The targets were 325-mesh powders (CERAC, typically 99.9% pure) pressed into shallow copper crucibles of 90 mm internal diameter.

The composition of the target was pure $PbTiO_3$ powder + 10 mol.% of PbO powder or a mixture of $PbTiO_3$, $PbZrO_3$, and 10 mol.% PbO powders. The PbO addition is intended to compensate for the loss of lead during sputtering deposition.

target	$PbTiO_3$ or PZT powder
sputtering gas	100 % Ar
sputtering pressure	0.8 - 1 Pa
substrate-to-target distance	60 mm
power	100 - 150 W
deposition rate	3 - 10 nm min^{-1}
post-deposition annealing	600°C or 800°C in oxygen flow

Table 1: Optimized preparation conditions

The substrates were not heated during deposition. After deposition the films were annealed in an oxygen flow of 10 to 100 sccm at 800°C for $PbTiO_3$ and 600°C for PZT. The thickness of the $PbTiO_3$ films was measured by profilometry (alpha-step, TENCOR instruments) and interferometry (Tolansky method). The crystal structure of the films was checked by X-ray diffraction (RIGAKU diffractometer, Cu $K\alpha_1$ radiation) and the morphology by scanning electron microscopy (CAMBRIDGE S-360 microscope). The dielectric constant and the dielectric loss factor were measured at 100 kHz by mean of an HP 4192 LF impedance analyzer. The D-E hysteresis loops were observed with a Sawyer-Tower circuit at 50 Hz excited by a periodic sawtooth signal and the resistivity and pyroelectric coefficient were measured with a KEITHLEY 617 electrometer.

For the pyroelectric measurements, the film was exposed to a periodic thermal radiation flux. The computerized control and acquisition system recorded the instantaneous temperature, its time-derivative and the current flowing across the film, allowing, after processing, to calculate the pyroelectric coefficient γ defined by

$$i_{pyro} = -\gamma A \frac{dT}{dt} \qquad (1)$$

where A is the overlapping area of bottom and upper electrodes and dT/dt is the temperature time-derivative.

RESULTS AND DISCUSSION

Structure

As already reported [22] and contrary to observations on films deposited on a heated substrate [4, 15, 18, 19], the as-deposited films are amorphous and become crystalline after a post-deposition thermal treatment [23]. The temperature required to obtain the ferroelectric structure depends on the chemical composition of the film. The annealing temperature lies between 800°C and 820°C for pure $PbTiO_3$ and is only 600°C to 650°C for PZT films. The composition determines the crystalline structure of the ferroelectric phase i.e. the structure is tetragonal (Fig. 1a) for pure $PbTiO_3$ and for

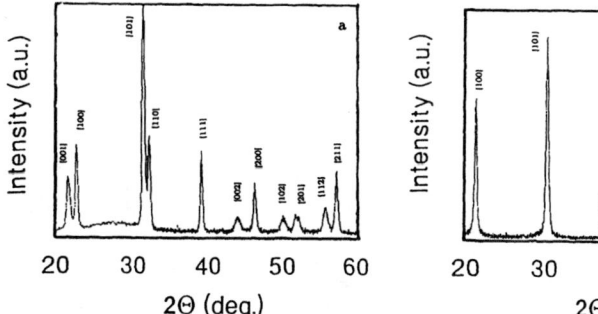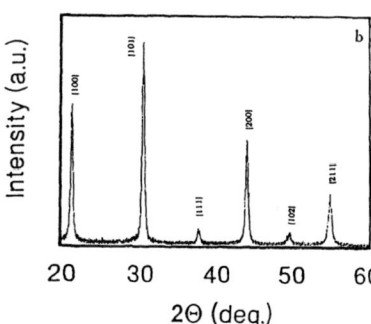

Figure 1: X-ray diffraction spectra at grazing incidence ($\theta = 5°$) of a) a film sputtered from a target of PbTiO$_3$ + 10 mol.% PbO powder and annealed at 800°C for 10 minutes in an oxygen flow of 10 sccm. The film has the tetragonal structure. b) a film sputtered from a target of PZT (53/47) + 10 mol.% PbO powder and annealed 5 minutes in an oxygen flow of 10 sccm. The film has the rhombohedral structure.

PZT with less than 40 mol.% zirconium. It is rhombohedral for PZT with more than 50 mol.% zirconium (Fig. 1b). In between lies the morphotropic phase boundary (MPB). In agreement with observations on films deposited at high substrate temperature [22], the presence of PbO in the target powder has little influence on the film structure except that films sputtered from PbO-enriched targets show smaller concentrations of undesirable phases, such as pyrochlore.

Electrical properties

The PbTiO$_3$ or PZT films have been deposited on top of 200 nm Pt dc-sputtered on a silicon substrate. During the early stage of the post-deposition annealing the platinum film reacts with the silicon substrate to form platinum silicide. This silicide layer was used as back electrode for the electrical measurements. On top of the film a platinum electrode was deposited, thus the electrical properties were measured across the film, along the normal-to-substrate direction.

The film has a dielectric constant of 950 at room temperature. It increases to 2000 at the Curie-temperature (Fig. 2). The permittivity vs. temperature shows a behavior characteristic of a diffuse ferroelectric-to-paraelectric transition. The broad peak is typical of mixed or multiphased materials.

The D-E hysteresis loop (Fig.3) characteristic of a ferroelectric material indicates that the film has a coercive field of 44 kV cm^{-1} and a remanent polarization of 18 μC cm^{-2} which is of the same order of magnitude as the values reported for highly-oriented films deposited on Pt/MgO or Pt/sapphire single-crystals [18, 24, 25].

Figure 4 demonstrates the pyroelectric properties of the film. It clearly shows that the response current is proportional to the time derivative of the temperature. The film of Fig. 4 has a pyroelectric coefficient of the order of 10^{-8} Ccm^{-2}K^{-1} which is comparable to that of ceramics of the same composition [26-28].

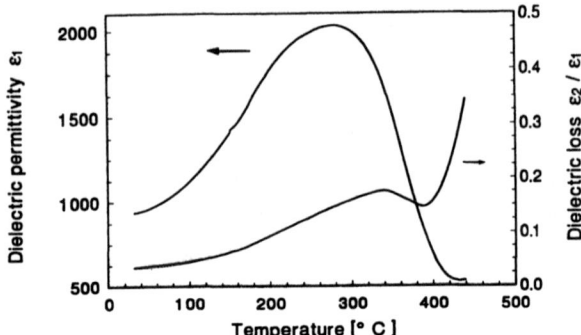

Figure 2: Dielectric permittivity ε_1 and dielectric loss $\varepsilon_2/\varepsilon_1$ at 100 kHz dependence on temperature for a (60/40) PZT thin film doped with 2 mol.% Mn. The film was deposited on platinum covered silicon and then annealed for 3 hours at 610°C. This curve shows a typical diffuse ferroelectric-paraelectric transition.

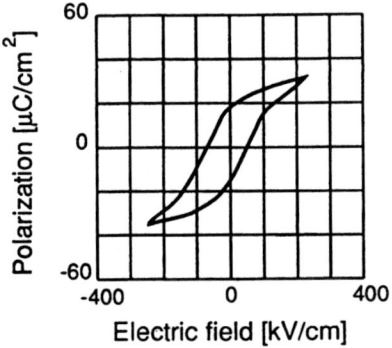

Figure 3: Polarization vs. applied electric field for a 1.5 μm thick PZT (53/47) film. The hysteresis loop is typical of a ferroelectric material. The film has not been poled before the measurement. The film has a spontaneous polarization of 24 μC cm^{-2}, a remanent polarization of 18 μCcm^{-2} and a coercive field of 44 kVcm^{-1}.

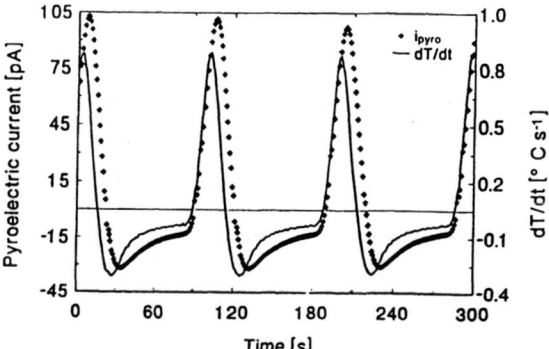

Figure 4: Pyroelectric response (***) of a 1.25 μm thick (60/40) PZT film doped with 2 mol.% Mn. The film was deposited on platinum covered silicon and was annealed for 3 hours at 610°C. The phase shift is due to the measurement geometry. The pyroelectric coefficient of the film is of the order of 10^{-8} Ccm^{-2}K^{-1}.

Although the films exhibited the ferroelectric structure after 5 to 10 minutes of post-deposition annealing, the electrical properties were improved by a longer anneal times of 3 to 5 hours.

CONCLUSION

Thin films of PbTiO$_3$ and PZT have been sucessfully deposited without substrate heating. The ferroelectric structure has been achieved by a post-deposition heat treatment. The structural, dielectric, ferroelectric and pyroelectric properties of the films are of the same order of magnitude as those of ceramics.

Aknowledgements

This work was supported by the National Research Program 19 of the Swiss National Foundation for Scientific Research.

REFERENCES

[1] K. Okamoto, Y. Nasu and Y. Hamakawa, IEEE Transactions on electron devices ED-28, 698 (1981).

[2] E. H. Putley, R. Watton and J. H. Ludlow, Ferroelectric 3, 263 (1972).

[3] H. P. Beerman, Ferroelectrics 2, 123 (1971).

[4] R. Takayama, Y. Tomita, K. Iijima and I. Ueda, J. Appl. Phys. 61, 411 (1987).

[5] R. Takayama, Y. Tomita, K. Iijima and I. Ueda, J. Appl. Phys. 63, 5868 (1988).

[6] K. Iijima, S. Kawashima and I. Ueda, Jpn. J. Appl. Phys 24-2, 482 (1985).

[7] M. Okuyama and Y. Hamakawa, Sensors and Materials 1, 13 (1988).

[8] J. F. Scott and C. A. Paz de Araujo, Science 246, 1400 (1989).

[9] R. B. Atkin, Ferroelectrics 3, 213 (1972).

[10] D. W. Chapman, J. Vac. Sci. Technol. 9, 425 (1971).

[11] L. K. Anderson, Ferroelectrics 3, 69 (1972).

[12] S. Y. Wu, W. J. Takei and M. H. Francombe, Ferroelectrics 10, 209 (1976).

[13] K. Sreenivas, M. Sayer and Garrett P., Thin Solid Films 172, 251 (1989).

[14] H. S. Potdar, S. D. Sathaye, A. B. Mandale and S. K. Date, Materials Letters 9, 71 (1990).

[15] X. Dengquan, X. Zhili, Z. Jumu, W. Derui, G. Huachong, X. Bizheng, and Y. Hong, Appl. Phys. Lett. 58, 36 (1991).

[16] *Ferroelectric Thin Films*, edited by E.R. Myers and A.I. Kingon (Mater. Res. Soc. Proc. 200 San Fransisco, 1990).

[17] H. Adachi, T. Mitsuyu, O. Yamazaki and K. Wasa, J. Appl. Phys. 60(2), 736 (1986).

[18] M. Adachi, T. Matsuzaki, T. Yamada, T. Shiosaki and A. Kawabata, Jpn. J. Appl. Phys. 26, 550 (1987).

[19] K. Iijima, R. Takayama, Y. Tomita and I. Ueda, J. Appl. Phys. 60(8), 2914 (1986).

[20] M. Okuyama, T. Ueda and Y. Hamakawa, Jpn. J. Appl. Phys. 24-2, 619 (1985).

[21] M. Okuyama, T. Ueda and Y. Hamakawa, Jpn. J. Appl. Phys. 24-3, 3 (1985).

[22] A. Pignolet, P. E. Schmid, L. Wang and F. Lévy, to be published in J. Phys. D: Appl. Phys., (1991).

[23] L. Wang, A. Pignolet and F. Lévy, Mat. Res. Bull. 25, 1495 (1990).

[24] K. Iijima, Y. Tomita, R. Takayama and I. Ueda, J. Appl. Phys. 60, 361 (1986).

[25] R. Takyama, A. Tomita and A. Abe, Ferroelectrics 95, 195 (1989).

[26] S. T. Liu, J. D. Heaps and O. N. Tufte, Ferroelectrics 3, 281 (1972).

[27] K. K. Deb, Ferroelectrics 88, 167 (1988).

[28] K. -H. Hellwege, *Landoldt-Börnstein*, New Series, Group III, vol. 16a (Springer-Verlag, Berlin, 1981), p 431.

THE EFFECTS OF LEAD-COMPENSATION AND THERMAL PROCESSING ON THE CHARACTERISTICS OF DC-MAGNETRON SPUTTERED LEAD ZIRCONATE TITANATE THIN FILMS

Vinay Chikarmane, Chandra Sudhama, Jiyoung Kim, Jack Lee and Al Tasch
Microelectronics Research Center, Electrical & Computer Engineering Department,
The University of Texas at Austin, Austin, TX 78712

ABSTRACT

The feasibility of the fabrication of thin film capacitors of Lead Zirconate Titanate (PZT) by reactive DC-Magnetron sputtering, with large switched charge and low leakage current densities for ultra-large scale integration Dynamic Random Access Memory (ULSI DRAM) applications has been demonstrated. As-deposited films were found to be predominantly pyrochlore; therefore, a subsequent phase transformation-inducing thermal processing step was key to obtaining device quality films. The importance of the thermal budget in optimizing the device characteristics of PZT films is discussed. The importance of the role of Pb compensation in lowering the required thermal budget and significantly enhancing device characteristics is shown.

INTRODUCTION

Limitations on dielectric thickness reduction due to tunneling, pinholes and reliability, and on cell structure complexity posed by processing technology, have recently led to considerable research activity in very high dielectric constant materials such as ferroelectrics for ULSI DRAM applications [1]. DRAM operation involves unipolar switching of the dielectric. Thus, it is necessary to maximize the switched charge density Qc' (defined as the difference between the maximum and the remnant polarization) for this application. The large switched charge density in capacitors fabricated from ferroelectrics such as PZT permits a reduction in device area, and is an important measure of film quality. These dielectrics must also have low leakage current in order to be candidates for use in semiconductor memories.

EXPERIMENTAL

PZT thin films of thickness 400nm were deposited at temperatures ranging from 32-500°C on Pt/TiO$_2$/SiO$_2$/Si multi-layered substrates by reactive DC-Magnetron sputtering from a multi-component Pb/Zr/Ti metal target in a pure oxygen ambient. The design of the target was similar to the one previously used by Croteau and Sayer [2]. The Pb content in the film was adjusted (compensated) by co-sputtering from a separate pure Pb metal target (r-f magnetron sputtering). Variations in the r-f power input to the pure Pb target were applied. The as-deposited films were annealed at different temperatures (300-900°C) in a tube furnace in a 100% nitrogen ambient. The top electrodes (also of Pt) were fabricated by a lift-off process and were 87μmx87μm square. Compositional analysis was performed using a Physical Electronics Scanning Auger Microscope with a single pass cylindrical mirror analyzer. X-Ray Diffraction (XRD) was performed on a Philips Diffractometer and Nomarski Microscopy on an Olympus BH-2 microscope. The Quasi-static capacitance-voltage (QSCV, sweep range +/-3V, ramp rate 0.3V/s), and leakage current density measurements were made on a HP 4140 pA meter interfaced to a computer. The switched charge density was obtained from the hysteresis loops, which were derived by integrating the area under the QSCV curves.

RESULTS AND DISCUSSION

The uncompensated PZT films (Zr/Ti::65/35) , deposited from the multi-component metal target alone, were deficient in Pb (compared to the Pb content in stoichiometric PZT) due to re-evaporation at all deposition temperatures above 32°C. The as-deposited uncompensated

Figure 1.(a), (b) and (c) Nomarski micrographs (200x) for Pb-compensated PZT films annealed at 650°C for 1h (T_{dep}= 200°C) for r-f power input to the Pb target of 20W,15W and 7W respectively. A steep increase in the perovskite-to-pyrochlore phase transformation rate with increasing Pb-compensation is observed.

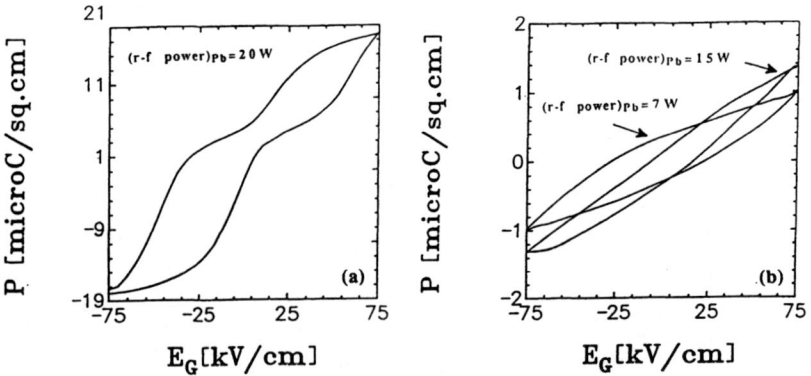

Figure 2. Hysteresis loops (+/- 3V sweeps) for Pb-compensated films in Figure 1 with r-f power input to the Pb target of (a) 20W and, (b) 15W, and 7W. Qc' for these films is 12.8μC/cm², 1.24 μC/cm², and 0.696 μC/cm² respectively.

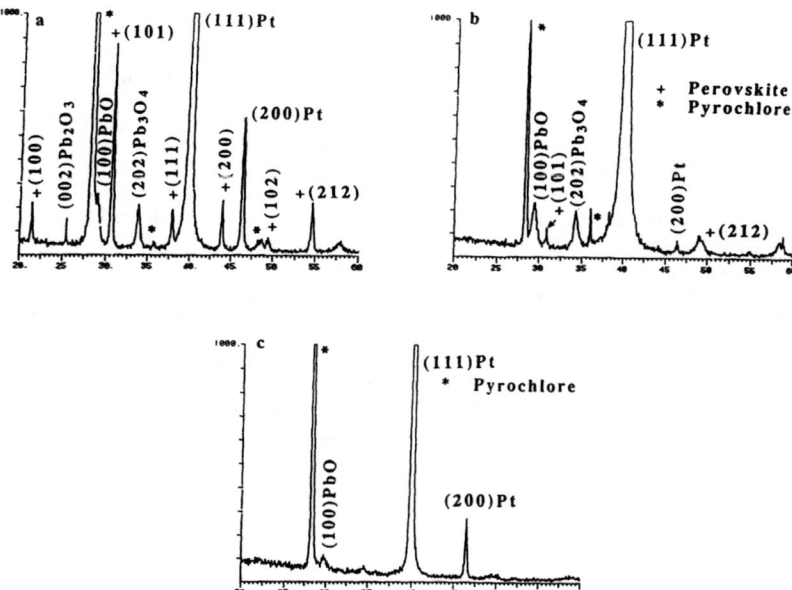

Figure 3. XRD spectra for the Pb-compensated films in Figure 1 with r-f power input to the Pb target of (a) 20W (b) 15W and, (c) 7W. An increase in the perovskite peak intensities with increasing Pb-compensation is observed.

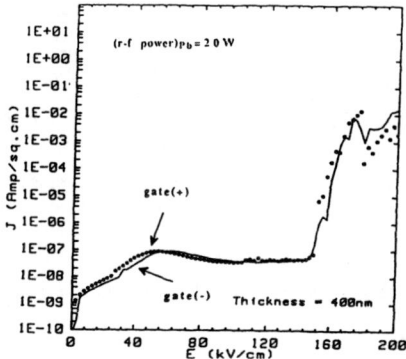

Figure 4. Typical leakage current characteristic for Pb-compensated film with r-f power input to the Pb target of 20W (T_{ann}=650°C, t_{ann}=1h, T_{dep}=200°C). The leakage current density at 3V is very low (9.05 x 10^{-8} & 7.28 x 10^{-8} A/cm^2 for (-) and (+) gate bias respectively), as desired.

films were rich in pyrochlore phase, and it was not possible to recover the perovskite phase by annealing for films deposited at high temperatures (300°C and higher) without causing the capacitor to short.

A deposition temperature of 200°C was selected as being optimum from the point of view of minimizing the thermal input in the as-deposited film while obtaining a fully crystalline structure as well as minimizing the Pb re-evaporation during deposition. It should be pointed out that the annealing time and temperature are crucial because although Q_c' increases, the leakage current density of the PZT thin film capacitors increases with increasing thermal budgets. There is clear evidence of the ferroelectric perovskite phase nucleation in uncompensated films deposited at 200°C and annealed at high temperatures (850°C, 1h), however Q_c' is low (2.09μC/cm^2), and inadequate for DRAM applications. Increasing the annealing temperature to 900°C in order to increase the perovskite content and therefore Q_c', causes the capacitors to short (i.e. have very large leakage current).

Pb compensation has been found to allow for lower thermal budgets, which is very significant from the point of view of integration of PZT into silicon IC's as well as obtaining superior device characteristics due to reduced thermal damage in these films. As-deposited Pb-compensated films contained the pyrochlore and (100) PbO phases, and no perovskite was detected by XRD. Appreciable pyrochlore-to-perovskite transformation was observed for compensated films annealed at 650°C for 1h. In Pb-compensated films, a large enhancement in the perovskite phase grain size with increasing Pb compensation (r.f power to the pure Pb target of 7W, 15W and 20W) is observed by Nomarski microscopy (Figure 1 (a), (b) & (c)). Since the annealing conditions are identical (650°C, 1h), this is due to a Pb-compensation-induced increase in the pyrochlore-to-perovskite phase transformation rate. This is in agreement with the hysteresis loops which shows that Q_c' registers a steep increase with increasing Pb compensation (Figure 2(a) & (b)).

The optical observations and electrical measurements are supported by the XRD spectra which confirm the increased formation of the ferroelectric perovskite phase with increasing Pb compensation (Figure 3(a) (b) & (c)). The leakage current density at 3V (~ 10^{-7} A/cm^2) is the best (lowest) reported so far for sputtered films (Figure 4).

CONCLUSIONS

Based on our results, PZT thin films with excellent electrical characteristics for ULSI DRAM applications have been fabricated by reactive DC-Magnetron sputtering from a multi-component (Pb/Zr/Ti) metal target with Pb compensation. The PZT capacitors showed large switched charge densities, very low leakage current densities. Uncompensated films are found to be Pb deficient resulting in a inhibition of the pyrochlore-to-perovskite phase transformation. However, the perovskite phase can be formed by annealing the films at high temperatures even in films highly deficient in Pb. Although this increases the pyrochlore-to-perovskite transformation, increasing the thermal budget also leads to a rapid degradation of the leakage characteristics due to the microcracking of the film. Pb compensation strongly enhances the pyrochlore-to-perovskite phase transformation rate and thereby allows for significantly less severe post-deposition thermal processing conditions. Large increases in switched charge densities can be obtained while achieving very low leakage current densities.

ACKNOWLEDGEMENT

This work was partially supported by SRC/SEMATECH through contract # 88-MC-505.

REFERENCES

[1] L.H.Parker and A.F.Tasch, IEEE Circuits and Devices magazine, pp.17-26 (1990).
[2] A.Croteau and M.Sayer, in Proceedings of the 6th IEEE International Symposium on Applications of Ferroelectrics, edited by V.E.Wood, pp.606-609 (1986).

NANOSTRUCTURE EVOLUTION DURING THE TRANSITION OF TiO_2, $PbTiO_3$ AND PZT FROM GELS TO CRYSTALLINE THIN FILMS.

Z.C. KANG[*+], A. GUPTA[#+], M.J MCKELVY[+], L. EYRING[*+] and S.K. DEY[#]
[*]Department of Chemistry; [+]Center for Solid State Science; [#]Chemical, Bio and Materials Engineering Department; Arizona State University, Tempe, AZ 85287

ABSTRACT

The nanostructure evolution of PZT, PT and T thin films has been studied by high-resolution electron microscopy (HREM) supported by other techniques such as thermal analysis, thermal mass spectrometric analysis and X-ray diffraction analysis. The evolution follows a common progression from amorphous film, to the development of condensed regions that develop crystalline order, to the final polycrystalline oxide thin film. If the precursor gel contains lead, the film develops fluctuating surface "blisters" that evolve to an oxide final product as well. Minor structural and compositional differences exist across the final oxide thin film.

INTRODUCTION

Ferroelectric thin films prepared by sol-gel processing are being paid considerable attention [1]. Processing involves drawing or spinning the thin-film gel followed by heat treatment to produce the crystalline thin, ferroelectric film. It is clearly of fundamental and technological value to understand the sequence of reactions and transformations that accompany gel to crystalline thin-film processing.

A brief report of the nanostructure evolution of a $Pb(Zr_{0.45}Ti_{0.55})O_3$ (PZT) thin-film gel has been made [2]. In this paper a systematic study of the structural evolution from alkoxy gel to PZT, $PbTiO_3$ (PT) and TiO_2 (T) utilizing high-resolution electron microscopy (HREM) together with thermal analysis (TGA and DTA), thermal mass spectrometric analysis (TMSA) and X-ray diffraction (XRD) is reported.

EXPERIMENTAL PART

The preparation procedures of the PZT thin-film gels for HREM studies was similar to that reported previously [2]. The specimens for the other techniques required powders made from dried gels. The precursor sol for PT was made from lead acetate, and titanium isopropoxide in methoxyethanol. The precursor sol for T was titanium isopropoxide solution. Electron-beam heating was used to age the thin-film specimens in the microscope.

RESULTS AND DISCUSSION

The amorphous state of the thin-film gels

As previously reported [2], the dipped and quickly dried PZT precursor film is amorphous. The same is true for the PT and T thin films as confirmed by the images and diffraction patterns of Fig. 1. The diffraction patterns all show a broad inner ring corresponding to nearest neighbor distances of 3.03 ± 0.03 Å and a weak second halo suggesting a continuous random network [3].

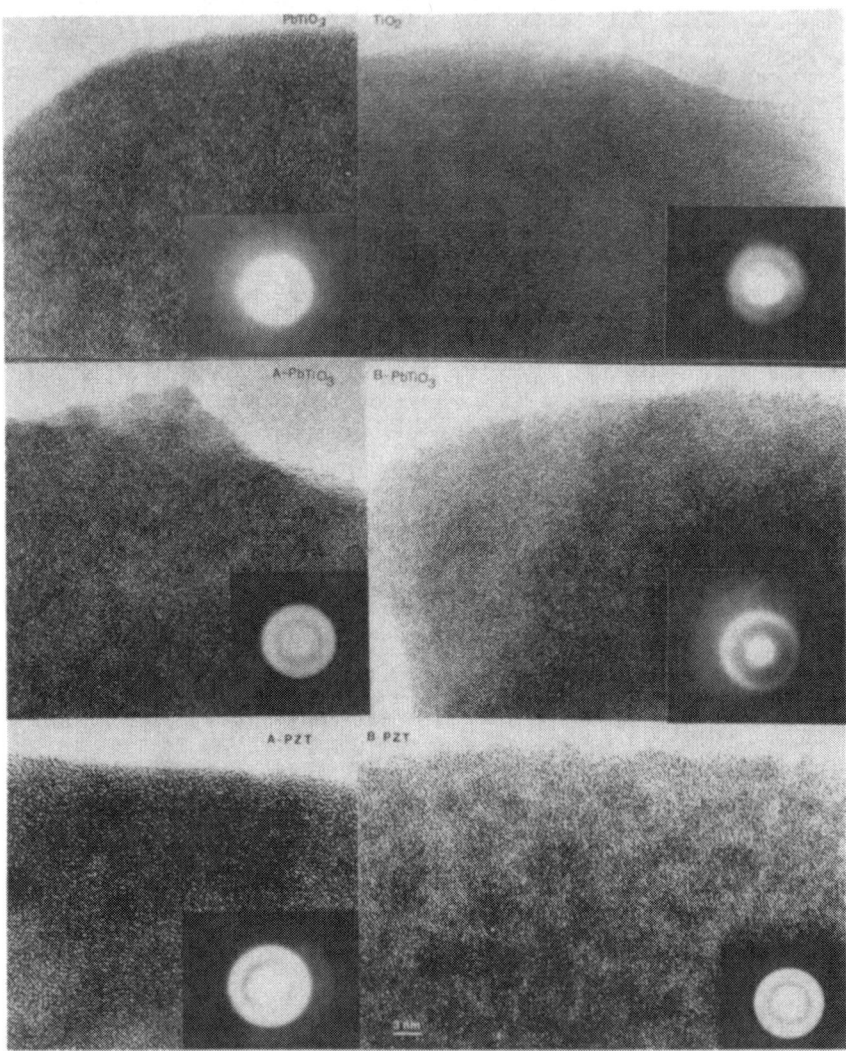

Figure 1. The amorphous state of T, PT and acid- and base-catalyzed PT and PZT.

All the images shown were taken about 10 minutes after beginning microscopic observation. There is no long-range order. However, if the beam for optical diffraction is small there are imaged regions that show patterns affirming a certain local point symmetry [3]. In the images of PZT and PT precursor films there are regions that begin to darken as observation is continued, while in the T film oriented linear structures appear (Fig. 2). The differences probably relate to the presence or absence of lead and the nature of condensation in the film.

Local crystallization and the emergence of blisters

After a few more minutes of beam irradiation the dark regions develop either one or two dimensional fringes that fluctuate in structural organization in a constrained way. These clusters are apparently inside the film. At the same time, growing blisters appear on the surface of the PZT and PT films (Fig. 2). The blisters are amorphous at first, but soon become polycrystalline with some inclusions that appear amorphous. The crystalline and amorphous regions fluctuate and interchange wildly during decomposition, displaying internal boundaries and faceted edges that change several times a second. The blisters are more or less hemispherical and rooted in the film. The pattern of behavior is similar for acid and base catalyzed gels except that it occurs during normal microscopal observation in the former but requires periodic, intense beam heating in the latter. This phenomenon is found only in samples that contain lead, hence, is absent in T.

The DTA results for acid catalyzed PZT and PT precursor gels under helium exhibit an exothermic peak over the temperature range ~220-290°C. In contrast, similar base catalyzed gels exhibit an endothermic response from 20-500°C. Vacuum TMSA indicates that primarily methoxyethanol, carbon dioxide and acetone are evolved from ~220-290°C for both the acid and base catalyzed PZT gels, suggesting acid vs. base catalysis has a significant impact on the local bonding arrangement in the gel. The presence of the latter two gases is apparently associated with acetate decomposition. The exotherm observed for the acid catalyzed gels is consistent with the lower gel decomposition temperature observed by HREM, since the exothermic process, once activated, can provide energy to further decomposition at lower external temperatures.

The nature of the crystallized thin-film

The early stages of film crystallization are shown in Fig. 3. The lattice of each dark spot has a different orientation suggesting no long-range intercluster order. The spaces between the clusters also show weak lattice fringes in some places. Optical diffraction patterns of the various crystalline dark regions indicate different c/a ratios implying slightly different compositions. The cluster sizes are less than 100Å and are observed to grow in size with increased irradiation time.

A model of the amorphous film

Under the experimental procedures used here the sols are incompletely hydrolyzed and the unadsorbed or unbound solvent is rapidly removed from the thin film, especially in the HREM experiments. This suggests that the amorphous film consists of a more or less continuous random network of partially hydrolyzed clusters with no translation symmetry relating the clusters.

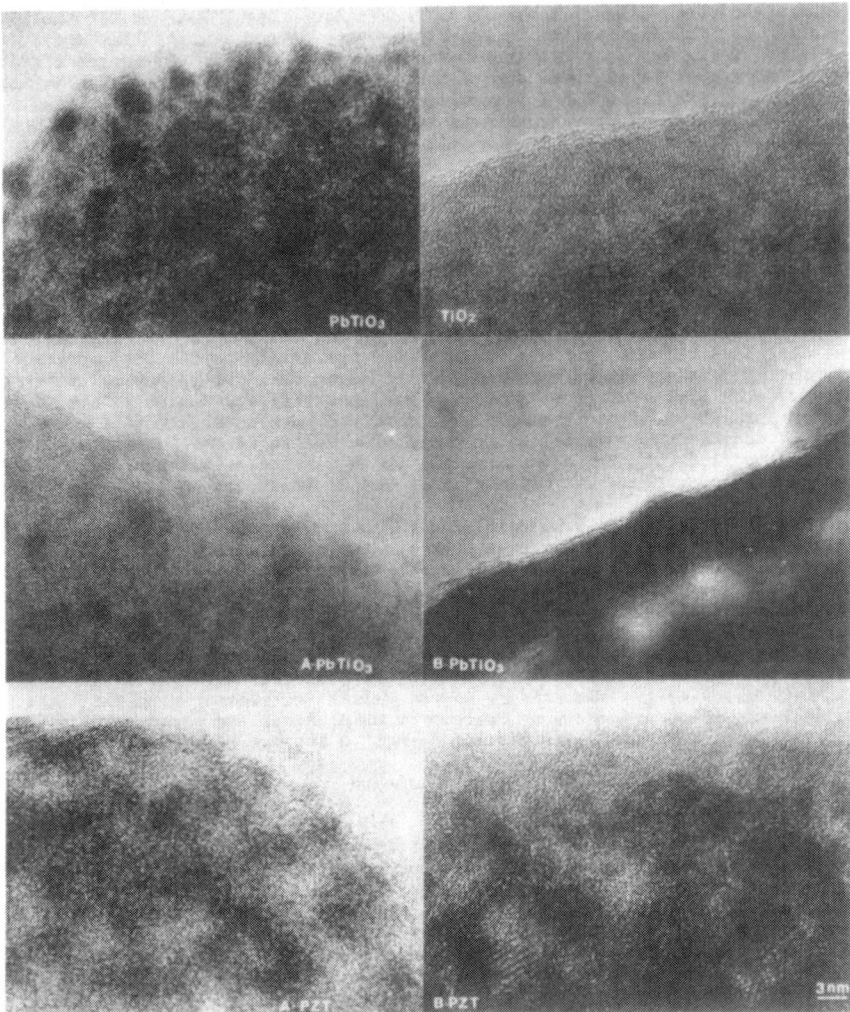

Figure 2. The dark spot and blister stage of PT and acid- and base-
catalyzed PT and PZT, and the oriented linear structure associated with T.

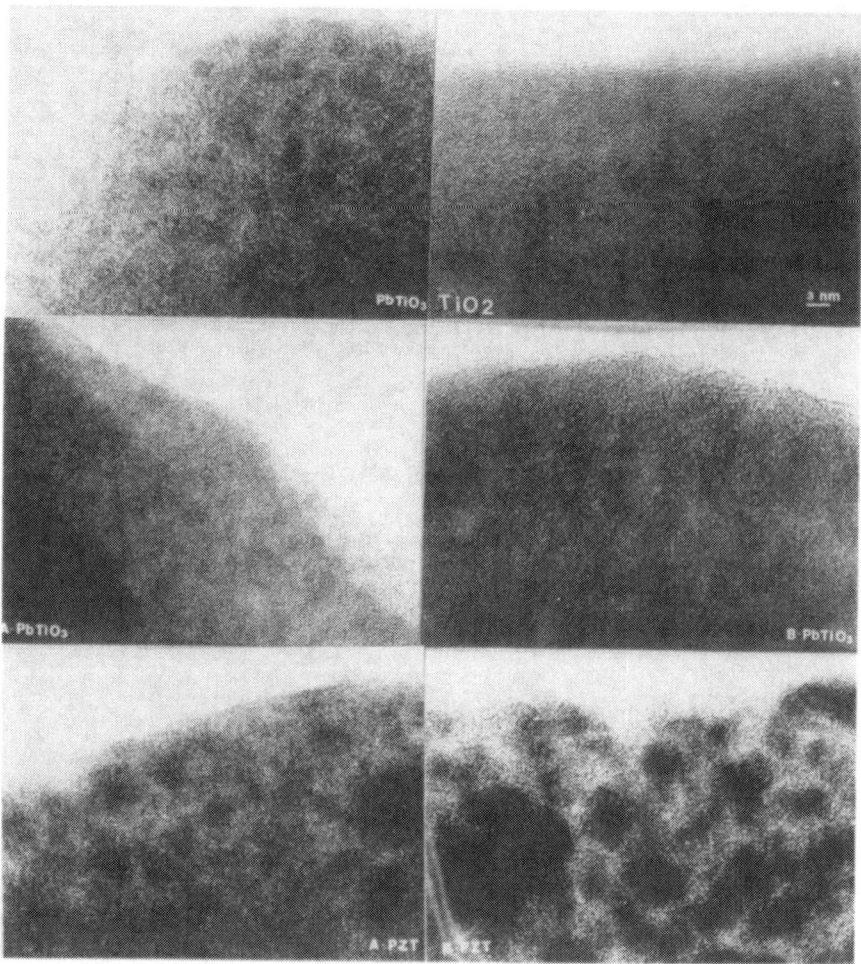

Figure 3. Polycrystalline thin-film stage of T, PT and acid- and base-catalyzed PT and PZT.

CONCLUSIONS

The structure of the PZT, PT and T thin-film gels studied here are also amorphous when dried. The lead-containing thin films, PZT and PT, evolve through stages where dark spots finally become crystalline and very active blisters form on the surface. The final film is polycrystalline with evidence of compositional inhomogeneity. The process occurs with a lower electron flux (and hence at a lower temperature) in the acid catalyzed gels than in the base catalyzed material. A full account of the structural and chemical studies of the gel-to-crystalline-film formation process will be published elsewhere.

ACKNOWLEDGEMENT

Support of the NSF through Research Grant DMR-8820017 is greatly appreciated. We wish also to thank the Center for Solid State Science for supported facilities and the Facility for High-Resolution Electron Microscopy (NSF Grant DMR-8913384).

REFERENCES

1. Z.C. Kang, S. K. Dey and L. Eyring, Mat. Res. Soc. Symp. Proc. Vol. 183 (1990) p. 291.
2. L. Eyring, A.K. Kang and S. K. Dey, to be presented at the American Ceramic Society Spring 1991 Meeting, Cincinnati, Ohio.
3. Richard Zallen, in The physics of amorphous solids, Wiley, New York, 1983, p. 74.

METASTABLE PYROCHLORE STRUCTURES IN SOL GEL SPIN COATED LEAD TITANATE THIN FILMS ON SILICON SUBSTRATE

JEON-KOOK LEE*, HYUNG-JIN JUNG* AND CHONG-HEE KIM**
*Div. Ceramics, Korea Institute of Science and Technology, Seoul 130-650, Korea
**Dept. Materials Science and Engineering, Korea Advanced Institute of Science and Technology, Seoul 130-012, Korea

ABSTRACT

High quality lead titanate thin films were fabricated by spin coating on a silicon substrate. The resulting dried gel layers were uniform in thickness through 2 X 2 cm² area, and polycrystalline perovskite structures developed almost crack free with a heat treatment above 500 °C in films with thickness above 0.36 μm.

Metastable pyrochlore structures were observed in films with thickness of 0.16 μm when heat treated at 500 and 600°C. But these structure did not appear in films with thickness of 0.36 μm. The thickness dependence in crystal structure of films was studied by varing the substrate condition and analyzing the interface between the film and substrate. In native oxide films on silicon substrates, amorphous dried gel layers were heterogeneously nucleated. Metastable cubic pyrochlore structure could be crystallized in amorphous native oxide (cubic property in random network structure).

INTRODUCTION

Much of the recent sol-gel ferroelectric thin film research has been directed to the PbO based perovskite compounds such as $PbTiO_3$, $Pb(Zr,Ti)O_3$. The microstructures and crystallizations of lead titanate thin films on a silicon substrate are highly dependent on the sol-gel processing variables such as substrate materials, film thickness, heat treating conditions.

The thickness dependence of several physical properties of ferroelectric films have been documented. The effective dielectric constant of PZT on silicon is very low and depends on the thickness of the PZT film and the annealing conditions [1]. At 500 °C, pyrochlore structure grew on glass plate when the thickness of the coated films was below 100 nm [2].

The objective of our work was to study the reason why the pyrochlore structure was formed in films on the silicon substrate with the thickness of 0.12 μm.

EXPERIMENTAL

Precursor Solution

Stock solutions of complex Pb-Ti alkoxide were prepared by reacting lead acetate trihydrate (99%, Alfa) with titanium isopropoxide (Alfa) in 2-methoxyethanol (99.9% HPLC grade, Aldrich) in a method similar to that reported by Blum [3]. The resulting Pb-Ti stock solution had 0.2 molar concentrations. The stock solutions were handled as moisture sensitive reagents and were stable.

The coating solutions were acidified with concentrated HNO3 to a resulting concentration of 0.1 molar concentration. Hydrolysis was initiated by the addition of an aqueous 2-methoxyethanol solution. The total hydrolysis water charge was 2 moles of H_2O per 1 mole of Pb.

Coating

Coatings were deposited on (100) p-type single crystal silicon wafer. The Pb, Ti coating solution was applied by spin coater utilizing a commercial photoresist spinner (Headway research EC 101D) operated at 3000 rpm. The wet films were allowed to hot plate heated at 215 °C for 5 min, after which they were transferred to an air atmosphere furnace for pyrolysis. The heating and cooling rates were 3 °C/min. with the exception of a cooling rate of 0.5 °C/min in the vicinity of the Curie temperature.

Characterizations

Microstructural and crystal structural characterizations of the films were accomplished by scanning electron microscopy and X-ray diffraction, respectively. In addition, overall film quality in metal oxide semiconductors was assessed by capacitance voltage plots.

RESULTS & DISCUSSION

Lead acetate trihydrates were partially methoxyethylated and 70% of the trihydrate were removed. Titanium isopropoxide underwent complete transesterifications by methoxyethanol. These results are similar to those of Ramamurthi[4]. Pb,Ti complex alkoxide are stable for coating.

Differential scanning calorimetries and weight losses of dried gels heated at 215 °C for 40 min. are shown in Fig.1. Weight loss of uncatalysed ones are more smoothly occured than that of nitric acid catalyzed ones above 200 °C.

In the differental scanning calorimetry curves, crystallization peaks at 500 °C of uncatalyzed case have a shoulder. But in the case of nitric acid catalyzed ones, this shoulder was not found. This suggests that two different phases nucleate during the crystallization of uncatalyzed gels.

Microstructures of singly coated films heat treated at 500 °C for 60 min. are shown in Fig.2. Nitric acid catalyzed films are more homogeneous microstructurally than uncatalyzed ones. Grains are small (below 0.1 μm) and grain boundaries are not clearly detected. Microstructural inhomogeneity of uncatalyzed films were composed by matrix and secondary phase. This formation of two phases was also observed in DSC data (Fig.1).

In cross sectional view, thickness was uniform up to 2 cm X 2 cm area and interfacial reaction between films and substrates was observed. Microstructural inhomogeneities of uncatalyzed films heat treated at 600 °C for 60 min. are more clearly observed in Fig.3. This bimodal microstructures is due to the incomplete hydrolysis and polycondensation because coating solutions were not catalyzed. Nitric acid catalyzed films were microstructurally homogeneous.

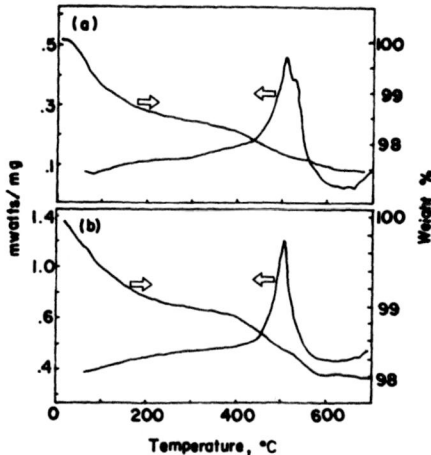

Fig. 1   Differential scanning calorimetry and weight losses of
hydrolyzed gels dried at 215 °C for 40 min.
(a) uncatalyzed   (b) nitric acid catalyzed

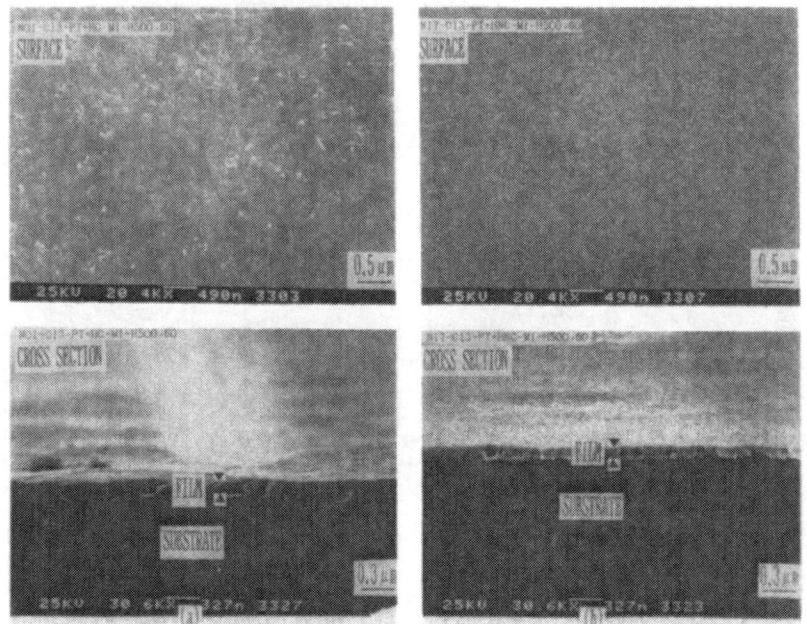

Fig. 2   Micrographs   of singly coated films heat treated
at 500 °C for 60 min.
(a) uncatalyzed   (b) nitric acid catalyzed

The sintering behavior of a ceramic film which is constrained by a rigid substrate is contrasted with the sintering behavior of a free film [5-7]. Constrained films show tensile stress in plane, and this stress makes microstructures of films bimodal phase (dense and porous parts).

These morphologies are shown in Fig.3(a), for the catalyzed case. But catalyzed ones were not shown. It is due to the difference of bonding strength (or reaction) between films and substrate.

Fig. 4 shows X-ray diffraction patterns of singly coated films heat treated at various temperatures. Uncatalyzed dried-gel films with thickness of 0.12 μm crystallized to the perovskite structures by heat treatment above 600°C. But nitric acid catalyzed dried-gel films cryatallized to the pyrochlore structures by heat treatment above 500°C. The pyrochlore structure is with the composition $Pb_2Ti_2O_6$. In films on silicon, titanium is easily substituted by silicon.

So pyrochlore structures were crystallized in nitric acid catalyzed films. It is suggested that microstructures are homogeneous but pyrochlore structures are crystallized by high reactivity between film and substrate in case of nitric acid catalyzed films. Microstructures of dried gel films with thickness of 0.36 μm (triple coating) were similar to that of 0.12 μm except that grains were grown, as shown in Fig. 5,6.

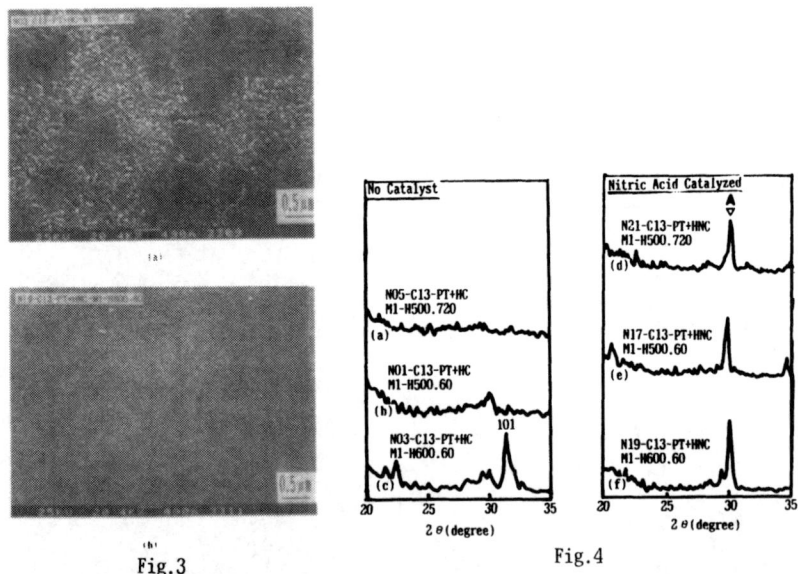

Fig.3   Fig.4

Fig. 3   Surface morphology of singly coated films heat treated 600°C for 60 min.

Fig. 4   X-ray diffraction patterns of variously heat treated films. Single coating, (a,b,c) uncatalyzed (d,e,f) nitric acid catalyzed (a) at 500°C for 720 min. (b) at 500 C for 60 min. (c) at 600°C for 60 min.

Surface morphologies of uncatalyzed films were more homogeneous by increasing heat treating temperature from 500 °C to 600 °C. Cross section of uncatalyzed films showed large grain through film thickness (0.36 μm), as shown in Fig. 5 (a). Cross section of nitric acid catalyzed films shown in Fig.5(b) was composed by 0.1 μm sized grains. Interfacial reaction between films and substrate can be found.

Fig.7 shows X-ray diffraction patterns of films using nitric acid catalyzed coating solutions. As increased film thickness, pyrochlore structures (mark A) were undetected in X-ray resolution. It is suggested that metastable cubid pyrochlore structures formed in interfacial region by reaction between film and substrate.

Fig. 8 shows capacitance voltage characteristics of films coated by various conditions and heat treated at various temperatures. In case of uncatalyzed films with thickness of 0.12 μm (see Fig. 8 (a) capacitance - voltage curves in depletion region become stiff by increasing heat treating temperature from 500 °C to 600 °C. Dielectric constants of film listed in Table I were slightly increase from 7.2 to 8.5 because perovskite structures are formed and microstructures are homogeneous. These low valués were due to the small grain [8].

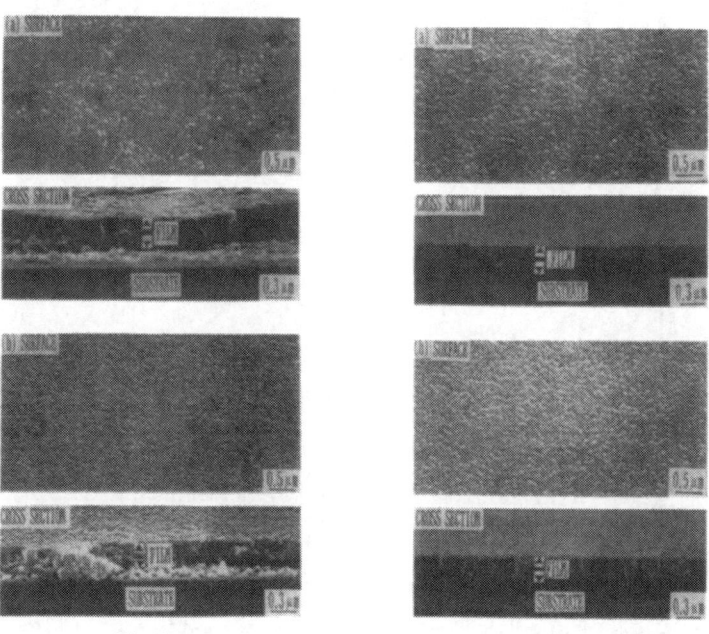

Fig.5   Fig.6

Fig. 5 Micrographs   of sol-gel derived thin films heat treated at 500 °C for 60 min. (triple coating) (a) uncatalyzed, (b) nitric acid catalyzed.

Fig. 6 Micrographs   of sol-gel derived thin films heat treated at 600 °C for 60 min. (triple coating) (a) uncatalyzed. (b) nitric acid catalyzed.

In the case of nitric acid catalyzed films with thickness of 0.12 μm (see Fig. 8 (b)), C-V curves were not changed by increasing heat treating temperatures from 500 °C to 600 °C because there is no change of crystal structure (pyrochlore structure). Dielectric constants of the catalyzed film were decreased from 3.2 to 2.8 and these values were lower than the case of the uncatalyzed film, because of reaction between the film and the substrate and formation of pyrochlore structures.

In case of uncatalyzed films with thickness of 0.36 μm (see Fig. 8 (c)), C-V curves in depletion region become stiff by increasing heat treating temperature from 500 °C to 600 °C. Dielectric constants of film listed in Table I were increased from 20.3 to 33.2. In case of nitric acid catalyzed films with thickness of 0.36 μm (see Fig. 8(d)), C-V curves also become stiff and dielectric constant was increased from 27.8 to 46. These values were due to the increase of thickness grain size and the decrease of interfacial phenomena by increasing film thickness. But in case of films with thickness of 0.65 μm,

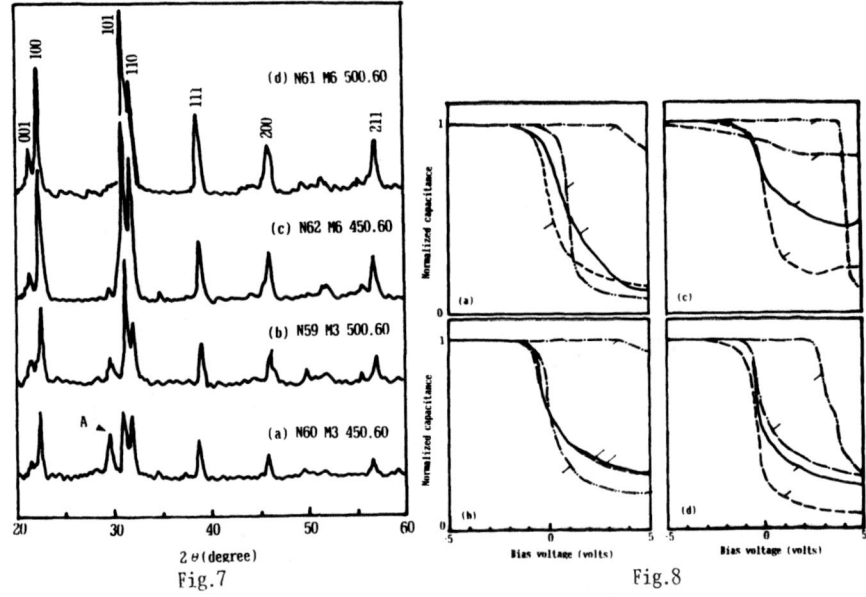

Fig.7

Fig.8

Fig. 7   X-ray diffraction patterns of films using nitric acid catalyzed coating solutions,
(a) triple coating, heat treated at 450 °C for 60 min.
(b) triple coating, heat treated at 500 °C for 60 min.
(c) 6 times coating, heat treated at 450 °C for 60 min.
(d) 6 times coating, heat treated at 500 °C for 60 min.

Fig. 8   Capacitance vs. voltage characteristics of films coated by the uncatalyzed or nitric acid catalyzed solutions. Heat treating at 500 °C for 60 min. (———). at 600 °C for 60 min. (————). at 500 °C for 720 min. (—·—·—·) and at 215 °C for 5 min. (—··—··—).
(a) Single coating, no catalyst, (b) Single coating, nitric acid catalyzed.
(c) Triple coating, no catalyst. (d) Triple coating, nitric acid catalyzed.

dielectric constant listed in Table I was 17.22 because of macro crack induced by stress from shrinkage.

CONCUSION

Metastable cubic pyrochlore structure could be crystallized in amorphous native oxide (cubic property in random network structure) and by the reaction between films and silicon substrate. These crystal layers were not detected in the film with thickness above 0.4 μm because pyrochlore structured layer was thin.

Table I   Capacitance Voltage Characteristics of Metal Oxide Semiconductors Devices for Various Coating Solutions. Multiplicity of Coating and Heat Treatment.

Multiplicity of coating	Catalyst	Heat Treating (60 min.)	Thickness (μm)	Dielectric constant
Single	No	500 °C	0.12	7.24
		600 °C	0.12	8.51
	Nitric acid	500 °C	0.12	3.24
		600 °C	0.12	2.81
Triple	No	500 °C	0.35	20.28
		600 °C	0.42	33.23
	Nitric acid	500 °C	0.44	27.83
		600 °C	0.36	46.00
6 times	Nitric acid	500 °C	0.65	17.22

REFERENCES

1. T.S. Kalkur, G. Argos and L. Kammerdiner, Mat. Res. Soc., Symp. Proc. Vol 200 (1990) 313
2. Y. Takahashi and K. Yamakuchi, J. Mater. Sci. 25 (1990) 3950
3. J.B. Blum and S.R. Gurkovich, J. Mater. Sci., 22 655 (1987).
4. S.D. Ramamurthi and D.A. Payne, J. Am. Ceram. Soc. 73 (8) 2547 (1990)
5. R.K. Bordia and R. Raj, J. Am. Ceram. Soc. 68 (6) 287 (1985).
6. G.W. Scherer and T.J. Garino, ibid 68 (4) 216 (1985).
7. T.J. Garino and H.K. Bowen, ibid 73 (2) 251 (1990).
8. V.P. Dudkevich and E.G. Fesenki, Ferroelectrics, 22 787 (1978).

METAL-FERROELECTRIC-SEMICONDUCTOR CHARACTERISTICS OF BaMgF$_4$ FILMS ON p-SILICON

T.S. KALKUR, J.R. KULKARNI*,R.Y. KWOR, L. LEVINSON** and L. KAMMERDINER***
Microelectronics Research Laboratories, Department of Electrical and Computer Engineering, University of Colorado at Colorado Springs, CO 80933
* Presently with National Semiconductors, Santa Clara, CA.
** Department of Physics, University of Colorado at Colorado Springs, CO 80933
*** Ramtron Corporation, Colorado Springs.

ABSTRACT

Capacitance-voltage characterstics of BaMgF$_4$ film deposited in an ion-assisted deposition system shows hysteresis and the direction of hysteresis corresponds to ferroelectric polarization. Electrical characterization of the films shows that these films can be used to implement non-destructive read-out non-volatile ferroelectric memories. These films were found to dissolve in water and other aqueous solutions. In order to overcome this problem, a suitable capping layer like zirconium oxide and amorphous silicon was deposited on BMF films. The shift in threshold voltage did not change significantly due to the incorporation of the capping layer. The shift in threshold voltage was found to be temperature dependent and this might be due to ionic conduction in fluorides.

INTRODUCTION

Recently, there has been tremendous interest in the use of ferroelectric materials like PZT and Bismuth Titanate for radiation hardened non-volatile memory applications (1-3). There are two approaches to implement non-volatile memories using ferroelectric materials. In one approach, which is commonly used for materials like PZT, the ferroelectric memory element is a simple metal-ferroelectric-metal capacitor. The memory read out in these capacitors is destructive because whenever the information is read, it has to be written so as to preserve the memory (1). This makes the memory element undergo fatigue as the read-write cycles are increased. In another approach, the ferroelectric material is incorporated as a gate insulator of the metal-insulator-semiconductor field effect transistor(MISFET) (4). This results in non-distructive read out of the information so that the fatigue of memory element can be minimised. The metal-ferroelectric-semiconductor characteristics of ferroecelectric materials like PZT and Bismuth Titanate have been studied (5-7).

Because of the interdiffusion of these ferroelectric materials with silicon, it was not possible to integrate them directly on silicon. The hysteresis observed in the capacitance-voltage characteristics of these devices was predominantly due to the injection of carriers at the interface rather than due to the field effect caused by ferroelectric behaviour. Recently, fluoride based ferroelectrics have attracted the attention of many investigators for the implementation of non-destructive read out ferroelectric MISFETs (8-10). In this paper, we are presenting the results of the characteristics of BMF fluoride MIS capacitors.

SAMPLE PREPARATION

The silicon wafers used for the study were of p-type with (100) orientation and had resistivities in the range from 6.8 to 10 ohm-cm. The BMF films were deposited using an ion-cluster beam deposition system as shown in fig.1. The source material was optical grade barium fluoride and magnesium fluoride mixed in equimolar concentrations and held in tantalum crucible. Initially, the ion-source temperature was increased so that both fluorides fused by melting. Later the source was cooled down to room temperature. During deposition, the wafer temperature was maintained at 400-450 C. The starting pressure in the vacuum system was 8×10^{-9} torr, and the pressure during the deposition of the fluoride film was approximately 1×10^{-7} torr. The deposition rate was approximately 1 A per sec. Some of the wafers were covered by 300 A of electron beam evaporated ZrO_2 films, and some by 200 A of electron beam evaporated Si films as the capping layer. The thicknesses of the deposited films were characterized by ellipsometry as well as with Dektak profilometer. Aluminum was used for both top and bottom electrodes for the MIS capacitor.

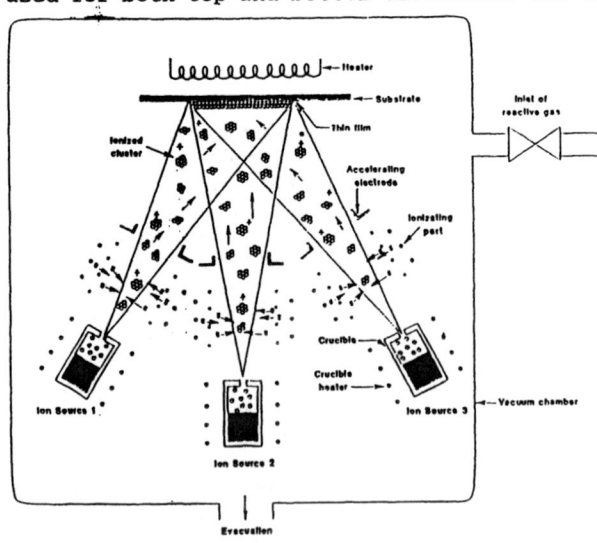

Fig.1 Ion Assisted Deposition System used for the deposition of BMF.

RESULTS AND DISCUSSION

The surface morphology of the as deposited BMF film was smooth, and no flaking or cracking of the structure was observed. The BMF film was dissolved with dilute nitric acid (20%) leaving a smooth Si surface, suggesting that there is negligible interaction of BMF film on silicon. Fig.2 shows the x-ray diffraction spectra of the BMF film (without cap and top electrode) and that of the charge in the crucible, determined by Scintag automated x-ray diffractometer. The structural analysis of the x-ray diffraction pattern of the film shows that the peaks correspond to orthorombic phase with a=4.130 A, b=5.819 A, and c=14.510 A and this was close to the value for charge in the crucible as well data reported in the literature (11).

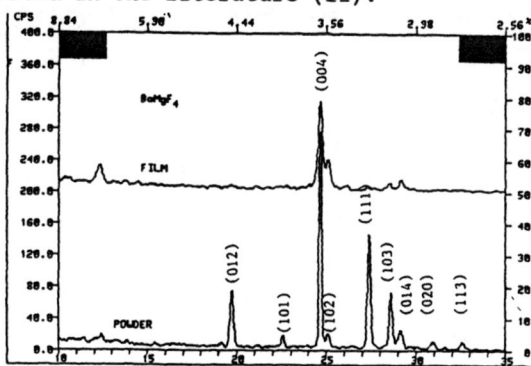

Fig.2 X-ray diffraction spectra of BMF. Film without cap and top electrode.

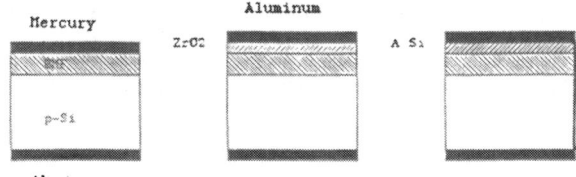

Fig.3 MIS device Structures used for the electrical characterization of BMF.

The device structures used for the electrical characterization of the ferroelectric film are shown in fig. 3. The capacitance vs voltage (C-V) characteristics of the MIS capacitor with mercury top electrode in Fig.4 clearly show regions of accumulation, depletion and inversion. The direction of hysteresis corresponds to ferroelectric polarization, and the shift in threshold voltage is 3.5V for bias voltage change from +5V to -5V.

Various top electrode materials like Au, Pt and Al for the MIS structures were deposited by vacuum evaporation. The adhesion of the electrode materials on these fluoride films was found to be poor. In addition to this, the top electrodes could not be patterned photolithographically on these films because they were found to dissolve in distilled deionized water as well as in positive photoresist developer. Therefore, various capping layers like ZrO_2 and amorphous silicon have been deposited on the BMF film. The capacitance-voltage curves of the MIS capacitors with ZrO_2 and amorphous Si as the capping layer is also shown in fig.4. The shift in threshold voltage did not change significantly due to the incorporation of a capping layer. The shift in the position of C-V characteristics for various device structures is due to the difference in the effective work function of the capping layers.

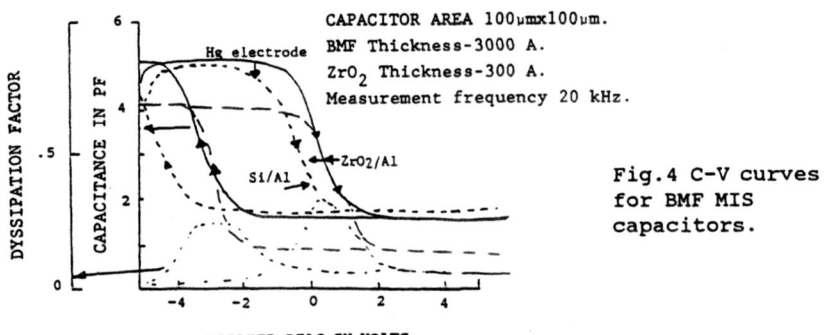

CAPACITOR AREA 100µmx100µm.
BMF Thickness-3000 A.
ZrO_2 Thickness-300 A.
Measurement frequency 20 kHz.

Fig.4 C-V curves for BMF MIS capacitors.

The shift in threshold voltage was found to depend on the bias voltage. Even for a bias voltage change of +3V to -3V, the threshold voltage change was very significant (2V). This shows that this material can be used for implementing non-volatile memory where information can be written at a voltage as low as 3v which is significantly lower than the present day MNOS based non-volatile memories.

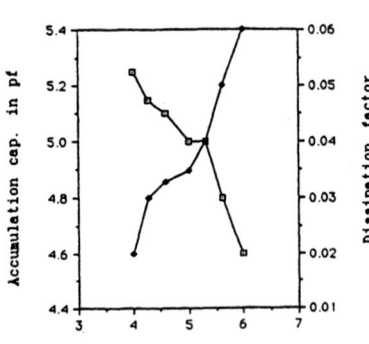

Fig.5 Variation of accumulation capacitance and dissipation factor with frequency.

Fig.5 shows the variation of the accumulation capacitance and dielectrc loss with frequency for BMF MIS capacitors. The accumulation capacitance decreases with increase in frequency and the diectric loss was found to increase with increase in frequency. The increase in dielectric loss might be due to various types of polarization taking place in the insulating layer. This effect has been observed in bulk single crystals of BMF (11).

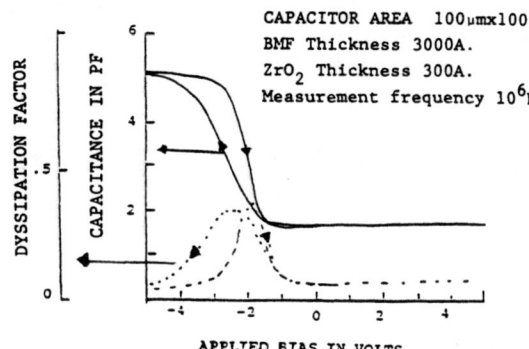

CAPACITOR AREA 100μmx100μm.
BMF Thickness 3000A.
ZrO_2 Thickness 300A.
Measurement frequency 10^6Hz

Fig.6 C-V curves for
Al/ZrO_2/BMF/ p-Si
MIS capacitors
at 80 C.

Fig.6 shows the C-V curves for Al/ZrO_2/BMF/p Si at a temperature of 80 C. With an increase in temperature, the shift in threshold voltage was found to decrease. This may be due to the loss of the polarization charges due to ionic conduction in fluorides. Fig.7 shows the current-voltage (I-V) curves for the MIS capacitors at various temperatures. The leakage current in the MIS capacitors was found to increase with increase in temperature. Preliminary analysis of the I-V curves shows that at low voltages the leakage current is due to Frenkel-Poole emission and at higher voltages, the leakage current is due to ionic conduction.

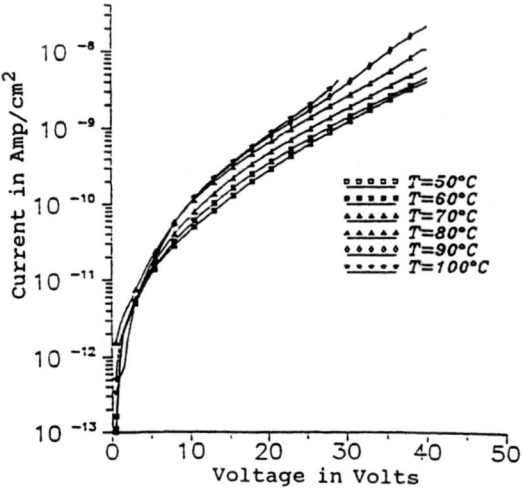

T=50°C
T=60°C
T=70°C
T=80°C
T=90°C
T=100°C

Fig.7 Current-voltage
characteristics of BMF
MIS capacitors at
various temperatures.
Amp/cm^2

CONCLUSIONS

The MIS characteristics of ion-assisted deposition of BMF films show hysteresis in the C-V characteristics, and the direction of hysteresis corresponds to ferroelectric polarization. The shift in threshold voltage does not change significantly due to the incorporation of a ceramic capping layer included to protect the BMF film. The hysteresis in the C-V characteristics depend on bias voltage as well as temperature.

ACKNOWLEDGEMENT

The authors are grateful to Mitsubishi Corporation for providing ICB facilities.

REFERENCES

1. S.S. Eaton, D.B. Butler, M. Parris, D. Wilson and H. McNeille, Proc. Intl. Solid State Circuits Conf., San Fransisco, CA, Feb. 1988.
2. J.F. Scott, L. Kammerdiner, M. Parris, S. Trayner, V. Ottenbachar, A. Shawbkeh and W.C. Oliver, J. Appl. Phys.64(2),787 (1988).
3. C.A. Araujo and G.W. Taylor, Ferroelectrics, vol.116, 215 (1991).
4. Yoshoi Nishi and Hisakazu IIzuka, Silicon Integrated Circuits, part A, 121 (1981).
5. T.S. Kalkur, George Argos and L. Kammerdiner, MRS Proceedings, vol.200, 313 (1990).
6.T.S. Kalkur, J.R. Kulkarni, Y.C. Lu, M.Rowe, W. Han and L. Kammerdiner, Ferroelectrics, vol.116, 135 (1991).
7. N. Shohata, S. Matsubara, Y. Miyasaka and M. Yonezawa, Proc. of 6th IEEE Intl. Symp. Appl. Ferroelectric, edited by V.E. Wood, New York, 580-584 (1986).
8. T.S. Kalkur, J.R. Kulkarni, R.Y. Kwor, L. Levinson and L. Kammerdiner, presented in 3rd Integrated Ferroelectrics Conference, Colorado Springs, 1991.
9.M.H. Francombe and S.V. Krisnaswamy, MRS Proceedings, vol.200, 179 (1990).
10. H. Buhay, S. Sinharoy, W.H. Kasner, M.H. Francombe, D.R. Lampe and E. Stepke, Proc. 7th Symp. on Applications of Ferroelectrics, University of Illinois at Urbana-Champion, June 6-8 (1990).
11. M. DiDomenico, Jr., M. Eibschutz, H.J. Guggenheim and I. Camlibel, Solid State Communications, 7 , 1119 (1969).

CHARACTERIZATION OF LEAD ZIRCONATE-TITANATE THIN FILMS PREPARED BY PULSED LASER DEPOSITION

C.K. CHIANG, W. WONG-NG, P.K. SCHENCK, L.P. COOK, M.D. VAUDIN

National Institute of Standards and Technology
Gaithersburg, MD 20899

P.S. BRODY, B.J. ROD, and K.W. Bennett

Harry Diamond Laboratories
Adelphi, MD 20783

ABSTRACT

Dense smooth lead zirconate-titanate thin films have been prepared by excimer laser deposition. The as-deposited films are amorphous as indicated by x-ray powder patterns. Differential scanning calorimetry studies show that the film has a glass transition at 301°C, and the amorphous to crystalline transformation takes place above 350°C to 650°C. Phase formation as a result of post-deposition heat treatment is described.

INTRODUCTION

For electrical and electronic applications of ceramic materials, we have applied pulsed laser deposition to produce various thin films, including barium titanate and lead zirconate-titanate (PZT).[1-3] The PZT films produced in our laboratory using a Nd/YAG laser showed ferroelectric hysteresis, indicating potential for application as non-volatile memories and related devices. However, these films had unsuitable microstructure, including large ejected particles and rough surfaces.[2] One cause of this morphology may be the thermal energy produced by the Nd/YAG laser during the sputtering. To improve these films we have since used an excimer laser, and in this paper we report the effect of post-depositional annealing on PZT films deposited by pulsed excimer laser deposition.

EXPERIMENTAL

The thin film deposition system was the same as reported earlier[1], except that we used an ArF excimer laser (193 nm). The deposition was done at a repetition rate of 10 Hz and nominal pulse width of 23 ns. The laser beam was focused to produce a fluence of typically 10-30 J/cm^2. A commercial PZT target (PZT 5A, 47% PbTiO$_3$ and 53% PbZrO$_3$) was used. During the deposition, the target was rotated and the laser beam was rastered across the target to maintain uniform material removal. A rotating substrate holder was used to hold substrates directly over the target surface at distances 3.0 cm from the target surface. The reaction chamber was continuously evacuated to <1 mTorr by oil free pumps. During deposition, oxygen was metered into the chamber to produce a background pressure of 100 mTorr. The growth rate of PZT film was about 50 nm/min. Typical thickness of the thin film was from 150 nm to 500 nm.

For electronic studies we used silicon < 100 > single crystal wafers as substrates which were coated with 500 nm platinum. For the materials processing studies, we used platinum substrates. After deposition, the films were annealed in air at various temperatures and times.

To determine the crystallization temperatures of the as-deposited PZT films differential scanning calorimetry (DSC) measurements were made. A platinum substrate similar to that upon which the film was deposited was used as the reference. DSC curves were measured at 20°C/min in flowing air. Each sample was measured for two cycles from 50°C to approximately 720°C. In one measurement, about 0.6 mg of as-deposited material was removed from the substrate mechanically and run in the DSC using an aluminum pan as the sample holder. The use of low mass aluminum pans for the sample holder and reference increased sensitivity and allowed us to detect the exact location of the glass transition temperature.

For an amorphous material, crystallization will not occur if it is annealed at a temperature less than its glass transition temperature. The DSC data provided us the guide for selecting the annealing temperature for the PZT film. We have annealed a series of the as-deposited amorphous PZT films in air for one hour at 350°C, 400°C, 450°, 500°, 550°C and 600°C.

The annealed films were studied by x-ray powder diffraction. The x-ray diffraction patterns were measured directly with the film in place on the platinum or silicon substrates, using Cu $K\alpha_1$ radiation. Because the x-ray measurements were done under the same conditions, the peak-heights could be used for a semi-quantitative analysis of the relative amounts of phases. We measured the peak-heights of selected major peaks to indicate variations in the ratio of the phases present as a function of annealing conditions. The different peak-heights were re-scaled relative to the intensity of the (101,110) peak for comparison.

Optical microscopy was used for examining the gross features of the films and scanning electron microscopy was used to examine the microstructure of the films. The hysteresis loop was measured using a balanced Sawyer-Tower circuit operated at 10 kHz. The electrode size for the hysteresis loop measurement was 5.0×10^{-4} cm^2. Extensive characterization on these films has been done, with details of the electrical characterization reported elsewhere.[3-4]

RESULTS

The as-deposited films were amorphous, as determined by their x-ray diffraction pattern which showed an amorphous hump with no sharp crystalline peaks. The films are of uniform thickness of approximately 400 nm over a region of 1 cm and have clear interference patterns indicating that they are transparent at this thickness. Optical microscopy of the films showed a smooth surface.

The DSC signal from the sample is very small due to the small amount of available material from the thin film samples. Since the amorphous to crystalline transition is a non-reversible transformation, only the initial heating trace contains the phase transformation information. The signals from the subsequent cooling trace or the second heating trace, or any other runs after that, showed only a smooth variation with temperature. The second heating trace of the same film reflects only the heat capacity of the film and the substrate. Hence, the second heating trace could be used as the background for analysis of the transition of the film. The background subtracted DSC traces of the as-deposited PZT film are shown in figures 1 and 2.

Figure 1 shows the initial heating trace of the PZT film on platinum. On heating at 20°C/min from room temperature to 720°C, the thin film on the platinum substrate showed a broad exothermal peak. This peak is due to re-crystallization, which transforms the amorphous film to crystalline form. The broadness of the peak is partially due to the heating rate which is high compared with the rate of crystallization. Nevertheless, the thermal process is completed at approximately 670°C. Examining figure 1 closely, one could identify T_g in figure 1, although it is not very clear.

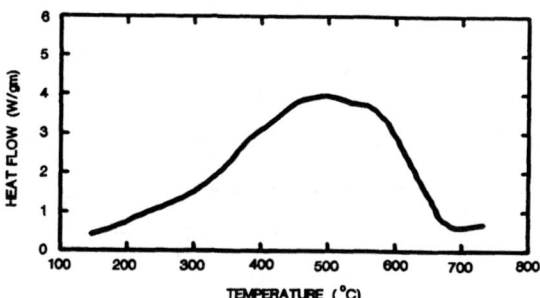

Figure 1   DSC trace of an as-deposited PZT film on a Pt substrate.

Figure 2 shows the trace obtained from the PZT film removed from the platinum substrate. The glass transition of the amorphous film is clearly shown as a step in the curve. It is identified with two parallel lines in the figure 2. From this data we determine the mid-point of the glass transition temperature of the film is 301°C.

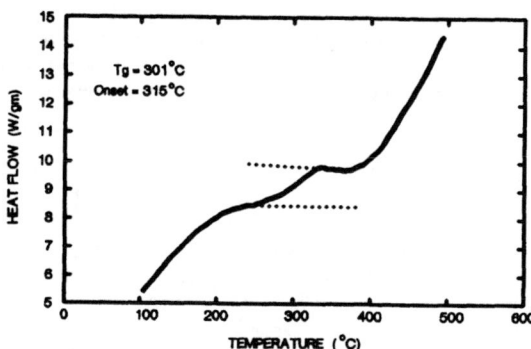

Figure 2   DSC traces of PZT material removed from an as deposited film. The parallel lines indicate the location of T_g.

The x-ray patterns of the PZT films after annealing at various temperatures are shown in figure 3. The 357°C pattern shows only an amorphous hump and the peaks from the platinum substrate. The patterns of 398°C, 450°C and 500°C shows that the crystalline peaks increase with increasing annealing temperature. The amorphous hump decreases at the same time.

These peaks are identified as the pyrochore phase of PZT. The volume fraction of the pyrochore phase reached a maximum at 500°C. At 550° and 602°C the x-ray patterns show nearly the same strength of the crystalline peaks. These patterns were identified as the perovskite phase of PZT[5]. At 500°C there is small amount of perovskite phase, but the pyrochore phase is not seen in the 550°C pattern.

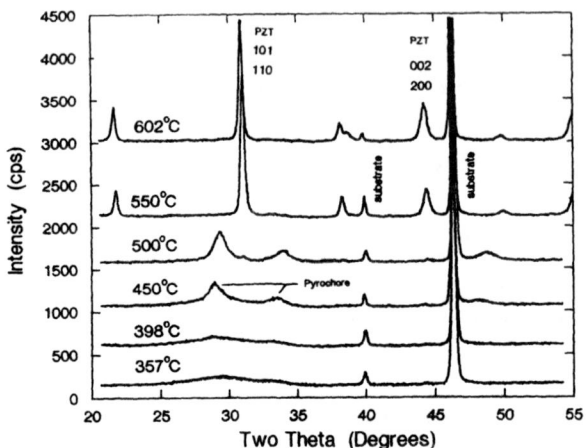

Figure 3 The x-ray diffraction patterns of the PZT films heat-treated at different temperatures.

Figure 4 shows the formation of the pyrochore and perovskite phases as a function of the annealing temperature. In the figure the formation of the pyrochore phase is plotted using the intensity of the strongest peak ($2\Theta = 29°$). The formation of perovskite phase is plotted using the intensity of both the $(110,101)$ and $(200,002)$ peaks (figure 3). This result indicates that there are two types of crystals formed in two temperature regions. The two regions separated by a relatively short temperature region of less than 50°. The two crystalline regions were not revealed in the DSC studies.

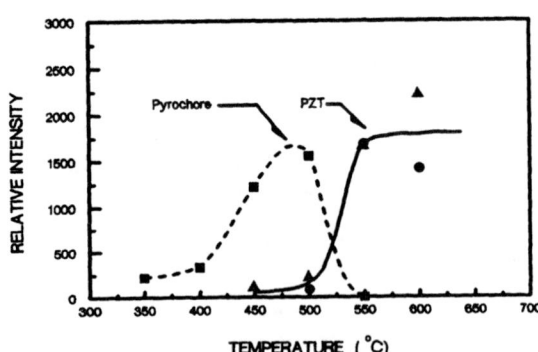

Figure 4 The formation of crystalline phases of PZT film as a function of isothermal heat-treatment.

These heat treated films have essentially no change in surface morphology examined with optical and electronic scanning microscopies. Figure 5 is a typical micrograph of the PZT film on platinum substrate. It can be seen that the smooth surface is decorated by many particles with diameters less then 1 μm.

Figure 6 shows a typical ferroelectric hysteresis loop obtained from an PZT thin film produced by excimer laser deposition. This film was about 180 nm in thickness. The maximum applied voltage was about 12 volts. The coercive field is approximately 80 kV/cm and the remanent polarization is approximately 14 μC/cm².

Figure 5   A SEM micrograph of a PZT film heat-treated at 550°C.

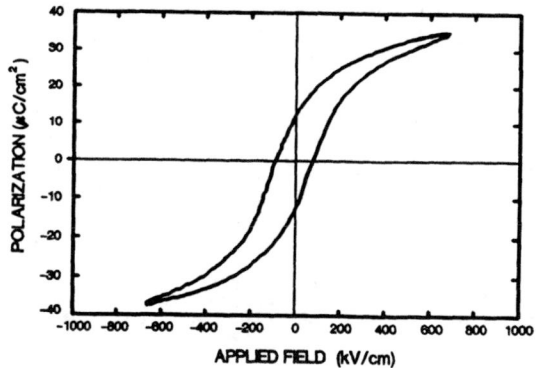

Figure 6   A typical hysteresis loop of a PZT film after post-depositional heat-treatment.

DISCUSSION AND CONCLUSION

The surface morphology of the excimer laser deposited films are significantly better than those of Nd/YAG laser[1,2]. Ejected particles are seen in both cases, but the number of particles seen in the excimer laser film is much less[4,6]. The ferroelectric properties of the eximer deposited film are also improved, the remanent polarization is almost a factor of two higher than in the Nd/YAG deposited film[2].

Using pulsed laser deposition to produce a PZT thin film, the as-deposited film is in an amorphous state if the substrate temperature is at ambient temperature. DSC data suggests that the film should remain in an amorphous state as long as the heat treatment temperature is less than its T_g or 301°C. The control of nucleation should be at a temperature slightly above T_g. Crystallization should occur if heat treated at a temperature higher than 400°C. However, the PZT processing should be designed to avoid the pyrochore phase which crystallized at temperature range between 400°C and 500°C. Also, the lead oxide in lead zirconate and its solid-solution is known to be volatile[5]. Care must be taken to avoid lead loss at temperatures higher than those reported here, especially at long annealing times. Because lower temperatures will be compatible with the processing of silicon substrates, it is concluded that the optimum processing temperature is in the vicinity of 550°C.

REFERENCES

1. P.K. Schenck, L.P. Cook, J. Zhao, E.N. Farabaugh, and C.K. Chiang, Proc. Symposium on Beam Solid Interaction: Physical Phenomena, ed. E.R. Myers and A.I. Kingon, Mat. Res. Soc., 157A, 587, (1990)

2. C.K. Chiang, L.P. Cook, P.K. Schenck, P.S. Brody, and J.M. Benedetto, "Ferroelectric Thin Films", Mat. Res. Soc., Vol. 200, 133, (1990).

3. P.S. Brody, B.S. Rod, J.M. Benedetto, K.W. Bennett, L.P. Cook, P.K. Schenck, C.K. Chiang and W. Wong-Ng, Proc. Seventh Intl Symp. on Applic. of Ferroelectrics, Univ. Ill, Urbana-Champaign, IL., (1990).

4. L.P. Cook, M.D. Vaudin, P.K. Schenck, W. Wong-Ng, C.K. Chiang, and P.S. Brody, p. 241-246 in C.V. Thompson, J.Y. Tsao, and D.J. Srolovitz, eds., Evolution of Thin-Film and Surface Microstructure, Mat. Res. Soc. Symp. Proc., vol. 202, (1991).

5. B. Jaffe, R.S. Roth, and S. Marzullo, J. Res. National Bur. Stds., 55, 239, (1955).

6. L.P. Cook, P.K. Schenck, J. Zhao, E.N. Farabaugh, C.K. Chiang and M.V. Vaudin, Ceramic Transactions: Ceramic Thin and Thick Films, ed. B.V. Hiremath, page 99, (1990).

RAMAN STUDIES OF STRESS-INDUCED PHASE TRANSFORMATIONS IN TITANIA FILMS

GREGORY J. EXARHOS AND NANCY J. HESS
Pacific Northwest Laboratory, PO BOX 999
MS K2-44, Richland WA 99352

ABSTRACT

Time-resolved micro-Raman spectroscopy is used to follow the amorphous to crystalline phase transformation in sol-gel deposited titania films induced thermally or through the action of applied hydrostatic pressure in a diamond anvil cell. Time-dependent phonon intensities intrinsic to the growing phase are related to the volume fraction of crystallite present at any time. The sigmoidally generated curves can be modeled in terms of modified Avrami ingrowth kinetics in which diffusion of the amorphous phase to the nucleation center is restricted by the morphology of the evolving phase. Phonon frequency and linewidth measurements during the course of the transformation probe changes in film stress and particle size which are used to understand the mechanistics of the transformation. Raman measurements also are used to derive a phase stability diagram for titania films.

INTRODUCTION

The kinetics and mechanistics of transformation phenomena in solids can be evaluated from inelastic light scattering measurements which probe not only the time dependent growth of the new phase, but also associated changes in localized chemical bonding both of which are driven by an imposed stress to the system. Previous studies of temperature- and laser-induced glass-crystal transformations in bulk oxide glasses demonstrated the utility of *in situ* Raman spectroscopy as a real-time diagnostic of the crystallization kinetics. [1,2] Raman measurements on thin film dielectrics provide an interesting challenge since the smaller amount of material present in the film and an increased signal from the substrate lead to greater difficulty in detecting the crystallizing phase. The availability of more sensitive detectors, and microprobe techniques which limit the probe depth of the excitation laser have made such measurements possible and allow time resolution on the order of milliseconds and spatial resolution on the order of micrometers. [3,4,5,6]

Laser Raman spectroscopy has been used to study thermally induced crystallization of amorphous TiO_2 films prepared using ion, electron beam, and sol-gel techniques. [7,8] These initial studies have established that deposition parameters and solution chemistry control both the rate of film crystallization and the phase of the crystallized material. Variations in processing conditions were proposed to influence the nature of the nucleation centers formed in the amorphous film during deposition. Thus, anatase, rutile, or mixed phases were found to evolve when samples were subjected to temperatures above 575 K. Film thickness, grain size, and inherent film stress also are predicted to influence the transformation dynamics. While thickness, and particle size effects on solid-solid phase transformations in thin films have been addressed [9,10], the stress dependence on the transformation and phase stability regions in T,P space are subjects which warrant further investigation.

Work reported here focuses on the influence of stress on the thermally-induced transformation of sol-gel deposited films. A spheroidal microstructure has been identified in such films following thermal annealing above 575 K, and measured rate constants for the transformation to anatase were found to depend upon solution mixing time and pH. [8]

Micro-Raman spectroscopy is used to follow the devitrification kinetics of amorphous titania films both free-standing and deposited on silica substrates. Raman intensities intrinsic to the growing crystalline phase obey a modified Avrami isotherm that incorporates diffusion constraints from which rate constants and critical exponents have been evaluated. Phonon frequencies and linewidths determined as a function of temperature and pressure for single crystal anatase hydrostatically constrained within a diamond anvil cell are correlated with measurements of the evolving anatase phase in isothermally annealed or high energy laser irradiated sol-gel films to evaluate transient stresses which develop during crystallization.

EXPERIMENTAL

Amorphous titania films (100 nm thick) were prepared from low pH alcoholic solutions containing titanium tetra-ethoxide using methods described previously. [11] Films were deposited on cleaned silica substrates using spin deposition methods or on silica fragments (50 micrometer diameter) which could be inserted into the diamond anvil cell for film evaluation at pressures to 100 Kbar. Following deposition, films were dried at room temperature in air. Films also were removed from the substrate for several measurements.

Isothermal growth of an anatase or rutile phase in sol-gel deposited films was followed by measuring vibrational line intensities at 143 cm^{-1} or 440 cm^{-1} respectively as a function of time. Measurements were performed using Spatially and Time Resolved Raman Spectroscopy (STRRS) to probe thin film samples subjected to four specific experimental conditions: (i) elevated temperature; (ii) elevated pressure; (iii) simultaneously applied temperature and pressure; and, (iv) high laser fluence irradiation at 1 bar and 298 K. All Raman spectra were excited using 50 mW of 514.5 nm excitation from an Ar^+ ion laser focused to a 20 micrometer diameter spot onto the sample by means of a 10X objective in the Raman microprobe (SPEX Model 1482). In case (iv) 100 mW of focused radiation was used. Backscattered light was directed to the entrance slits of a 0.5 m triple spectrometer (SPEX Model 1877) equipped with 1200 and 600 lines/mm gratings in the dispersive and filter stages respectively. Dispersed radiation was detected by means of a liquid nitrogen cooled CCD (PI Model 1433-C) detector. Spectra were acquired periodically while the amorphous films crystallized.

Experiments at elevated pressure were performed on free films and coated silica fragments using a Merrill-Bassett diamond anvil cell [12] containing KCl as a pressure medium. Some measurements were performed without a pressure medium for correlation with previous work where isothermal crystallization was followed only as a function of temperature. [8] Initial measurements using the standard hydrostatic pressure medium (4:1, methanol-ethanol mixture) resulted in solvent-film photo-induced chemical reaction even at relatively low laser powers. The pressure-induced fluorescence frequency shift of Sm:YAG was used as the pressure sensor for experiments in the diamond anvil cell. [13,14]

RESULTS

Raman spectra of the amorphous, anatase, and rutile phases of TiO_2 films deposited on silica substrates are illustrated in Figure 1. The rutile and anatase phases are easily distinguishable from one other; the Raman spectrum of amorphous TiO_2 exhibits features similar to that of the rutile phase, however, linewidths have increased by a factor of three and the line intensity has decreased by an order of magnitude. Correlation of this data with spectra of amorphous TiO_2 films at different temperatures and pressures is used to establish stability regions in P,T space for phases which evolve from the sol-gel deposited films.

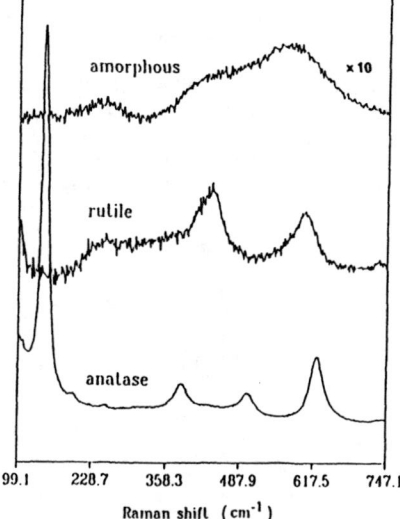

Figure 1. Raman spectra of
TiO$_2$ phases.

Raman spectra of a single crystal of anatase obtained from the Grao Mogol locality in Minas Gerais, Brazil were acquired as a simultaneous function of temperature and pressure in a heated diamond anvil cell. The frequency and linewidth variation in T,P space is illustrated in Figure 2 for the low frequency E$_g$ mode at 143 cm^{-1}. In this T,P regime both frequency and linewidth increase with temperature and with pressure. The anomalous increase in resonance frequency with temperature could be associated with the anatase to rutile transformation which proceeds above 1075 K. These results for the single crystal can be correlated with measured variations in frequency and linewidth which accompany ingrowth of the crystalline phase during isothermal crystallization of the amorphous film. Based upon this correlation, transient changes in film stress can be evaluated.

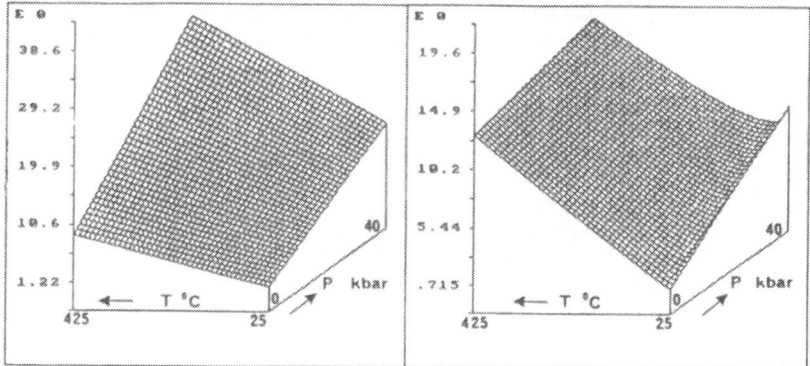

Figure 2. Variation in linewidth and frequency for single crystal anatase as a function of both temperature and pressure for the 143 cm^{-1} mode.

Previous work [8] summarized in Figure 3 demonstrates how Raman line intensity measurements are used to follow the crystallization kinetics of sol-gel films. Here, time-dependent intensities are related to the fraction of material transformed. The sigmoidal crystal ingrowth behavior is well represented by an empirical Avrami equation having a critical exponent of about 2. This relatively low value for three-dimensional crystal growth can be understood in terms of a mass-limited diffusion model in which the dense crystal nuclei pull-away from the amorphous matrix during growth thereby restricting the transport of matrix material to the nucleation center. [15] Rate constants determined at different temperatures indicate an activation energy for crystallization approaching 142 kJ/mole.

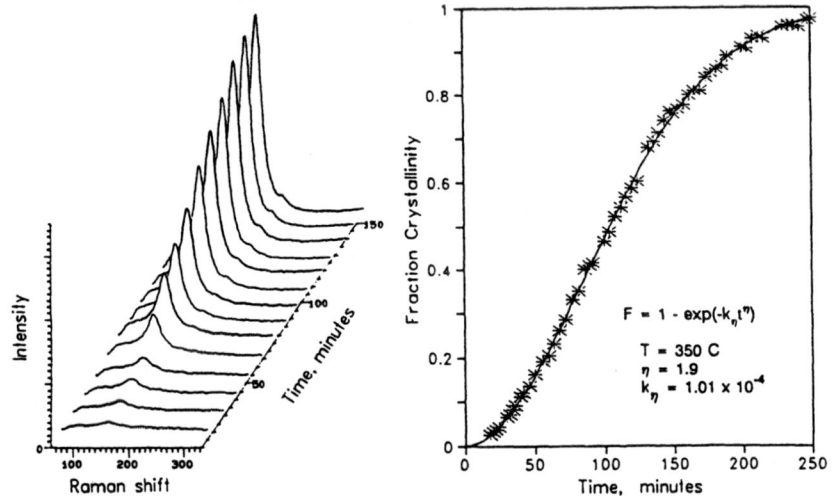

Figure 3. Isothermal crystallization kinetics of sol-gel TiO_2 films at 625 K.

Amorphous sol-gel films exposed to an applied hydrostatic pressure in a diamond anvil cell transform irreversibly to anatase or rutile phases depending upon the temperature and the magnitude of the applied pressure. Figure 4 illustrates regions of phase stability for these films in P,T space determined from the measured Raman spectra. Amorphous films were found to crystallize spontaneously upon application of pressure in excess of 15 kbar.

Figure 4. Regions of phase stability for sol-gel deposited titania films on silica.

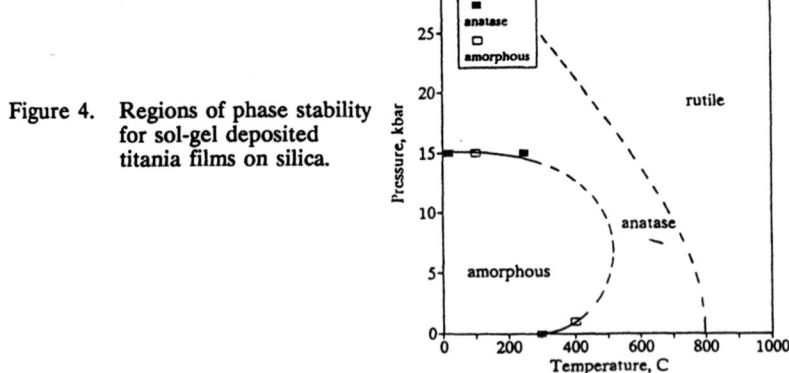

CW laser irradiation of amorphous sputter-deposited titania films is known to induce crystallization to anatase, rutile or mixed phases dependent upon the laser power density and the exposure time. [7] Based upon the phase stability map shown in Figure 4, irradiation at moderate power levels (*ca* 100 mW) should be sufficient to induce the amorphous → anatase transformation in these stressed films. Time-resolved Raman spectra of a sol-gel film exposed to a moderately high incident laser power have been measured. Figure 5 depicts the intensity variation of the 143 cm^{-1} anatase feature with time along with measured changes in the mode frequency and linewidth. The data presented in Figure 2 for single crystal anatase can be used to interpret the thin film measurement. During the first few minutes of irradiation, the film temperature increases and crystallization is initiated. Observed changes in the vibrational frequency and linewidth correlate with film heating which soon reaches a steady state condition. At later times, both frequency and linewidth decrease indicating that as the transformation is driven to completion the initial stress in the film relaxes. This example demonstrates how Raman spectroscopy is used to follow both the ingrowth of crystalline phases in amorphous films and to evaluate changes in the magnitude of film stress as it transforms to the denser phase.

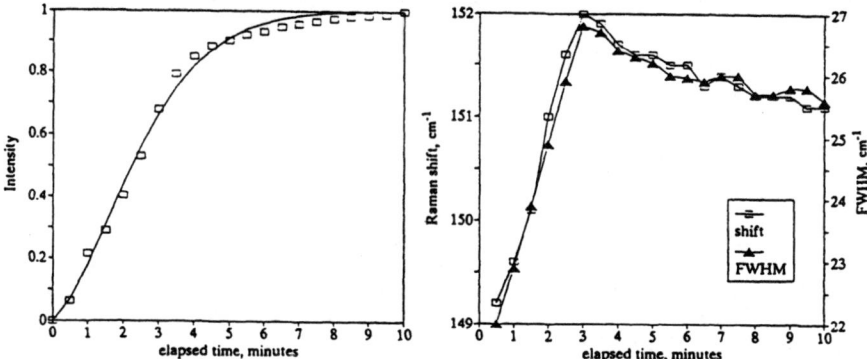

Figure 5.   Time dependent variation in mode intensity, frequency and linewidth for the 143 cm^{-1} E$_g$ mode intrinsic to anatase which evolves during laser-induced crystallization of an amorphous sol-gel titania film.

DISCUSSION

In situ measurements of the transformation dynamics in amorphous sol-gel films are readily performed by means of laser Raman spectroscopy. In addition to characterization of TiO$_2$ phases and determination of the time-dependent growth of a particular crystalline phase, variations in vibrational mode frequencies and linewidths can be used to evaluate transient changes in the inherent film stress. Measurements also are used to establish phase stability regions in P,T space. The stability boundaries in this system are very much dependent upon film deposition parameters. She, *et.al.* have shown for sputter-deposited amorphous titania films that the nature of the crystalline phase which evolves during thermal annealing of these films depends upon the technique used to deposit the film. Contrasting behavior of films deposited using different methods was explained by the presence of specific nucleation centers introduced into the film during deposition. The nucleation centers were thought to specifically nucleate either the anatase or rutile phase, and their relative number would control the final phase composition of the crystallized film.

Work reported here shows that intrinsic film stress also can be a controlling factor in the isothermal crystallization of amorphous titania films. Heterogeneity in the stress distribution can lead to growth of either anatase or rutile phases in the temperature regime where these experiments have been performed. At temperatures near 1075 K, rutile is the thermodynamically stable phase and persists even upon cooling. [4] Relative differences in the growth rates for these two crystalline phases and stress inhomogeneity can account for the presence of mixed phases in thermally processed films. Raman microprobe spectroscopy is an effective method by which to quantify phase inhomogeneity in thin films and the influence of localized stress on solid state transformations.

ACKNOWLEDGEMENTS

This work has been supported by the Materials Sciences Division Basic Energy Sciences, U.S. Department of Energy. Pacific Northwest Laboratory is operated by Battelle Memorial Institute for the U.S. Department of Energy under Contract DE-AC06-76 RLO 1830.

REFERENCES

[1]   G.J. Exarhos and Wm. M. Risen, Jr., J. Am. Ceram. Soc. **57(9)**:401 (1974).
[2]   E.I. Kamitsos, M.A. Karakassides, A.P. Patsis, and G.D. Chryssikos, J. Non-Crys. Sol. **116**:115 (1990).
[3]   G.J. Exarhos, Mat. Res. Soc. Proc. **48**:461 (1985).
[4]   G.J. Exarhos, and P.L. Morse, Pro. Soc. Photo-Opt. Instrumen. Eng. **540**:460 (1985).
[5]   C.Y. She, and L.S. Hsu, in *Laser Induced Damage in Optical Materials:1986*, Ed. H.E. Bennett, A.H. Guenther, D. Milam, and B.E. Newnam, NBS Sp. Pub. **746**:383 (1988).
[6]   G.J. Exarhos, and D.M. Friedrich, Microbeam Anal. **23**:147 (1987).
[7]   L.S. Hsu, R. Rujkorakarn, J.R. Sites, and C.Y. She, J. Appl. Phys. **59(10)**:3475 (1986).
[8]   G.J. Exarhos, and M. Aloi, Thin Solid Films **193/194**:42 (1990).
[9]   V.D. Das, and D. Karunakaran, J. Phys. Chem. Solids **46(5)**:551 (1985).
[10] J.R. Sambles, J. Phys. Chem. Solids **46(5)**:525 (1985).
[11] W.S. Frydrych, G.J. Exarhos, K.F. Ferris, and N.J. Hess, Mat. Res. Soc. Symp. Proc. **121**:343 (1988).
[12] L. Merrill, and W. Bassett, Rev. Sci. Instru. **45**:290 (1974).
[13] N.J. Hess, and G.J. Exarhos, High Press. Res. **(2)**:57 (1989).
[14] N.J. Hess, and D. Schiferl, J. Appl. Phys. **68(5)**:1953 (1990).
[15] S.Z.D. Cheng, and B. Wunderlich, Macromol. **21**:3327 (1988).

OXIDATION KINETICS OF YBa$_2$Cu$_3$O$_{7-x}$ THIN FILMS IN THE PRESENCE OF ATOMIC OXYGEN AND MOLECULAR OXYGEN BY IN-SITU RESISTIVITY MEASUREMENT

K. Yamamoto
Central Reserch Laboratories, Kanegafuchi Chemical Industry, Co. Ltd.
2-8-,1-chome, Yoshida-cho, Hyogo-ku, Kobe652, Japan
B.M. Lairson, J.C. Bravman and T.H. Geballe
Stanford University, Stanford CA94305

ABSTRACT

The kinetics of oxidation in YBa$_2$Cu$_3$O$_{7-x}$ thin films in the presence of molecular and atomic oxygen ambients have been studied. The resistivity of c-axis, a-axis, and mixed a+c axis oriented films, deposited in-situ by off-axis magnetron sputtering, was measured as a function of time subsequent to a change in the ambient conditions. The oxidation process is shown to be thermally activated and can be characterized by a diffusion model with an activation energy which varies from approximately 1.2eV in the presence of molecular oxygen to 0.6eV for a flux of 2x10^{15} oxygen atoms/cm^{2}sec. In both cases, diffusivity is found to be insensitive to oxygen stoichiometry, but the rate of oxidation is found to be sensitive to the microstructure and orientation of the films.

1 Introduction

To study oxygen diffusion in YBa$_2$Cu$_3$O$_{7-x}$ (hereafter referred to as YBaCuO or YBaCuO$_{7-x}$, where x is the oxygen deficiency), various relaxation monitoring techniques, such as conductivity[1], thermogravimetry[2] and SIMS profiles of tracers[3] have been used. From oxygen tracer diffusion studies of both polycrystalline and single crystal YBaCuO[4], it has been proposed that oxygen diffusion is highly anisotropic, as might be expected from the crystal structure, with diffusion in the a-b plane much faster than diffusion parallel to the c-axis. In contrast to the case for bulk materials, there are very few reports of oxygenation in thin films, especially in-situ films. Some results for the oxidation of thin films at low temperature using an oxygen plasma have been reported. In these processes, atomic oxygen plays an important role[5]. The importance of atomic oxygen during growth is also reported by several workers[6]-[9].

To investigate the oxygenation process, we report in this paper systematic in-situ conductivity measurements similar to those reported by Tu, et al.[1] Several films, including those with a-axis, mixed a+c-axis and c-axis orientations deposited on several substrates (MgO, LaAlO$_3$, yttria stabilized zirconia (YSZ)), prepared by off-axis magnetron sputtering, were used to study both the orientation dependence and microstructural dependence of the oxygenation process. Both molecular oxygen and atomic oxygen were employed to understand the various processes involved in the oxidation of thin films.

2 Experimental Results and Discussion
2-1 Experimental

In-situ YBaCuO films were deposited on several single crystal substrates by off-axis magnetron sputtering[10]. Some properties of films used in this study are listed in Table 1. The general characteristics of YBaCuO films deposited in this manner have been established by many techniques, details of which are published elsewhere[11].

In-situ conductivity measurements during oxidation were performed using 4-point DC transport. Electrical contacts were made either by using gold wire with Ag paste (with an anneal at 500°C) or by wire bonding onto evaporated gold. The contacts were tested for ohmic behavior before the in-situ measurements were made. A schematic of the general experimental arrangement is shown in Figure 1.

Oxidation of the films was achieved by exposing the films to either ambient molecular oxygen or an oxygen plasma at a fixed substrate temperature. Increments in the

Table1 (In-situ grown films used in this study)

Substrate	T_c	Texture	Thickness	Comments
MgO	~86	c-axis	3000	(1)
"	~85	c-axis	2000	(1)
"	~89	c-axis	4000	(1)
"	~87	c-axis	8000	(1)
LaAlO3	~76	a-axis	4000	(2)
"	~84	a(50%)+c	3000	
"	~86	c-axis	3000	
YSZ	~81	c-axis	3000	(3)

1) Sharp resistive transition (width < 1.5K). 2) Broad resistive transition (90K-76K).
3) Large mosaic spread measured by XRD (~2°). Broad resistive transition(85K-81K).

ambient oxygen pressure were achieved by changing the pumping speed; typically the oxygen pressure could be changed from 10mTorr to 100mTorr in 10-20 sec.

A compact ECR plasma apparatus was used to produce atomic oxygen, as shown schematically in Figure 1. The atomic oxygen flux was measured by the oxidation rate of Ag on a standard quartz thickness monitor, where measuring the initial oxidation of the silver determines a lower limit on the atomic oxygen flux. For these conditions, an atomic oxygen flux of 2×10^{15} atom/cm^{2}sec was measured, which is a typical flux for measurements reported below. The details of this method for determining atomic oxygen flux are discussed elsewhere[12],[13].

2-2 Kinetics of Oxygenation in Molecular Oxygen

The kinetics for oxygenation of the sample were monitored as follows. The thin film sample was initially brought to a fixed temperature-pressure point T_0-P_0 in the YBaCuO phase diagram[2]. The sample was allowed to equilibrate at this point, as monitored by the conductance of the sample. After the conductance of the sample was saturated, the pressure was incremented to a new value, P_1, at time t_0. A typical pressure increment for this experiment was one-half or one decade of pressure. The conductivity was then measured at fixed time intervals, typically until the conductance had again saturated. The temperature was held fixed at T_0.

The observed changes in conductivity are attributed to changes in the oxygen content of the films, as has previously been reported in ceramic YBaCuO[1] and in single crystals[4]. The speed with which the film reaches a new equilibrium oxygen content is controlled by the kinetic limitations in the system.

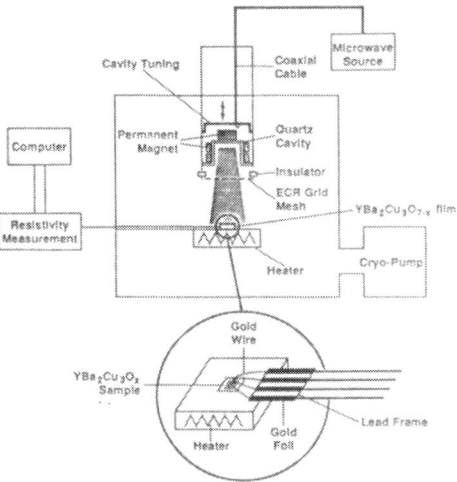

Figure 1: A schematic of the general experimental arrangement for the vacuum system.

In bulk ceramics[1] and in single crystals[4], the limiting kinetic process has been taken to be the bulk diffusion of oxygen. We assume here that, for changes in the ambient oxygen pressure, diffusion of oxygen inside the films is similarly the rate limiting process.

The simplest model for diffusion which might be applicable in the case of these thin films is one-dimensional diffusion governed by Fick's Law, with the film assumed to be homogeneous, but not necessarily isotropic. If the chemical diffusivity D is approximately constant vs. oxygen content C in the film, the diffusion equation can be solved by imposition of the appropriate boundary conditions.

For small changes in the applied molecular oxygen pressure, we assume a locally linear relationship between oxygen content and the conductivity. (We consider larger changes in the oxygen content in the presence of a plasma[14].) For a locally linear relationship, the solution[15] for the sample conductance S becomes

$$S(t) = S_\infty - \Delta S \sum_{n=0}^{\infty} \frac{1}{(2n+1)^2} \exp\left\{\frac{-D(2n+1)^2\pi^2 t}{4\ell^2}\right\}$$

(1)

In this equation, S_∞ is the conductance of the sample after infinite time, ΔS is the difference between the conductance at t_0 and the saturation value, and D is a measure of the relaxation time, interpreted as the diffusivity, and ℓ is the thickness of the film. We use the first two terms in the sum in Equation 1 to perform least-squares fitting to the data.

Figure 2 illustrates a fit of Equation (1) to experimental data for a 3000Å c-axis film on MgO at 480°C, for a pressure change at time t_0 from 10mTorr to 100mTorr. In this case, for $D=3.8\times10^{-14}$, the fitting is satisfactory, which is representative of fits to data at temperatures >450°C.

Figure 3 illustrates D vs. oxygen pressure for a film at two different temperatures. The pressure on the x-axis is taken as the final value of the oxygen pressure based on the phase diagram[2]. The difference between initial and final oxygen pressures is on the order of the spacing between the points. The oxygen deficiency measured ranges from x=0.27 to x=0.48 at 450°C and from x=0.45 to x=0.77 at 560°C. As can be seen in the figure, very little dependence of the diffusion constant on oxygen pressure is observed in the case of these thin films. This agrees with data from Rothman[3] showing that the diffusion coefficient is not a strong function of oxygen partial pressure from 3×10^3 to 1×10^5 Pa at 600°C (with $D=1\times10^{-10}$cm²/sec). The result shown by Figure 3 indicates that we can largely neglect the actual oxygen stoichiometry of the films when measuring diffusivity at various temperatures. The values of D vs. temperature reported below are therefore based on measured relaxation at a variety of oxygen pressures, with no particular attention to the exact equilibrium oxygen stoichiometry of the film, save that all of the conditions are for the region of the phase diagram in which YBaCuO is stable.

Diffusion coefficients vs. temperature were determined for c-axis oriented films on LaAlO₃, YSZ and MgO, for a-axis films on LaAlO₃, and for a mixed a+c oriented film on LaAlO₃, for film thicknesses between 2000Å and 8000Å. Figure 4(a), (b), (c) illustrates diffusivity vs. temperature for various pure c-axis oriented films on MgO and on YSZ(poorer microstructure), an a-axis film on LaAlO₃, and mixed a+c oriented film on LaAlO₃, respectively.

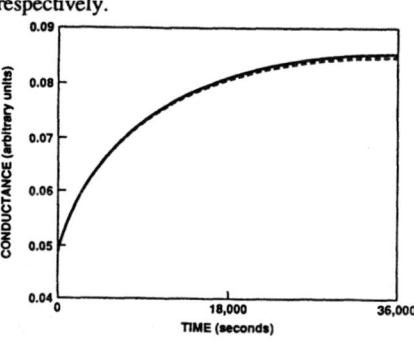

Figure 2: The time dependence of the conductance (dotted line) and a fit of Equation 6 to experimental data (solid line) for a c-axis oriented film on MgO at a constant temperature of 480 °C.

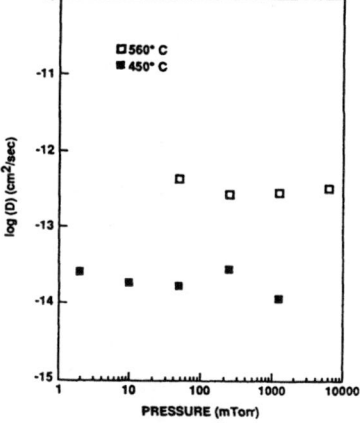

Figure 3: Diffusivity vs oxygen pressure at 560 °C (open squares) and 450 °C (filled squares). Each diffusivity was calculated for a molecular oxygen pressure increment of half decade of pressure. The pressure given on the x-axis is the final value of the oxygen pressure.

Data for c-axis films on LaAlO₃ and MgO lie very close to one another, while the diffusivity of the c-axis oriented film on YSZ is observed to be much higher. The dotted lines in the figure represent the measured diffusivity in the c-axis and in the a-b axis of single crystals[4]. The diffusivity of an a-axis film is observed to be approximately 20 times higher than that observed in c-axis oriented films on MgO. It is not certain how directly comparable the diffusivity values obtained from this film are to a-axis diffusivity in pure bulk material or to the c-axis films. The a-axis films of the type discussed here are somewhat (several percent) rich in barium, and are observed to have broad superconducting transition widths (~10K). The measured diffusivity of mixed a+c oriented film lies very close to that determined for the a-axis oriented films and the c-axis oriented film on YSZ.

A notable feature in the data for the various films is that the activation energy determined from D vs. 1/T is very similar for all of the films, approximately 1.2 eV, and that this value is quite similar to the values reported for diffusion in bulk ceramics[1,3]. This indicates that the mechanism for oxygen uptake subsequent to a change in the oxygen pressure is the same in these two systems. Also, the values determined for D appear very sensitive to the microstructure and orientation of the films. One very straightforward interpretation of these observations is that the diffusion is very anisotropic in the films, as has been reported in single crystals[4]. The values for D_0 would then be representative of the defect structure of the films, while the activation energy E_0 would be the same for the films, representing diffusion in the a-b plane. For example, in c-axis oriented films, D_0 would depend on the density of sites at which oxygen could move easily through the thickness of the film, presumably along planar defects, while E_0 would be the activation energy for motion along a-b planes. The appropriate value for ℓ, the diffusion length, would in this case be approximately the distance between these planar defects. (Even for films of pure orientation, it is known that these films contain many low-angle grain boundaries.[16])

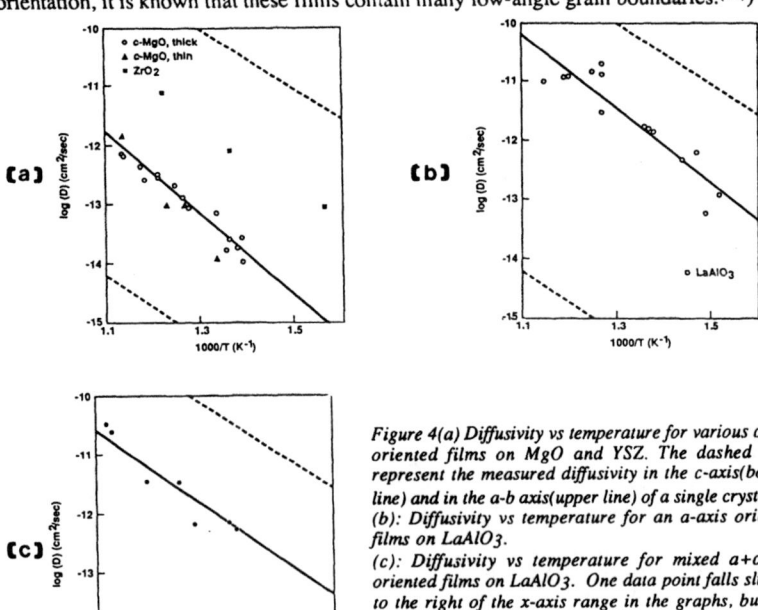

Figure 4(a) Diffusivity vs temperature for various c-axis oriented films on MgO and YSZ. The dashed lines represent the measured diffusivity in the c-axis(bottom line) and in the a-b axis(upper line) of a single crystal[4].
(b): Diffusivity vs temperature for an a-axis oriented films on LaAlO3.
(c): Diffusivity vs temperature for mixed a+c-axis oriented films on LaAlO3. One data point falls slightly to the right of the x-axis range in the graphs, but was included to obtain the best fit line shown.

2-3 Kinetics of Oxygenation in Atomic Oxygen

The time dependence of the conductivity was investigated for various films, in the same manner as with molecular oxygen above. The original oxygen pressure was set at 0.2mTorr in almost every case. The atomic oxygen flux was ~2x10^{15} atom/cm^{2}sec. After the conductivity (resistivity) reached equilibrium in molecular oxygen, the ECR plasma was initiated. In Figure 5 the conductance of a c-axis oriented YBaCuO film on MgO in the presence of atomic oxygen (solid line) and molecular oxygen (dashed line) are shown as a function of time at a constant temperature of 480°C. The original resistivity in each case is the equilibrium value, for applied oxygen pressures of 0.2mTorr and 10mTorr, respectively. The conductivity change in the presence of molecular oxygen is for a pressure change from 10mTorr to 100mTorr. A clear difference in time dependence of the conductance is observed between the two cases. This difference is greatest in c-axis oriented films. For the film shown in Figure 5, saturation occurs approximately 20 times faster in the presence of atomic oxygen that in molecular oxygen.

We quantify the time dependence of the conductivity in the presence of atomic oxygen using the diffusion model discussed in above section. Since the ratio of conductance from the initial to saturated value is large, the logarithm of the conductance was used for fitting, as shown in the reference (14). The same procedure was performed at different temperatures to determine the activation energy in the c-axis oriented films. Reasonably good fits can be obtained over the entire temperature range. Figure 6 shows the Arrhenius plot of log(Diffusivity) versus temperature for c-axis oriented films on MgO substrates. For comparison the Arrhenius plot for this sample in molecular oxygen is also plotted in Figure 6. The activation energy is approximately 0.65eV for atomic oxygen and 1.2eV for molecular oxygen, showing a clear activation energy difference between these cases. The small activation energy for atomic oxygen can explain the occurrence of low temperature plasma oxidation of YBaCuO films.

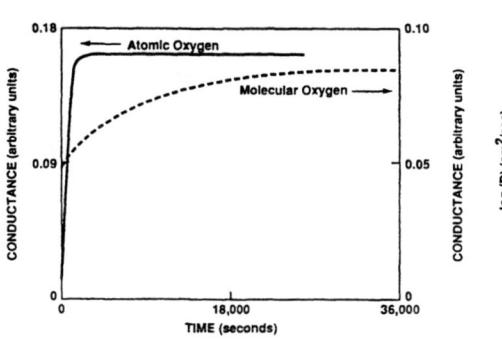

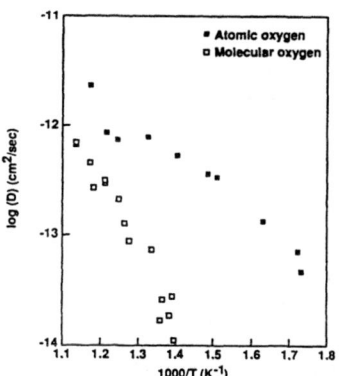

Figure 5: The conductance of a c-axis film on MgO in the presence of atomic oxygen (solid line) and molecular oxygen (dashed line) as a function of time at a constant temperature of 480°C.

Figure 6: The Arrhenius plot of diffusivity versus temperature for a c-axis oriented films on MgO, in the presence of the ECR plasma and in the presence of molecular oxygen.

3 Conclusion

We have modeled the changes in conductivity with time using a 1-dimensional diffusion equation to extract values for the diffusion coefficient D in the presence of various changes in applied oxygen pressure and atomic oxygen flux.

The values we extract for D are insensitive to the oxygen stoichiometry, provided there are no large gradients in oxygen concentration. D is found to be sensitive to the microstructure and orientation of the films, in the presence either of atomic or molecular oxygen. The activation energy we find in the presence of molecular oxygen, ~1.2eV for films of all orientations, is comparable to the values previously reported for single crystals and ceramics of YBaCuO. This indicates that the mechanism for diffusion is the same for these systems, probably representing migration in the a-b plane. We find a much lower activation energy in the presence of atomic oxygen, ~0.6eV. (We also find for plasma oxidation that the diffusion is insensitive to oxygen stoichiometry [14]) The balance of our data indicates that this low activation energy for diffusion in the presence of atomic oxygen is the result of the large gradients in steady-state oxygen content which occur at lower temperatures when the plasma is started, as opposed to enhanced migration over a surface barrier. The large gradients cause variations in unit cell dimensions which can possibly cause stress enhanced diffusion. Further work could test this hypothesis.

Acknowledgements

The authors would like to thank D. L. Keith and W. Holmes for their help and the use of the vacuum facility. This work has been supported in part by the Air Force Office for Scientific Research, by the Stanford Center for Materials Research under the NSF-MRL program, by EPRI contract RP8009-11, and by the Kanegafuchi Chemical Industry Co., Ltd.

References

(1) K.N. Tu, N.C. Yeh, S.I. Park, and C.C. Tsuei, Phys Rev. B, **39**, 304(1989)
(2) K. Kishio, K. Suzuki, T. Hasegawa, T. Yamamoto and K. Kitazawa, J. Solid State Chemistry, **182**, 192 (1989).
(3) S. J. Rothman, J. L. Roubort, and J. E. Baker, Phys. Rev.B, **40**, 13(1989).
(4) S. J. Rothman, J. L. Roubort, J.Z. Liu, J.W. Downey, L. J. Thompson, Y. Fang, D. Shi, J.E. Baker, J.P. Rice, D.M. Ginsberg, P.D. Han, and D.A. Payne, Proceedings of the 1989 TMS Fall Meeting Symposium on Atomic Migration and Defects in Materials.
(5) H. Tamura, A. Yoshida, S. Morohashi, and S. Hasuo, Appl. Phys. Lett. **53**,(1988) 618.
(6) N. Missert, R. H. Hammond, J. E. Mooij, V. Matijasevic, P. Rosenthal, T. H. Geballe, A. Kapitulnik, M. R. Beasley, S. Laderman, C. Lu, E. Garwin and R. Barton, IEEE Trans. Magn. MAG **25**, 2418 (1989).
(7) T. Aida, A. Tsukamoto, K. Imagawa, T. Fukazawa, S. Saito, K. Shindo, K. Takagi and K. Miyauchi, Jpn. J. Appl. Phys. **28**, L635 (1989).
(8) E.S. Hellman, D.G. Schlom, N. Missert, K. Char, J. S. Harris,Jr., M.R. Beasley, A. Kapitulnik, T.H. Geballe, J. N. Eckstein, S.L.Weng and C. Webb, J. Vac. Sci. Technol. **B6**, 799(1988).
(9)J. Kwo, M. Hong, D. J. Trevor, R. M. Fleming, A. E. White, J. P. Mannaerts, R. C. Farrow, A. R. Kortan and K. T. Short, Physica C **162-164**, 623 (1989).
(10) C. B. Eom, J. Z. Sun, K. Yamamoto, A. F. Marshall, K. E. Luther, S. S. Laderman and T. H. Geballe, Appl. Phys. Lett., 54, 595 (1989).
(11) C.B. Eom, et al., Physica C 171(1990)354.
(12) K.Yamamoto, et al., to be published in J. Vac Sci. Tech.
(13) V. Matijasevic, E.L. Garwin, R.H. Hammond, Rev. Sci. Instrum. **61**,1747(1990).
(14) K. Yamamoto, B.M. Lairson, J.C. Bravman and T.H. Geballe, to be published in J.Appl. Phys.
(15) J. Crank, "The Mathematics of Diffusion," Clarendon, Oxford, 1975.
(16) S.K. Streiffer, B.M. Lairson, C.B. Eom, J.C. Bravman, and T.H. Geballe, Proc. MRS **183**, 363(1990)

Nucleation and Growth kinetics of Cu$_2$O During Reduction of CuO Thin Films

Jian Li*, K.N.Tu** J.W. Mayer*
*Dept. of Materials Science and Engineering, Cornell University, Ithaca, NY 14853
**IBM, T.J. Watson Research Center, Yorktown Heights, NY 10598

Abstract:

The combination of $^{16}O(\alpha,\alpha)^{16}O$ oxygen resonance measurement and transmission electron microscopy (TEM) provides an unique and effective method to study the kinetics of nucleation and growth of Cu$_2$O phase during reduction. *In situ* TEM observation showed that isolated and large Cu$_2$O grains emerge from the small CuO grain matrix and the growth of Cu$_2$O grains is linear with time. We propose that the discontinuous morphology of grain growth of Cu$_2$O is due to the migration of the Cu$_2$O-CuO phase boundary induced by oxygen out-diffusion along the moving phase boundary. Based on the classical analysis of phase transformation by Johnson, Mehl and Avrami, the activation enthalpy of nucleation of Cu$_2$O phase in the CuO matrix has been deduced as ΔE_n=2.3 eV. The specific interfacial energy between CuO and Cu$_2$O phases has been estimated as 0.5 eV/atom.

I. Introduction:

Copper is widely used in electronic packaging and in integrated circuit technology [1]. Cu is readily oxidized at low temperature, leading to poor adhesion between Cu and a solder pad. Thus, the degradation of the package may occur. Studies on the phase transformation and stability of copper oxide thin films are of importance for device fabrication. The investigation of oxygen in and out diffusion induced phenomena in pure copper oxides can contribute to our understanding of adhesion failure mechanism, e.g. in tape automate bonding (TAB) technology.

Oxygen diffusion in the copper oxides leads not only to oxygen concentration variations but also to microstructural changes. In this study, we combine the transmission electron microscopy (TEM) and oxygen resonance, an extended Rutherford backscattering technique to study the nucleation and growth kinetics of Cu$_2$O phase during reduction of CuO thin films in vacuum.

II. Experimental:

Thin films of CuO were prepared by reactive cathodic sputtering of Cu in oxygen atmosphere onto Si(100) substrate and thin layer of SiO$_2$ covered NaCl substrate, followed by an annealing at 300°C for 30 minutes in ambient oxygen. The base pressure in the sputtering chamber was 2×10^{-7} Torr. A Rutherford backscattering (RBS) was employed to measure the film thickness at 70 nm and the composition as CuO. Self-supporting CuO films were floated off the NaCl substrates and were placed on transmission electron microscopic grids for *in situ* observation. A thin SiO$_2$ membrane acts as a passivation layer to prevent oxygen loss from one side. A hot stage was used in the electron microscopy upon vacuum annealing. CuO thin films on Si(100) substrate were isothermally annealed at different temperatures in a quartz tube furnace with a vacuum level 2×10^{-7} Torr. A Tandetron accelerator with a 1.7 MV terminal was used to conduct $^{16}O(\alpha,\alpha)^{16}O$ scattering resonance measurements for quantifying the oxygen concentration of the thin film copper oxide on Si substrate.

III. Results and discussions:

The growth of Cu₂O phase in CuO matrix: CuO thin film sample with a thin layer of SiO₂ covered on one side was annealed *in situ* in the hot stage of electron microscope. Figure 1 shows the bright field images before and after vacuum annealing. The initial grains which are identified as pure CuO phase are randomly oriented with a mean grain size of 15 nm. After the annealing in vacuum to 510°C, grain growth occurs as a function of time, 15, 30 and 45 minutes corresponding to the bright field images in Fig.1 (b), (c) and (d). Examining the diffraction patterns of the samples with isolated and large grains randomly distributed in the fine grain matrix of CuO phase, these larger grains are Cu₂O phase. The growth kinetics of Cu₂O grains were measured by isothermal annealing in vacuum. A linear growth of Cu₂O grains is found (see fig.2).

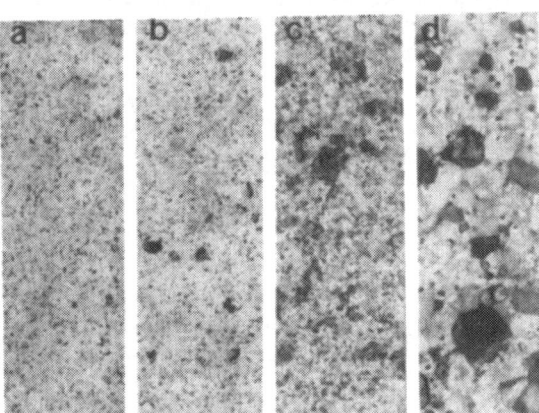

200nm

Fig.1. Bright field TEM images of copper oxide films (a) As-preapred CuO; and 510°C in vacuum (b)15 min.;(c)30 min.and (d) 45 min.

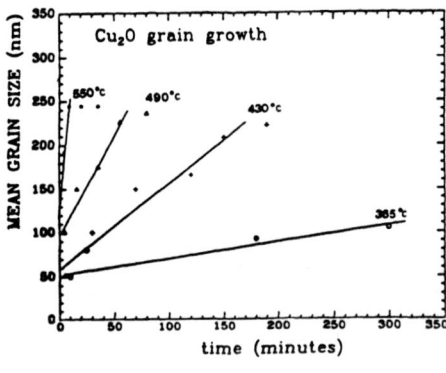

Fig.2. Isotherma anneal curves of CuO films.

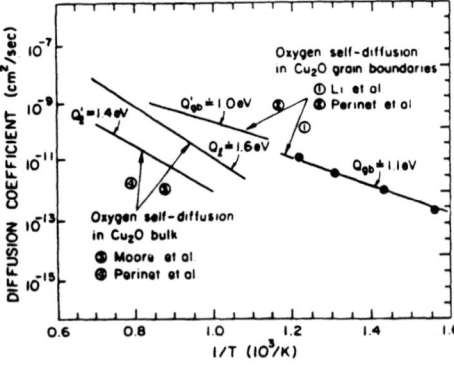

Fig.3. Arrhenius plot of oxygen diffuion coefficients in both bulk and thin copper oxide samples.

The unique morphological changes due to oxygen diffusion in some cubic oxide materials have been connected with diffusion-induced grain boundary migration (DIGM) [2,3]. In this study, the question of driving force for the Cu_2O grain growth in isolation at temperature about 0.3 of the melting point of copper oxide arises as to whether the curvature of the boundary drives the transformation or the free energy change of the transformation moves the boundary. We propose that the discontinuous morphology of grain growth of Cu_2O is due to the migration of the Cu_2O-CuO phase boundary induced by oxygen out-diffusion along the moving phase boundary [4]. Based on this theory, the oxygen diffusion coefficient along phase boundaries can be estimated as:

$$D_{gb} = \frac{Z^2 V(T)}{\delta} \tag{1}$$

where Z, V(T) and δ are the half of the film thickness, mean grain boundary migration velocity and phase boundary thickness, respectively. Based on this calculation, the values of D_{gb} are shown in Fig.3, and the activation energy Q_{gb} has been determined to 1.1 eV, which is the characteristics of diffusion induced phenomenon. The comparison of oxygen lattice diffusion and grain boundary diffusion in bulk copper oxide is made in Fig.3 [5].

(b). The nucleation kinetics of Cu_2O in CuO matrix: The grain growth of Cu_2O phase in the CuO matrix during reduction can best be studied by using a thin film sample in an electron microscopy. However, the nucleation kinetics of the Cu_2O phase is not readily observed and quantified. $^{16}O(\alpha, \alpha)^{16}O$ elastic nuclear resonance near 3.05 MeV has the sensitivity and offers an effective means of extending Rutherford backscattering technique (RBS) to the quantification of oxygen concentrations in the oxide thin films [6]. By employing this technique, the fractional volume change of transformed phase such as Cu_2O phase in the CuO matrix can be deduced.

When the Cu_2O phase forms in the CuO matrix during reduction, a phase boundary is created. Since both phases have a very different atomic configuration, the phase boundary forms a incoherent interfaces. Very little is known about the detailed atomic structure and interfacial energy of this kind of incoherent interfaces. In this study, we report the measurement of the oxygen loss rate during the reduction of CuO thin film by means of oxygen resonance $^{16}O(\alpha, \alpha)^{16}O$. By combining the electron microscopy observation of Cu_2O grain growth, we have deduced the activation enthalpy of nucleation, which enables us to estimate the interfacial energy between CuO and Cu_2O phases.

CuO thin films on Si(100) substrate were also isothermal annealed in vacuum at different temperatures. $^{16}O(\alpha, \alpha)^{16}O$ resonance has been employed to measure the oxygen concentration change during the reduction of the CuO thin film sample. Figure 4 shows that the oxygen variation near surface of the CuO sample has been monitored at 3.05 MeV as a function of time at 550°C. By subtracting the Si background, the oxygen peaks have been simulated as x=0.82, 0.72, 0.58 and 0.52 for the form of CuO_x corresponding to 15, 30, 45 and 60 minutes annealing at 550°C. The oxygen concentration in the annealed copper oxide thin films vs. annealing time has been plotted in Fig.5(a), indicating the oxygen loss rates at temperatures ranging from 375 to 550°C during reduction.

Since the two phases CuO and Cu_2O can be distinguished by their morphology, i.e. the nature of grain size during reduction of CuO thin films, we also can quantitatively deduced the atomic fraction X_t of the corresponding transformed phases by measuring the overall concentration of the oxygen in the form of CuO_x containing both CuO and Cu_2O phases; in other words, the overall oxygen concentration comes from the contributions of both the oxygen bonding to the Cu^+ state and the oxygen bonding to the Cu^{2+} state. The fraction X_t of Cu_2O is given by

$$CuO_x = X_t Cu_2O + (1 - X_t)CuO \tag{2}$$

By knowing the overall concentration of oxygen x, the fraction X_t can be deduced. Here, we ignore the contribution of free state oxygen content since the solubility of oxygen in the compounds and the deviation from the stoichiometry are quite small [7]. Based on the oxygen resonance measurements of oxygen concentration change near the surface of the copper oxide sample in Fig.5 (a), we plot the atomic fraction of transformed Cu_2O phase versus the annealing time at different temperatures in Fig.5 (b).

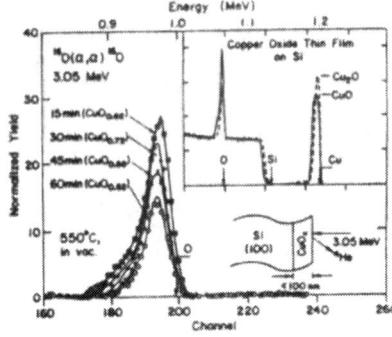

Fig.4. Oxygen resonance measurements of copper oxide films during reduction at 550°C for different times in vacuum.

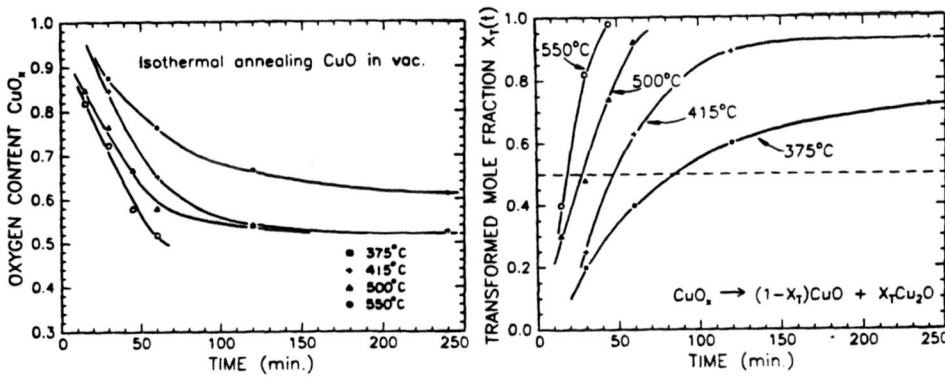

Fig.5a. Oxygen content in CuO_x vs. time during isothermal annealing.

Fig.5b. Transformed mole fraction of Cu_2O vs. time at different temperatures.

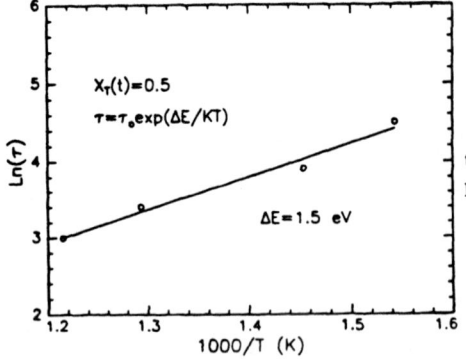

Fig.6. Lagarithm of time required to transform 1/2 of the CuO film into Cu_2O phase as a function of $1/kT$.

To describe the nucleation and growth of Cu_2O, we take the classical analysis of phase transformation by Mehl, Johnson and Avrami [8]. Two assumptions are set up for the validity of the Avrami equation: (1) the nucleation of Cu_2O phase is random with a constant rate; and (2) the growth is isotropic and linear with time. The nucleation and growth behavior of Cu_2O satisfies the condition described by Johnson, Mehl and Avrami mode of phase transformation. The transformed fraction X_{ext} of Cu_2O which includes impingements is related to the actual area fraction X_t by using the Avrami equation:

$$X_t = 1 - exp(-X_{ext}) \tag{3}$$

Since the Cu_2O grains can grow up to 250 nm in diameter, more than 3 times larger than the film thickness, the nucleation and growth can reasonably treated as a two-dimensional process. Based on the two assumptions mentioned above, we can express the X_{ext} as:

$$X_{ext} = \int_{\tau=0}^{\tau=t} \pi I G^2 (t - \tau)^2 d\tau = \frac{\pi I G^2 t^3}{3} \tag{4}$$

Here, I and G stand for the nucleation and growth rates, which have been assumed to follow Boltzmann's distribution:

$$I = I_o exp(-\frac{\Delta E_n}{kT}) \tag{5}$$

$$G = G_o exp(-\frac{\Delta E_g}{kT}) \tag{6}$$

and ΔE_n and ΔE_g are the activation enthalpies of nucleation and growth, respectively. By combining Eq.(3) and (4) and taking logarithm at a constant value of X_t=0.5 ($t = \tau$), we have:

$$-\frac{\Delta E_n + 2\Delta E_g}{kT} + 3ln\tau = constant \tag{7}$$

In Fig.5 (b), we draw a horizontal line at $X_t = 0.5$ to intercept the four annealing curves so that the corresponding annealing times for them can be decided from the horizontal axis. The relation between time τ and temperature T can be obtained. Through the Arrhenius relation given by

$$\tau = \tau_o exp(\frac{\Delta E}{kT}) \tag{8}$$

The activation energy of the CuO to Cu_2O transformation is found to be

$$\Delta E = \frac{\Delta E_n + 2\Delta E_g}{3} = 1.5 eV \tag{9}$$

Fig. 6 shows the results obtained.

The activation energy of phase transformation of CuO to Cu_2O and activation enthalpy of growth of Cu_2O have been independently obtained by using $^{16}O(\alpha, \alpha)^{16}O$ resonance measurement and electron microscope, respectively. From Eq.(9), we deduce the activation enthalpy of nucleation ΔE_n=2.3 eV. By knowing ΔE_n of Cu_2O phase, we can estimate the interfacial energy of between CuO and Cu_2O by using classical nucleation theory [9]. To form a nucleus of N atoms, the energy change ΔE_n is

$$\Delta E_n = -aN\Delta E_h + bN^{2/3}\gamma_{int} \tag{10}$$

Where ΔE_h and γ_{int} are heat of phase transformation and specific interfacial energy formed between CuO and Cu_2O phases, respectively. For the critical nucleus where

$$\frac{\partial \Delta E_n}{\partial N} = 0 \tag{11}$$

By assuming unity of the geometrical constants a and b, we obtain

$$\gamma_{int} \simeq \Delta E_n^{1/3} \Delta E_h^{2/3} = 0.5 eV/atom \tag{12}$$

Here, ΔE_h has been measured to be 0.2 eV/atom [10].

From an interfacial energy standpoint of view, it is favorable for a newly formed phase to be surrounded by low-energy coherent interfaces. In general coherent interfacial energies range from 0.001 eV/atom to 0.1 eV/atom [11]. There are, however, usually no other planes of good matching such as between the CuO (monoclinic structure) and Cu_2O (cubic structure). The Cu_2O phase must consequently be bounded by high-energy incoherent interfaces. The interfacial energy obtained in CuO/Cu_2O system is at least 5 times larger than that in other systems with coherent interfaces.

The error of the estimation of the interfacial energy is directly related to the validity of the two dimensional model based on the Avrami equation and classical nucleation theory. The nucleation rate of Cu_2O phase may not be constant all the time during the nucleation and growth, i.e. the nucleation rate may slow down when the grain size of Cu_2O phase is larger than the film thickness. In our experiment, the mode parameter n which is related to the mechanism of nucleation and growth morphology in the CuO/Cu_2O system has been determined from the plot $ln ln[1 - X_T(t)]^{-1}$ vs. $ln t$. The value of n was found to be around 2.5, which is less than 3 for the ideal two- dimension case. The role of film surface to the nucleation rate and oxygen diffusion are still unclear.

In summary, Combining the transmission electron microscopy with oxygen resonance techniques, we study the kinetics of Cu_2O grain growth, and deduce the activation enthalpy of nucleation of Cu_2O phase as $\Delta E_n = 2.3$ eV. The specific interfacial energy between CuO and Cu_2O phase has been estimated as 0.5 eV/atom based on classical nucleation theory.

We acknowledge discussions with G.Vizkelety, G.Amsel, J.Siejka, F.Abel and C.Ortega (Univ. Paris). This work was supported in part by the U.S. Defense Advanced Research Projects Agency.

Reference

1. D.P.Seraphim, R.Lasky and C.Y.Li, Principles of Electronic Packaging (McGraw-Hill, New York,1989).
2. J.W. Cahn, J.D. Pan, and R.W. Balluffi, Scr. Metall. 32, 29, (1984).
3. T.A. Parthasarathy and P.G. Shewmon, Acta Metall. 32, 29 (1984).
4. Jian Li, S.Q. Wang, J.W. Mayer and K.N. Tu, Phys.Rev. B39, 12367 (1989).
5. F.Perinet, J.Le, Duigou and C.Monty, in Non-stoichiometric Compounds ed by J.Nowotny and W.Weppner (Kluwer city, 1989) p387.
6. Jian Li, G.Vizkelety, P.Revesz and J.W.Mayer, to be published in J.Mat. Res.
7. Jie Xue and R.Dieckmann, to be published.
8. M.Avrami, J.Chem.Phys. 9,177 (1941).
9. D. Turnbull, Solid State Physics, 3, 226 (1956).
10. G.V. Samsonov, in The oxide handbook, (IFI/PLENUM, NY,1982),ch.2.
11. D.A. Porter and K.E. Easterling, Phase transformations in metals and alloys, (Van Nostrand Reinhold Co. England, 1982), p.144.

NUCLEATION AND ABNORMAL GRAIN GROWTH OF ALPHA-Al$_2$O$_3$ IN GAMMA-ALUMINA MATRIX

T. C. Chou and T. G. Nieh
Lockheed Missiles and Space Company, Research and Development Division, O/93-10, B/204, Palo Alto, CA 94304

ABSTRACT

The microstructures of reactive sputter-deposited alumina films have been studied by transmission electron microscopy. The as-deposited films contained γ-Al$_2$O$_3$ phase in an amorphous alumina matrix. Annealing of the films at 1200°C for 2 h resulted in nucleation and concurrent anomalous grain growth of α-Al$_2$O$_3$ in a polycrystalline γ-Al$_2$O$_3$ matrix which exhibited a layered microstructure and was strongly textured along [001]. The grain sizes of α-Al$_2$O$_3$ varied from 3 to 20 μm, while the average grain size of γ-Al$_2$O$_3$ was only about 50 nm. It appears that the nucleation kinetics of α-Al$_2$O$_3$ was slow. As a result, the abnormal grain growth of α-Al$_2$O$_3$ proceeded by consuming surrounding γ-Al$_2$O$_3$ grains. An atomic model is presented to explain the origin of layered structure in γ-Al$_2$O$_3$. The nucleation mechanism of α-Al$_2$O$_3$ in γ-alumina matrix is suggested. Orientation relationships between γ- and α-Al$_2$O$_3$ are reported. The anomalous grain growth of α-Al$_2$O$_3$ is discussed in terms of γ/α interface boundary migration.

INTRODUCTION

Aluminum oxide, in either bulk or thin film coating forms, finds many important applications as a component for structural, tribological, and electrical usages. For those applications, the thermodynamically-stable phase α-Al$_2$O$_3$ (commonly known as corundum) is desired because of its high melting point (due to strong bonding), high temperature stability, extreme hardness, and a high dielectric constant.

Aluminum oxide is known to exist in various metastable polymorphs in addition to α-Al$_2$O$_3$ [1]. The polymorphs include: gamma (γ), delta (δ), theta (θ), eta (η), kappa (κ), chi (χ), beta (β), and iota (ι). It has been noted that the starting phase of Al$_2$O$_3$ polymorphs may vary, depending upon the processing techniques and conditions, and phase transformation sequence can take place by various routes during post-processing treatments [1-9]. By the physical vapor deposition technique, a commonly reported transformation route is: amorphous $\rightarrow$ γ $\rightarrow$ θ + δ $\rightarrow$ α, in which the transformation products at each stage may be a mixture of several phases.

Recently, thin alumina films (50 nm to 1.2 μm in thickness) were produced by r.f. reactive sputtering deposition [10], and their microstructures were studied after annealing at temperatures ranging from 800 to 1200°C. A few striking results were observed in the films annealed at 1200°C: nucleation and concurrent explosive grain growth of α-Al$_2$O$_3$ took place in a textured, polycrystalline γ-Al$_2$O$_3$ matrix; γ- and α-Al$_2$O$_3$ phases exhibited special orientation relationships. In this paper, we summarize the experimental results and offer explanations for the concurrence of nucleation/abnormal grain growth of α-Al$_2$O$_3$.

EXPERIMENTAL PROCEDURES

Alumina films were deposited from a hot pressed Al$_2$O$_3$ target (99.99%) by r.f. magnetron reactive sputtering deposition in a plasma containing argon and oxygen. Prior to deposition, the substrates were cleaned by reverse bias sputter etching at 400 V for 5 min. The oxygen concentration in the gas, the r.f. power on the target, and the bias voltage for the substrates were varied to optimize quality of the films. Typical deposition parameters used in this study were: 4% oxygen, 2 kw r.f. power, and 200 V r.f. bias voltage. The base pressure prior to deposition was 8 x 10^{-8} torr and the processing gas pressure during deposition was 5 mtorr. No substrate heating was applied during the deposition; the substrate temperature was measured by heat sensitive tapes and found to be no more than 60°C. The average deposition rate of alumina was about 1.2 nm/min.

The Al_2O_3 films (50 nm in thickness) were deposited onto water-soluble NaCl substrates. The NaCl substrates were <100>-oriented with 99.9% purity grade. Free standing films were produced by dissolving away the NaCl substrates in water, and the floating films were retrieved by 400-mesh Ni grids (3 mm in diameter) and dried in air. The films were then subjected to annealing in vacuum (2 x 10^{-6} torr) at temperatures ranging from 800 to 1200°C for different times. The composition and microstructure of the films were characterized by Auger electron spectroscopy (AES) and JEOL 2000FX transmission electron microscopy (TEM), respectively.

RESULTS AND DISCUSSION

The as-deposited films were first examined by AES to confirm the presence of alumina. The low and high energy peaks related to aluminum in the film occurred at 54, 1338, and 1389 eV (the corresponding peaks for elemental Al are at 68, 1345, and 1396 eV). Depth concentration profiles indicated that the compositions of Al and O in the films were uniform throughout the thickness.

According to TEM studies, the as-deposited alumina films were predominantly amorphous with some γ-Al_2O_3 nanocrystalline phase embedded therein. By high resolution lattice imaging, the average grain size of γ-Al_2O_3 was determined to be about 10 nm. Based on electron diffraction [10], annealing of the films between 800 and 1100°C for various times gave rise to crystallization of amorphous matrix (into γ-Al_2O_3), development of structural texture (in γ-Al_2O_3), and various degrees of $\gamma \rightarrow \alpha$ phase transformation. It is important to point out that, however, no α-Al_2O_3 grains were ever detected in the films annealed at or below 1100°C. The average grain sizes of γ-Al_2O_3 did not change appreciably as the annealing temperature increased; they varied from 10 to 20 nm.

Annealing of the films at 1200°C for 2 h resulted in explosive grain growth of a few α-Al_2O_3 grains in a γ-Al_2O_3 matrix. The grain size of α-Al_2O_3 varies from 3 to 20 μm, while the average grain size of γ-Al_2O_3 is only about 50 nm. Fig.1 shows a TEM micrograph of an area containing a large α-Al_2O_3 grain embedded in a γ-Al_2O_3 matrix. Those anomalously grown α-Al_2O_3 grains were predominantly single crystals, manifested by the continuous extension of extinction contours within each grain. Electron diffraction analysis indicated that most of the α-Al_2O_3 grains were [0001]-oriented.

Fig.2 shows a TEM centred dark field image, using the $(400)_\gamma$ reflection, from an area containing a large α-Al_2O_3 grain and a number of γ-Al_2O_3 grains. The γ-Al_2O_3 grains are characterized by a layered microstructure, while the α-Al_2O_3 grain is featureless. Figs.3 (a) and (b) show a typical electron diffraction pattern and its indices, respectively, from the γ-Al_2O_3 matrix. Close examination of the diffraction pattern indicates that the pattern consists of a single crystalline (or a highly-textured)

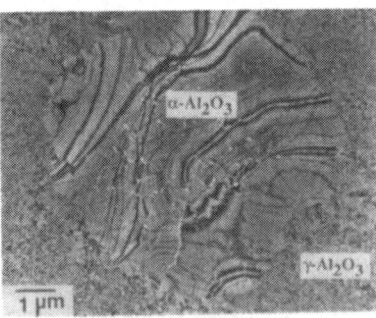

Fig.1 A TEM micrograph showing the coexistence of a large α-Al_2O_3 grain with polycrystalline γ-Al_2O_3 grains.

Fig.2 A TEM centred dark field image showing the microstructural characteristics of γ-Al_2O_3 (left) and α-Al_2O_3 (right).

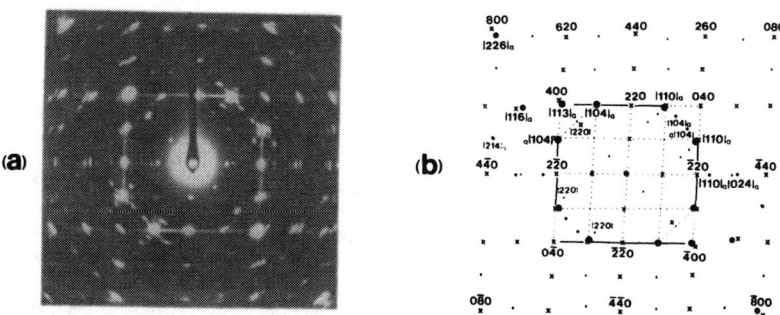

Fig.3 (a) An electron diffraction pattern and (b) an indexed pattern obtained from the polycrystalline γ-Al_2O_3 matrix.

γ-Al_2O_3 spot pattern superimposed by a polycrystalline α-Al_2O_3 ring pattern along specific orientations. Most importantly, the intensities of the α-Al_2O_3 diffraction pattern are enhanced only in certain sections in the neighborhood of the γ-Al_2O_3 diffraction spots. For example, the diffraction patterns from the {400}, {310}, and {800} planes of γ-Al_2O_3 overlapped with those from the {113}, {110}, and {226} planes, respectively, of α-Al_2O_3. It is intriguing to note that although the γ-Al_2O_3 matrix is polycrystalline in nature, electron diffraction pattern indicates that it is strongly textured along the {400}, {440}, and {800} crystallographic planes and has a [001] preferred orientation. Furthermore, heavy streaking is noted to occur along the 220 and 110 reciprocal lattice (R.L.) rows of the γ-Al_2O_3 pattern. The emergence of a polycrystalline α-Al_2O_3 diffraction pattern in a γ-Al_2O_3 matrix indicates that $\gamma \rightarrow \alpha$ nucleation process is undergoing, although no individual α-Al_2O_3 grains were observed.

The appearance of streaks in the γ-Al_2O_3 diffraction pattern is a result of the Fourier synthesis of a faulted, layered microstructure formed in γ-Al_2O_3. Their occurrence along the 220 and 110 R.L. rows indicates that the layers are aligned on the {$\bar{1}$10} and {220} planes. It was found that the presently formed layered structure exhibits a certain periodicity (or superstructure) in some regions. In one area as shown in Fig.4(a), 5 extra spots between origin and $\bar{2}$20, and between 110 and $\bar{1}$30 are observed in the convergent beam diffraction pattern, see Figs.4(b) and 4(c). This indicates that the superstructure has an interlayer spacing equivalent to 16.8 Å which is consistent with the value directly measured from the bright field image. In other words, the superstructure exhibits a periodicity every 6 layers of the {220} planes (d= 2.8 Å) of γ-Al_2O_3.

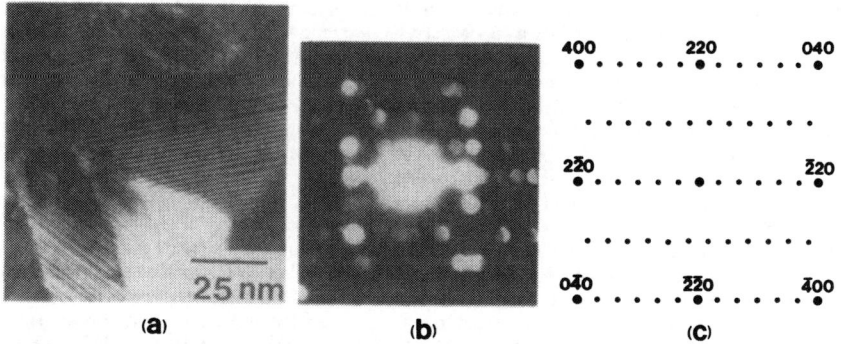

Fig.4 (a) A TEM bright field image showing the superstructure formed in γ-Al_2O_3. (b) A convergent beam electron diffraction pattern and (c) an indexed pattern obtained from the area shown in (a).

Layered microstructures have been observed in many materials, such as SiC, TiSi$_2$, oxide superconductors, etc. The most common explanation for their formation is the change of stacking sequence of atomic layers along one dimension. Such a change can be accomplished through either a purely atomic arrangement (called polytypism) or a small fluctuation of composition (called polytypoidism).

In the following we present an atomic model to demonstrate the feasibility of forming one-dimensional structures in γ-Al$_2$O$_3$ along {220} planes. By viewing the (110) and (220) planes in a γ-Al$_2$O$_3$ cell, see Figs.5(a) and 5(b), respectively, it is noticed that a simple translation of the (220) plane by 1/2 [001] will yield the same atomic arrangement as the (110) plane. In other words, the atomic structure of γ-Al$_2$O$_3$ can be retained if change of stacking sequence of atomic layers occurs on the {220} planes. The order of stacking of atomic layers can follow each other in any of these positions so that there is a family of polytypic structures which is based on the stacking along <220> or <110> directions. In view of the prevalent formation of layered structure in γ-Al$_2$O$_3$, the stacking fault energy along {220} and {110} planes may be low.

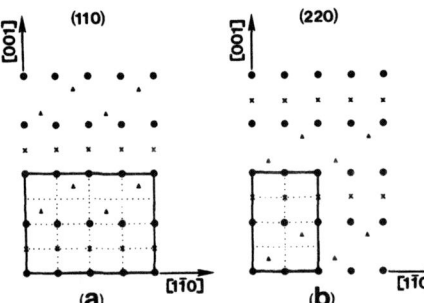

Fig.5 Schematic drawings showing the atomic arrangement on (a) the (110) and (b) the (220) planes of γ-Al$_2$O$_3$. A simple translation of a (220) plane by 1/2 [001] can exactly reproduce a (110) plane, thereby retaining the atomic structure of γ-Al$_2$O$_3$. (● : oxygen ion sites, x : octahedral cation sites, ▲ : tetrahedral cation sites).

It is intriguing to note in the present results that, on one side, the $\gamma \rightarrow \alpha$ phase transformation (i.e., nucleation of α-Al$_2$O$_3$) is in progress, while completely transformed α-Al$_2$O$_3$ grains are formed on the other side and they exhibit explosive grain growth. Since no α-Al$_2$O$_3$ grains were ever observed to form in the transformation-pending γ-Al$_2$O$_3$ matrix, the abnormal grain growth of α-Al$_2$O$_3$ and the accompanying $\gamma \rightarrow \alpha$ phase transformation are believed to proceed by consuming the γ-Al$_2$O$_3$ grains mediated through γ/α interface boundary migration. It is important to contrast the present results with the commonly reported secondary grain growth in thin films [11]. In the later case, the abnormal grain growth occurs in a homogeneous, single phase solid solution, and is accomplished by consuming smaller grains of the same kind through grain boundary migration. In the former case, however, the grain growth of α-Al$_2$O$_3$ is operated by consuming its surrounding smaller γ-Al$_2$O$_3$ grains through γ/α interface boundary migration. Most importantly, the former one involves phase transformation in the regions which are to be swept by the interface boundary. As a result, an additional term (i.e., the change of free energy of formation due to phase transformation) is introduced in the driving forces for the grain growth. A detailed analysis on the driving forces for secondary grain growth involving phase transformation has been provided elsewhere [12]. Based on calculations therein, the driving force from the change of free energy of formation (from γ- to α-Al$_2$O$_3$) is 3 to 5 orders of magnitude greater than those from surface energy anisotropy and grain boundary energy.

The concurrence of nucleation and explosive grain growth of α-Al$_2$O$_3$ strongly suggests that the nucleation kinetics of α-Al$_2$O$_3$ is slow and may be a rate limiting step for $\gamma \rightarrow \alpha$ phase transformation. Since neither composition change nor long range diffusion is required as the parent γ-Al$_2$O$_3$ converts into the α-Al$_2$O$_3$, it is speculated the nucleation of α-Al$_2$O$_3$ in γ-alumina matrix may be controlled by a structurally-dominated mechanism. In other words, it is believed that there exists certain orientation relationships between γ- and α-Al$_2$O$_3$, which allow the nucleation of α-Al$_2$O$_3$ taking place along preferred crystallographic planes of γ-Al$_2$O$_3$. This hypothesis is supported by the previously shown electron diffraction that the overlapping pattern from α-Al$_2$O$_3$ is intensified in the neighborhood of the spot pattern from γ-Al$_2$O$_3$.

To understand the structural correlation between γ- and α-Al$_2$O$_3$, electron diffraction patterns were taken from areas in the vicinity of a γ/α interface boundary. Figs.6(b) and 6(c) show electron diffraction patterns obtained from the polycrystalline γ-Al$_2$O$_3$ region and the neighboring single crystalline α-Al$_2$O$_3$ region from an area shown in Fig.6(a). It is noted in Fig.6(c) that the α-Al$_2$O$_3$ grain is [0001] oriented, and its electron diffraction pattern superimposes with that from the polycrystalline γ-Al$_2$O$_3$ along specific orientations. According to Fig.6(b), the polycrystalline γ-Al$_2$O$_3$ is also strongly textured along [001]. Fig.6(d) shows a schematic drawing illustrating the relationship of the diffraction patterns between γ- and α-Al$_2$O$_3$. Diffraction spots from the $(\bar{2}110)/(2\bar{1}\bar{1}0)$ and $(\bar{3}030)/(30\bar{3}0)$ planes of α-Al$_2$O$_3$ were found to coincide with those from the $(3\bar{1}0)/(\bar{3}10)$ and $(4\bar{4}0)/(\bar{4}40)$ planes, respectively, of γ-Al$_2$O$_3$, although small mismatches were noted to occur among $(11\bar{2}0)_\alpha$-$(\bar{1}30)_\gamma$, $(\bar{1}\bar{1}20)_\alpha$-$(1\bar{3}0)_\gamma$, $(\bar{3}300)_\alpha$-$\{\bar{4}40\}_\gamma$, and $(3\bar{3}00)_\alpha$-$\{440\}_\gamma$.

Based on the electron diffraction patterns obtained from the polycrystalline γ-Al$_2$O$_3$, layered structure is oriented primarily along the $\{220\}$ and $\{110\}$ planes. Since no layered structure is formed in γ-Al$_2$O$_3$ without the presence of completely transformed α-Al$_2$O$_3$ grains in its vicinity and fully transformed α-Al$_2$O$_3$ does not contain layered microstructure, it is likely that the layered structure formed in the present γ-Al$_2$O$_3$ thin films is characteristic of an intermediate state during $\gamma \rightarrow \alpha$ phase transformation. If this hypothesis is valid, the nucleation of α-Al$_2$O$_3$ may start from the $\{220\}$ or $\{110\}$ planes of γ-Al$_2$O$_3$.

In view of the fact that epitaxial films can be grown virtually on any single crystalline substrates of various orientations, it is believed that lattice match between the crystallographic planes of γ- and α-Al$_2$O$_3$ plays a vital role in determining the nucleation sites. According to the present

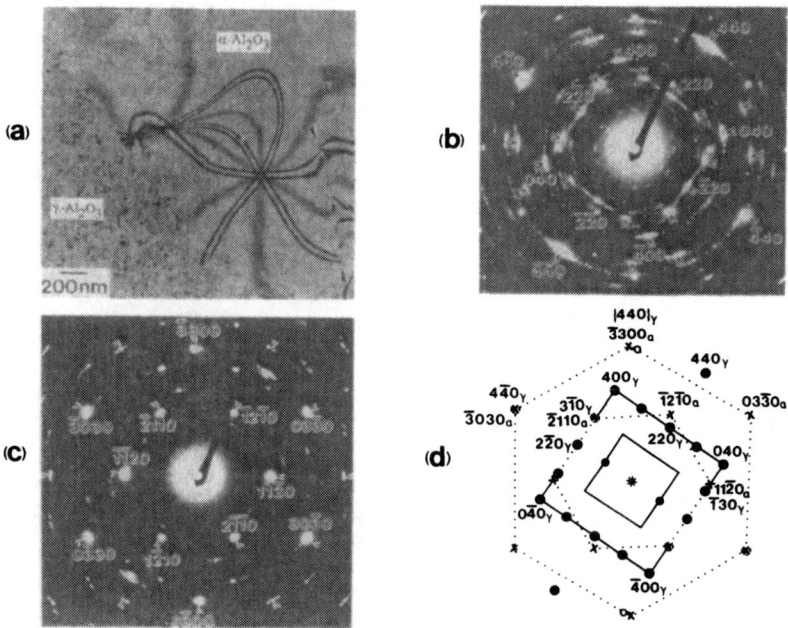

Fig.6 (a) A TEM bright field image showing the coexistence of a large α-Al$_2$O$_3$ grain and fine polycrystalline γ-Al$_2$O$_3$ grains. A low angle grain boundary is indicated by arrows in (a). Electron diffraction patterns obtained from the polycrystalline γ-Al$_2$O$_3$ region and the single crystalline α-Al$_2$O$_3$ region are shown in (b) and (c), respectively. A schematic drawing illustrating the orientation relationships between γ- and α-Al$_2$O$_3$ is shown in (d). $\{440\}_\gamma$ indicates diffraction generated from the $\{440\}$ planes of polycrystalline γ-Al$_2$O$_3$.

results, it is postulated that the most likely nucleation sites for α-Al_2O_3 reside on the {220} or {110} planes of γ-Al_2O_3 because of the presence of a high population of structural defects. Moreover, polycrystalline γ-Al_2O_3 films were found to exhibit a strong texture along the {400} and {440} planes and the d-spacings of both crystallographic planes closely match with those of {113} and {300} planes, respectively, of α-Al_2O_3 [10]. The close lattice match between γ- and α-Al_2O_3 may promote collective, massive growth of α-Al_2O_3 along preferred orientations, thereby resulting in its abnormal grain growth.

SUMMARY

As-deposited, r.f. reactive sputtered alumina films contained γ-Al_2O_3 in an amorphous alumina matrix. The γ-Al_2O_3 grains had an average grain size of about 10 nm. Annealing of the films between 800 and 1100°C gave rise to crystallization of the amorphous alumina matrix, evolution of texture, and various degrees of $\gamma \rightarrow \alpha$ phase transformation. In the films annealed at 1200°C for 2 h, nucleation and concurrent abnormal grain growth of α-Al_2O_3 took place in a polycrystalline γ-Al_2O_3 matrix. The grain size of α-Al_2O_3 varied from 3 to 20 μm, while the polycrystalline γ-Al_2O_3 had an average grain size of 50 nm. Those large α-Al_2O_3 grains were predominantly [0001]-oriented. The polycrystalline γ-Al_2O_3 matrix was strongly textured along the [001] orientation; its microstructure was characterized by the global formation of layered structure. The layered structure was aligned mainly along the {110} and {220} planes. While most of the layers showed one-dimensional, disordered arrangement, a superstructure with 16.8 Å periodicity was identified. An atomic model is presented to explain the origin of the formation of layered structure. According to the model, a simple translation of the {220} planes by 1/2 [001] can reproduce the same atomic structure as the {110} planes, thereby giving rise to the formation of one-dimensional layered structure along the {220} planes. The high propensity of introducing stacking faults on the {220} and {110} planes of γ-Al_2O_3 is believed to facilitate the nucleation of α-Al_2O_3. Because of close lattice match between selected crystallographic planes of γ- and α-Al_2O_3, cooperative growth of α-Al_2O_3 along some preferred orientations is possible, which in turn gives rise to anomalous grain growth of α-Al_2O_3. The orientation relationships between γ- and α-Al_2O_3 were determined to be $[001]_\gamma$ // $[0001]_\alpha$, $(4\bar{4}0)_\gamma$ // $(\bar{3}030)_\alpha$, and $(3\bar{1}0)_\gamma$ // $(\bar{2}110)_\alpha$. The abnormal grain growth of α-Al_2O_3 is suggested to mediate through γ/α interface boundary migration.

ACKNOWLEDGEMENTS

The authors are grateful to J. Mardinly, D. Adamson, and H. S. Hu for their technical assistance. This work was supported by the Lockheed Independent Research Program.

REFERENCES

1. J. A. Thornton and J. Chin, Ceram. Bulletin 56, 504-12 (1977).
2. B. A. Movchan and A. V. Demchishin, Z. Metal. Metalloved. 28, 653-60 (1969).
3. R. G. Frieser, J. Electrochem. Soc. 113, 357-60 (1966).
4. C. A. T. Salama, J. Electrochem. Soc. 117, 913-17 (1970).
5. I. H. Pratt, Solid State Technol. 12, 49 (1969)
6. A. L. Dragoo and J. J. Diamond, J. Am. Ceram. Soc. 50, 68-574 (1967).
7. K. G. Khakhanashvilli, F. N. Tavadze, O. P. Shalamberidze, and E. R. Kuteliya, Problemy Spetsial'noi Elektrometallurgii 3, 48-54 (1987).
8. B. Lux, C. Colombier, H. Altena, and K. Stjernberg, Thin Solid Films 138, 49-64 (1986).
9. J. Skogsmo, P. Liu, C. Chatfield, and H. Norden, 12th International Plansee Seminar V.3, (Metallwerk Plansee Gmbh, Reutte, Tirol, Austria, 1989), pp.129-42.
10. T. C. Chou and T. G. Nieh, Thin Solid Films (in press).
11. C. V. Thompson, J. Floro, and H. I. Smith, J. Appl. Phys. 67, 4099 (1990).
12. T. C. Chou and T. G. Nieh, submitted to J. Mat. Res.

INTERACTION BETWEEN DISLOCATIONS AND NiFe2O4 PRECIPITATES IN A NIO MATRIX

SCOTT R. SUMMERFELT* AND C. BARRY CARTER**
Dept. of Materials Science and Engineering, Bard Hall, Cornell University, Ithaca, NY 14853

ABSTRACT

Three different types of dislocation interactions with $NiFe_2O_4$ (spinel crystal structure) precipitates in a NiO matrix have been studied. In the first, the movement of dislocations introduced by room temperature deformation is impeded by the spinel precipitates. Glide dislocations in the NiO with $1/2<011>$ Burgers vectors and $\{0\bar{1}1\}$ glide planes cannot pass through the spinel precipitates without forming stacking faults because the perfect NiO dislocations are partial dislocations in $NiFe_2O_4$. Many dislocation loops but no stacking faults were observed in the deformed samples indicating that the gliding dislocations formed the loops when they moved past the precipitates. In the second type of interactions, cusps were formed in the spinel-NiO interface at close to the dislocation loops when the sample was heat treated; the cusps indicate preferential dissolution of the spinel. In the final interaction, the dislocations were shown to act as preferential nucleation sites when spinel was precipitated from the NiO matrix. At slow nucleation rates, $NiFe_2O_4$ precipitated only on the dislocations; when the nucleation rate was increased, precipitation occurred both on and away from the dislocations. Precipitates which form at a dislocation may contain a stacking fault extending from the partial dislocation to a cusp in the spinel-NiO interface. When this occurred, the stacking faults were observed to be faceted parallel to either $\{111\}$ or $\{011\}$ planes.

INTRODUCTION

The spinel-wüstite system is particularly suitable for a precipitation study because both structures have face-centered-cubic oxygen sublattices. In the absence of misfit dislocations, precipitate growth can therefore occur by the movement of cations alone (if misfit dislocations are present movement on both ion sublattices must occur). In the $NiFe_2O_4$-NiO system, the small misfit in the oxygen sublattice (less than 0.2% at room temperature) can result in the growth of large coherent precipitates [1]. The dislocations in NiO and perfect dislocations in spinel have similar $1/2<011>$ Burgers vectors and $\{011\}$ glide planes [2]. Since the lattice parameter of $NiFe_2O_4$ is almost twice that of NiO, a perfect dislocation in NiO is a partial dislocation in $NiFe_2O_4$ [3]. Previous studies of $MgFe_2O_4$ precipitates in MgO have shown that precipitation hardening and preferential nucleation on dislocations occur [4-7].

In the present study, the interaction between dislocations and $NiFe_2O_4$ precipitates in a NiO matrix has been studied. The $NiFe_2O_4$-NiO system was chosen because precipitation [1,3], particle coarsening [1], and nucleation and growth in thin-film specimens [1] have been studied previously. The effect of the spinel precipitates on dislocation movement was studied by precipitating spinel in NiO and then introducing the dislocations by deformation. The effect of the dislocations on the spinel precipitate morphology was studied by reannealing a transmission electron microscope (TEM) specimen which already contained dislocations. Finally, the effect on spinel nucleation was studied by precipitating $NiFe_2O_4$ in an Fe-doped NiO TEM sample which contained dislocations.

EXPERIMENTAL

Disks of polycrystalline Fe-doped NiO which were chemically homogeneous were prepared by hot pressing high purity powders. The disks which were heat treated to precipitate TEM samples were prepared by mechanically thinning and polishing the disks to 120 μm, dimpling and ion milling to perforation using 5keV Ar ions. The samples were characterized by bright-field (BF) and centered-dark-field (CDF) imaging with a JEOL 1200EX operated at 120keV. Energy dispersive spectroscopy (EDS), performed in a JEOL 200CX TEMSCAN on some of the samples showed that they contained approximately 4 cation-% Fe. The solution temperature for 4 cation-% Fe in NiO is ~1200°C in air [8].

In the first study, a disk was cooled from above the solvus temperature at 100°C/hr to 800°C and then quenched. This heat treatment produces coherent dendritic $NiFe_2O_4$ [1]. A TEM sample was prepared from the disk using the procedure described above; the sample was dimpled to less than ~8μm and ion milled from one side. The electron-transparent regions of the sample were, therefore, close to the original polished surface.

In the second study, a disk was cooled at 100°C/hr from above the solvus temperature and then prepared as a TEM sample using the same procedure used in the first study. The TEM sample was placed in a nickel ferrite container and subjected to a series of heat treatments: 500°C for 40 min; 700, 750, 800, 825, 850 and 875°C for 30min; 900°C for 30, 60 and 120min; and 920°C for 30min, in order to study particle coarsening and shape transformation.

In the third study, a disk was made into a TEM sample using the standard procedure except that this disk was dimpled to perforation and then ion milled from both sides for only a short time. The TEM sample was enclosed in a Fe-doped NiO container which had exactly the same composition as the TEM sample and then subjected to a series of heat treatments. Images were recorded after heat treatments A7 and A12 which involved heating the sample to 1025°C for 10min, and then cooling to 850°C (A7) or 815°C (A12) and holding for 30min. The 1025°C anneal dissolves all of the existing spinel particles; reprecipitation occurs with a slow rate at 850°C or at a faster rate at 815°C.

RESULTS AND DISCUSSION

In the first study, dislocation motion was impeded by the spinel precipitates such that the dislocations formed loops around the precipitates. Fig. 1 is a weak-beam image recorded with $g = (0\bar{1}1)$ NiO near the [011] zone axes. Both the spinel precipitates and the NiO matrix appear dark in the image although there is enough contrast to distinguish the location and shape of the spinel precipitates. The morphology of these dendritic precipitates is characterized by branches in the <011> directions which are bounded by {111} and [011} planes [1].

The bright lines in Fig. 1 are dislocations, most of which are located at the spinel-wüstite interface. Some of the dendrites have several dislocations looped around them. Dislocations within the NiO grains were only observed in samples which were ion milled less than ~5μm from both sides. These dislocations were, therefore, introduced by mechanical deformation at room temperature during TEM specimen preparation. This deformation layer is generally removed during ion milling. The perfect NiO dislocations are only partial dislocations in the spinel and hence cannot glide through the precipitates without forming stacking faults. Instead, the dislocations bend around the precipitates forming loops.

Analysis of weak-beam images such as those shown in Fig. 2 are consistent with the dislocations having $1/2$<011> Burgers vectors and {0$\bar{1}$1} glide planes. Fig. 2 shows two weak-beam images of the same region recorded near the [011] zone axis using (A) $g= (\bar{1}\bar{1}1)$ NiO and (B) $g = (1\bar{1}1)_{NiO}$. In Fig. 2A, the dislocations appear as bright lines, usually next to a precipitate, while the dislocations are invisible in Fig. 2B. The dislocations are nearly parallel to the [11$\bar{1}$] direction and are not end on. These observations are consistent with the expected Burgers vectors and glide planes for glissile dislocations in NiO. No stacking faults were observed in this sample indicating that the formation of loops around the precipitates is favored over cutting of the precipitates.

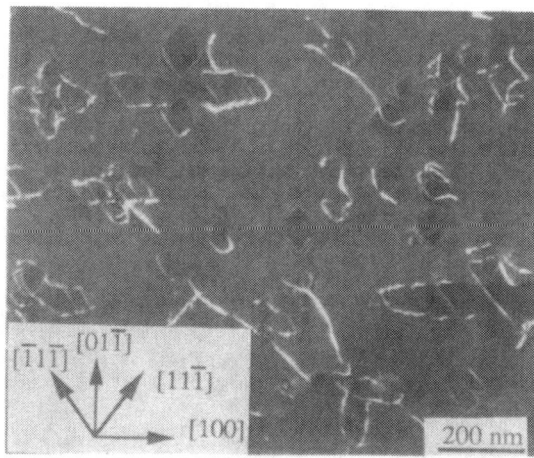

Fig. 1 Weak-beam image of dislocation loops around spinel precipitates. The sample is viewed from near the [011] direction and was formed using the $\mathbf{g} = (0\bar{2}2)_{NiO}$ reflection in the g,3.1g diffraction condition.

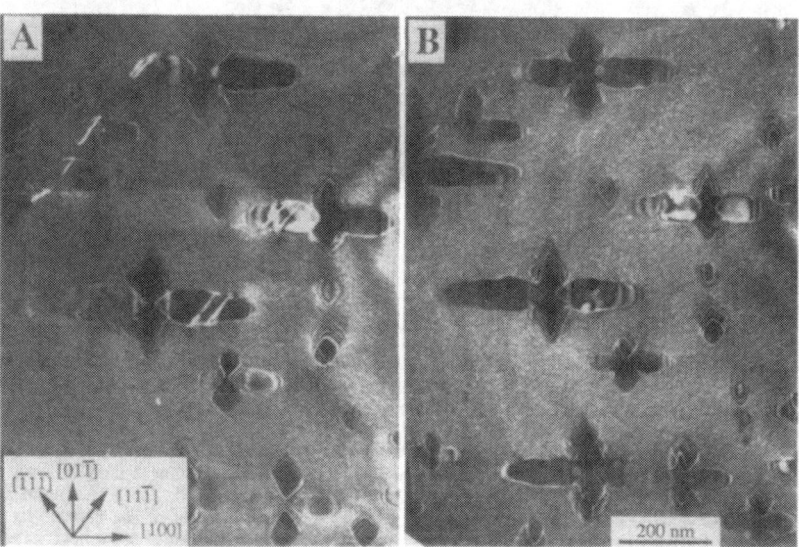

Fig. 2 Weak-beam images of dislocation loops around spinel precipitates. The sample is viewed from near the [011] direction using the NiO reflections (A) $\mathbf{g} = (\bar{1}\bar{1}1)$, (B) $\mathbf{g} = (0\bar{1}1)$.

354

In the second study, cusps were formed in the spinel-NiO interface at the dislocation loops when the sample was heat treated such that shape transformation occurred. Figs 3A-B are CDF images recorded near the [011] zone axes using the $(0\bar{2}2)_{spinel}$ reflection. The same region was characterized after heat treatments of A) 500°C for 40 min, and B) 800°C for 30min. The precipitates in Fig. 3A have a dendritic morphology. After further heat treatment, cusp formed in the spinel-wüstite interfaces (Fig. 3B) of precipitate N. Dislocations encircle two of the branches on this particle (arrowed). After an additional heat treatment, these branches and the dislocations circling them both disappeared. If the dislocation is completely wrapped around the spinel particle, then its line tension exerts a force to reduce the length of the dislocation. A cusp in the spinel particle forms at the dislocation such that the dislocation's line tension is balanced by the additional interface energy.

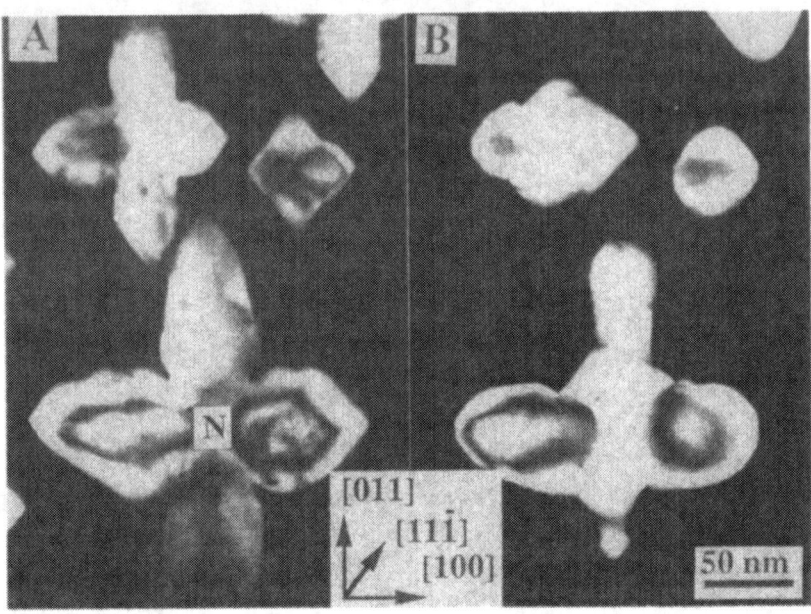

Fig. 3 Series of heat treatments of TEM sample containing spinel precipitates and dislocations. Images were recorded near the [011] zone axis using the $(0\bar{2}2)$ spinel reflection and the same area is recorded after heat treatments of (A) 500°C for 40 min and (B) 850°C for 30 min.

In the third study, NiO dislocations were shown to act as preferential sites for the nucleation of spinel precipitates. Fig. 4 shows two CDF images recorded near the [011] zone axis using the $(0\bar{2}2)$ spinel reflection. The images are from the same area of the TEM sample after heat treatments A7 and A12. In heat treatments A7 and A12, abrupt heating of the sample to above 1025°C dissolved the initial spinel particles so that nucleation and growth of spinel occurred after the sample was cooled to 850°C (A7) or 815°C (A12) [1]. Four dislocations in Fig. 4 are almost end-on and traverse the thin foil. Preferential nucleation of spinel has occurred on three of these dislocations as shown in Fig. 4A near the centers of particles L, M and N. As shown in Fig. 4B, many precipitates do not nucleate on dislocations when the nucleation rate is faster. These dislocations were originally perfect NiO dislocations but, once the spinel particles grew around them, the dislocations became spinel partial dislocations. Stacking faults, therefore, extend from the partial dislocation to the spinel-wüstite interface. These stacking faults appear as dark lines extending from the centers of the particles L, M and N (Fig. 4A) and L, M, N and O (Fig. 4B) to cusps in the spinel-wüstite interface. Since the stacking faults are very narrow in these images they are most likely edge-on. The stacking faults appear to be faceted parallel to ((0$\bar{1}$1), (1$\bar{1}$1) and ($\bar{1}$1$\bar{1}$) planes. The stacking fault in particle L (Fig. 4A) appears to have an overall (100) orientation, but closer inspection reveals that it is actually composed of stacking faults with alternating (1$\bar{1}$1) and ($\bar{1}$1$\bar{1}$) planes.

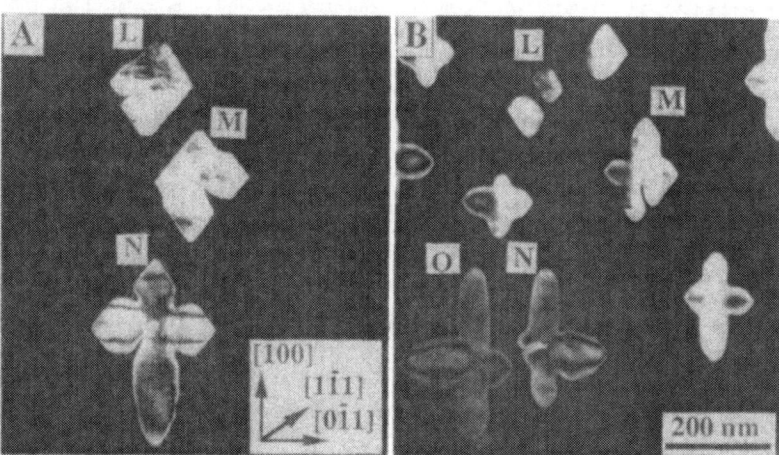

Fig. 4 Series of heat treatments on Fe-doped NiO TEM sample in order to study precipitation of spinel on NiO dislocations. The images were recorded in the same region from near the [011] zone axis using the (0$\bar{2}$2) spinel reflection. (A) heat treatment A7 and (B) heat treatment A12.

CONCLUSION

Three types of interactions between dislocations and precipitates in a NiO matrix have been studied. The movement of dislocations introduced by room-temperature deformation was impeded by the spinel precipitates. The dislocations formed loops around the precipitates rather than cutting the precipitates. Weak-beam images are consistent with the dislocations having $^1/_2$<011> Burgers vectors and {011} glide planes. Dislocations looped around the precipitates resulted in the formation of cusps in the spinel precipitates during shape transformation. The cusps are the result of the dislocation line tension. The dislocations also act as preferential sites for spinel nucleation. The spinel precipitates which grew around NiO dislocations contained stacking faults which extended from the dislocation core to the spinel-wüstite interface. The stacking faults form because the perfect NiO dislocations are only partial dislocations in the spinel. Cusps in the precipitates containing dislocations also occurred at the intersection of the stacking faults with the spinel-wüstite interface. The stacking faults were faceted along {011} and {111} planes.

ACKNOWLEDGMENTS

The authors would like to thank Dr. Lisa Tietz for discussions, Mr Ray Coles for careful maintenance of the electron microscopes and Ms Margaret Fabrizio for photographic work. The electron microscope Facility is supported, in part, by NSF through the Materials Science Center at Cornell. This research has been supported by the NSF grant No. DMR-8901218.

REFERENCES

*   Present address: Texas Instruments, Inc., Dallas. TX 75243
**   Present address: Department of Chemical Engineering and Materials Science, University of Minnesota, Minneapolis, MN 55455
1.   S. R. Summerfelt, Ph.D. Thesis, Cornell University (1990).
2.   J. P. Hirth and J. Lothe, Theory of Dislocations Second Edition, John Wiley & Sons. New York (19812).
3.   K.M. Ostyn, C.B. Carter, FM. Koehne, H. Falke and H. Schmalzried, J. Am. Ceram. Soc. 67, 679 (1984).
4.   G. W. Groves and M.E. Fine, J. Appl. Phys. 35, 3587 (1964).
5.   R.W. Davidge, J. Mater. Sci. 2, 339 (1967).
5.   R.W. Davidge, J. Mater. Sci. 2, 339 (1967).
6.   B. Reppich and H. Knoch, Deformation of Ceramic Materials, edited by Bradt and Tressler, Plenum Press (1975).
7.   G.R. Pulliam, J. Amer. Cer. Soc. 46, 202 (1963).
8.   F. Schneider and H. Schmalzried, Z. Phys. Chem. Neue Folge, 166, 1 (1990).

EFFECT OF ANNEALING ON THE DIELECTRIC PROPERTIES
AND MICROSTRUCTURE OF TANTALUM OXIDE THIN FILMS

CHANG HWAN CHUN*, GEUN HONG KIM*, AND KYOUNG-SOO YI**
* ADD, TAEJEON, KOREA ; ** ETRI, TAEJEON, KOREA

ABSTRACT

Effects of heat treatments on the dielectric properties of tantalum oxide thin films(250Å) deposited on the p-Si substrates by RF reactive sputtering were investigated. The leakage current density was considerably reduced from 10^{-9} to $10^{-12} A/\mu m^2$ at an electric field of 2MV/cm after rapid thermal annealing in O_2 at 1000°C , while little leakage reduction was observed after annealing at 500°C.
The structural changes of tantalum oxide thin film after annealing were examined using high resolution transmission electron microscopy. The leakage reduction after annealing can be attributed to crystallization and reoxidation of the amorphous tantalum oxide thin film.

INTRODUCTION

There have been continuing requirements to scale down the thickness of capacitor dielectric in dynamic random access memories (DRAMs). Recently, conventional SiO_2 or Si_3N_4/SiO_2 film is close to a physical limit of dielectric strength.
The use of high dielectric constant insulators will eliminate this problem. Tantalum pentoxide with a dielectric constant of about 25, six times larger than 3.8 of SiO_2, is one of the promising materials.
In spite of its high dielectric constant, tantalum oxide films deposited by various methods have been reported to be too leaky to be used in practice. [1-3] Some workers who have tried to reduce the leakage current have reported that annealing the tantalum oxide films in oxygen or ozone significantly reduced the leakage current.[4,5]
This paper examines the effects of sputtering and subsequent annealing conditions on the leakage current and microstructure of tantalum oxide films deposited by rf-reactive magnetron sputtering.

EXPERIMENT

Nominal resistivity of 6-9Ω cm, 5" diameter p-type Si substrates of <100> orientation were used for this study. Tantalum oxide films of 25-30 nm in thickness were deposited via RF magnetron sputtering in Ar and O_2 mixtures on chemically cleaned Si substrates. Si substrates were cleaned in a $H_2SO_4/$ H_2O_2(4:1) solution for 10 minutes at 110°C and native oxide was removed in 10% HF solution. As a sputtering target, 99.99% high purity Ta_2O_5 was used. Reactive sputtering was done using oxygen partial pressures of 0, 5, 10 and 15%. During sputtering the substrates were not intentionally heated.
In order to investigate annealing effects on the properties of tantalum oxide films, as-deposited samples were subjected to thermal treatments in O_2. Furnace annealing was carried out for 1 hr at 500°C while rapid thermal annealings(RTA) for 1 min at temperatures of 800-1000°C.
Al of 1μm thickness was sputtered on the deposited tantalum oxide film and photolithographically patterned as rectangular electrodes for capacitors with area of $100\times100\mu m^2$. The electrodes were annealed in N_2/H_2 mixture for 30 min at 420°C. Leakage currents were examined in the accumulation region by HP

Mat. Res. Soc. Symp. Proc. Vol. 230. ⸀1992 Materials Research Society

4140B pA meter applying a negative voltage to Al electrode.

Chemical compositions of as-deposited tantalum oxide films were analysed by Auger Electron Spectroscopy. Microstructures of tantalum oxide films and interface layers between Si substrates and the tantalum oxide films were examined in a JEOL 4000FX TEM. The specimen preparation techniques for TEM analysis was that developed by J. Benedict et al.[6]

RESULTS AND DISCUSSION

Dielectric properties of as-deposited samples

During RF magnetron reactive sputtering, the deposition rates of tantalum oxide films depend upon oxygen partial pressure in Ar/O$_2$ mixture. In the case of pure Ar, deposition rate is 10.1 nm/min which decreases to 2.7, 1.9, and 1.7 nm/min for oxygen partial pressures of 5, 10 and 15% respectively.

Oxygen partial pressure also affects chemical composition of tantalum oxide film. In spite of the fact that the sputtering target was Ta$_2$O$_5$ of 99.99% high purity, oxygen atoms in the as-deposited film sputtered in pure Ar are so deficient that the average atomic ratio of oxygen to tantalum was less than 1:1. This ratio continues to rise slightly with increasing oxygen partial pressure in Ar/O$_2$ mixtures. But it is almost saturated to 1:1 at oxygen partial pressure of 10% or above. Therefore 10% oxygen partial pressure was selected to deposit tantalum oxide films which were heat-treated subsequently to examine annealing effects.

Fig.1 shows typical results of I-V measurement for as-deposited tantalum oxide films. The figure presents the leakage current density as a function of applied electric field. N$_1$ denotes the specimen deposited in pure Ar(0% oxygen partial pressure) and N$_2$, N$_3$, and N$_4$ denote the specimens deposited in 5, 10, and 15% oxygen partial pressure respectively. For N$_1$ the leakage curve is always an order of magnitude higher than the ones for N$_2$, N$_3$, and N$_4$, which are almost equal.

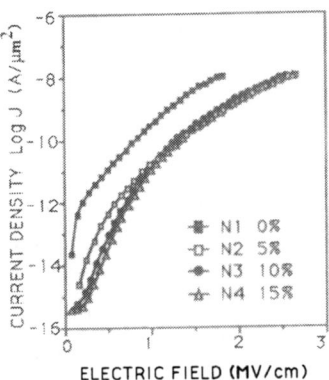

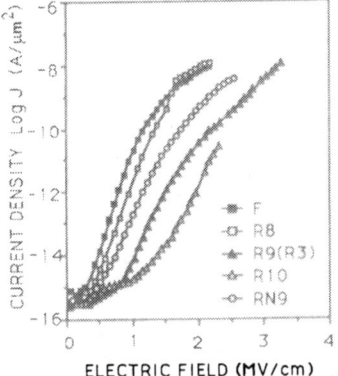

Fig. 1. Current Density-Electric Field characteristics of tantalum oxide films deposited in various oxygen partial pressures, before annealing.

Fig. 2. Current Density-Electric Field characteristics of tantalum oxide films deposited in 10 % O$_2$ partial pressure and subsequently annealed at various conditions.

The microstructures of the as-deposited tantalum oxide films, examined by TEM, are found to be entirely amorphous.

Annealing Effect

For the tantalum oxide films which were deposited on the Si substrates in 10% oxygen partial pressure, the effects of thermal treatments on the dielectric properties and microstructure were examined. Fig.2 shows the relationship between the leakage current density and the applied electric field after annealing. F denotes the furnace annealing at 500°C for 1hr; and R8, R9, and R10 denote RTAs at temperatures of 800, 900, and 1000°C respectively. Except for F, RTA reduced considerably the leakage current as shown. It is not the holding time but the temperature of annealing that mainly affects the dielectric property. It is observed that up to 1000°C the higher the heat treatment temperature, the lower the leakage current.

For RN9, which was heat-treated under the same condition as R9 but for using N_2 atmosphere instead of O_2, the leakage curve is nearly about one order of magnitude higher than the one for R9. So it is also recognized that reoxidation of oxygen deficient tantalum oxide film during annealing contributes to reduce the leakage current. This reoxidation was confirmed by TEM analysis on microstructure to be discussed later.

Fig.3 shows TEM images of tantalum oxide/Si for F. While 5-10Å of SiO_2 layer, which is considered as a remainder of native SiO_2, is found at the interface, other structural changes such as crystallization are not found after 500°C annealing. Electron diffraction patterns for the tantalum oxide film in plan view also show no evidence of crystallization. These results are consistent with the little enhancement of dielectric properties after the 500°C furnace annealing.

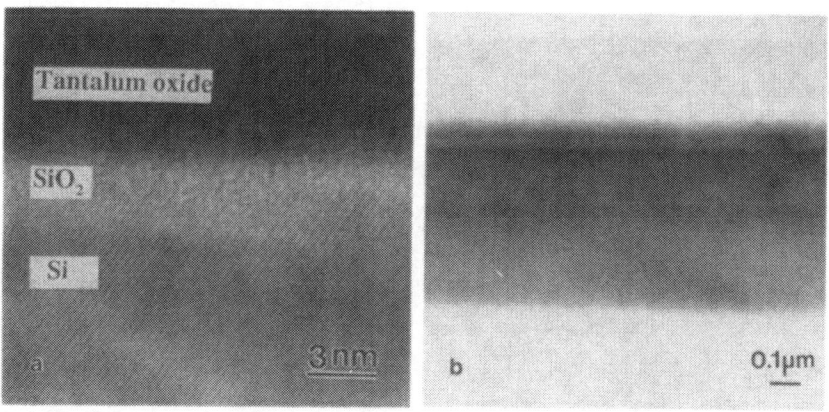

Fig. 3. Transmission electron micrograph images of tantalum oxide film on silicon O_2 annealed at 500 °C for 1 hr: (a) cross-sectional view including the interfacial SiO_2 layer (5 - 10 Å), (b) tilted view of (a).

Among the RTA specimens which were heat-treated at high temperatures R9 was selected to investigate the relationship between the microstructural change and improvement of dielectric property after RTA. Even in a time period as short as 1 min, 900°C is high enough to crystallize the amorphous tantalum oxide film completely.

Fig. 4 shows the micro-beam diffraction patterns from various kinds of products such as Ta_2O_5, TaO_2 and Ta_2Si. Recalling the fact that the average atomic ratio of oxygen to tantalum in as-deposited film is nearly 1:1 or below, it is evident that crystallite phases such as Ta_2O_5, TaO_2 are resulted from the reoxidation and densification of film during RTA in O_2. It is considered that in spite of the existance of grain boundaries the decrease of the leakage current is due to the reoxidation and densification of the tantalum oxide film.

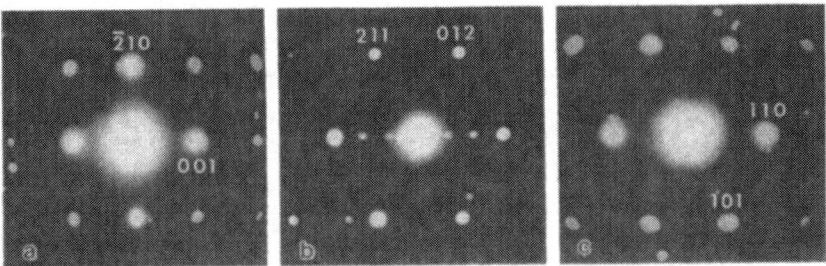

Fig. 4. Micro-beam diffraction patterns of tantalum oxide film on silicon O_2 annealed at 900 °C for 1 min: (a)Ta_2O_5 [120], (b)Ta_2Si [1$\overline{4}$2], and (c)TaO_2 [1$\overline{1}\overline{1}$].

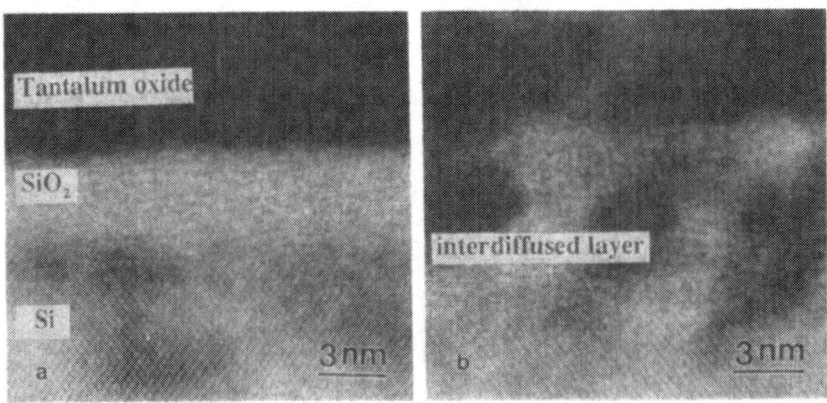

Fig. 5. Cross-sectional transmission electron micrograph images of tantalum oxide film on silicon O_2 annealed at 900 °C for 1 min including: (a) the interfacial SiO_2 layer (40 - 50 Å), (b) the interdiffused layer (80 - 100 Å).

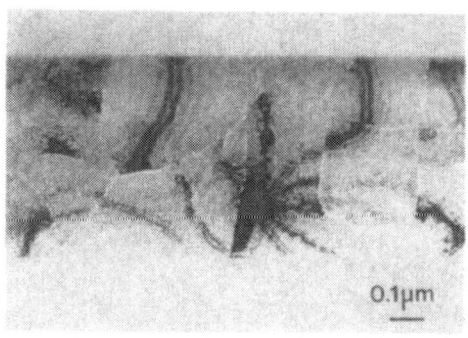

Fig. 6. Transmission electron micrograph images of entirely crystallized tantalum oxide film on silicon O_2 annealed at 900 °C for 1 min.

The presence a Ta_2Si in the tantalum oxide film implies that Si atoms diffuse into the tantalum oxide film during RTA. Since this would cause deteriorating effects on the dielectric properties, it might be necessary to devise a proper method to prevent interfacial diffusion during RTA in further study.

High resolution cross-sectional TEM images of tantalum oxide/Si for R9 are shown in Fig.5. It is shown not only the thickness of the interfacial SiO_2 layer was increased to be 40-50Å but also interdiffused layer of about 80Å thickness, consisted of randomly oriented microcrystallites of Ta_2Si and amorphous phases, was formed during RTA.

While micro-beam diffraction patterns illustrated the crystallization, it was very difficult to indentify the grain boundaries of each crystalline in bright or dark field images in plan view. Fig.6 is an bright field image of tantalum oxide film for R9, which was obtained by increase of the beam path through film with tilting the cross-section view specimen about 40° around an axis parallel to the cross-section and to the interface simultaneously. The entirely crystallized tantalum oxide film is found to be composed of grains of 1500-2000Å size.

CONCLUSION

In the rf-reactive magnetron sputtering, the optimum atmosphere to deposit tantalum oxide film with reasonable dielectric properties is 10% O_2 and 90% Ar mixture.

The leakage current density of tantalum oxide film is reduced from 10^{-9} to 10^{-12} A/μm^2 at an electric field of 2 MV/cm after rapid thermal annealing at 1000℃ in O_2. TEM examinations of the microsturctural change after annealing show that this reduction of the leakage current is attributed to the crystallization and reoxidation of amorphous tantalum oxide film during annealing.

REFERENCES

1. S. Seki, T. Vnagami, O. Kogure, and B. Tsujiyama, J. Vac. Sci. Technol.
    A, 5, 1771 (1987)
2. S. Banerjee, B. Shen, I. Chen, J. Bohlman, G. Brown, and R. Doering,
    J. Appl. Phys., 65(3), 1140 (1989)
3. M. Saitoh, T. mori, and H. Tamura, Tech. Digest of '86 IEDM, 680 (1986)
4. H. Shinriki, M. Nakata, Y. Nishioka and K. Mukai, Digest of Symp. on
    VLSI Technology, 25~26 (1989)
5. C. Isobe and M. Saitoh, Appl. Phys. Lett., 56(10), 907~909 (1990)
6. J.P. Benedict, S.J. Klepeis, W.G. Vandygrift, and Ron Anderson, EMSA
    Bulletin 19:2 November, 74~79 (1989)

Author Index

Subject Index

CPSIA information can be obtained at www.ICGtesting.com
Printed in the USA
LVOW06s1017220514

386805LV00011B/448/P